Advances in Experimental Medicine and Biology

Volume 1498

Advances in Experimental Medicine and Biology provides a platform for scientific contributions in the main disciplines of the biomedicine and the life sciences. This series publishes thematic volumes on contemporary research in the areas of microbiology, immunology, neurosciences, biochemistry, biomedical engineering, genetics, physiology, and cancer research. Covering emerging topics and techniques in basic and clinical science, it brings together clinicians and researchers from various fields.

Advances in Experimental Medicine and Biology has been publishing exceptional works in the field for over 40 years, and is indexed in SCOPUS, Medline (PubMed), EMBASE, Reaxys, EMBiology, the Chemical Abstracts Service (CAS), and Pathway Studio.

2023 CiteScore: 5.9

Sally C. Pias • Qi Wang
Joseph C. LaManna
Editors

Oxygen Transport to Tissue XLVI

Springer

Editors
Sally C. Pias
Department of Chemistry
New Mexico Institute of Mining
and Technology (New Mexico Tech)
Socorro, NM, USA

Joseph C. LaManna
Department of Physiology and Biophysics
Case Western Reserve University
School of Medicine
Cleveland, OH, USA

Qi Wang
Department of Chemistry
New Mexico Institute of Mining
and Technology (New Mexico Tech)
Socorro, NM, USA

Present Address: Department
of Chemistry and Biochemistry
Presbyterian College
Clinton, SC, USA

International Society on Oxygen Transport to Tissue
isott, isott 2024, ISOTT, Socorro, NewMexico
2024-6-9 – 2024-6-14
https://www.isott2024.com/

ISSN 0065-2598 ISSN 2214-8019 (electronic)
Advances in Experimental Medicine and Biology
ISBN 978-3-032-10388-8 ISBN 978-3-032-10389-5 (eBook)
https://doi.org/10.1007/978-3-032-10389-5

This Springer imprint is published by the registered company Springer Nature Switzerland AG
The registered company address is: Gewerbestrasse 11, 6330 Cham, Switzerland

In Memoriam: Duane Frederick Bruley (August 3, 1933–August 23, 2024)[1]

***Duane Frederick Bruley** had many talents but remained humble and caring as a scientist, researcher, professor, academic administrator, author, athlete, and coach. Following an international "Workshop on Oxygen Supply" in Dortmund/Germany 1971 (attended by D.F. Bruley, M.H. Knisely, P. Vaupel), Duane and Haim Bicher founded the **International Society on Oxygen Transport to Tissue (ISOTT)** in 1973, in honor of Duane's mentor Melvin H. Knisely, a physiologist and Chairman of the Department of Anatomy at the Medical College of Charleston, SC, USA. Duane gave his last video presentation at the 2023 ISOTT meeting in Tokyo, Japan. He will be missed by all in the ISOTT community and beyond—one of many lasting legacies from a life well spent and fully lived.*

[1] Some portions of the In Memoriam have been taken or adapted with permission from an obituary published in the Chippewa Herald on September 3–4, 2024.

***Duane's involvement in ISOTT** was considerable. He served as President Elect 1981–82; President 1983; International Executive Committee Member 1973–1985; Society Member 1973–Present; Editor or Co-Editor of 19 volumes of "Oxygen Transport to Tissue" (Springer* Advances in Experimental Medicine and Biology *series); and ISOTT Historian 2008–2024. The formalization of ISOTT took place at the Charleston-Clemson meeting in 1973. The Society now has about 200 members representing essentially every major tissue oxygen research laboratory in the world. ISOTT meets alternate years in North America, Europe, and Asia. At the meeting 2023 in Tokyo, ISOTT celebrated its 50th Anniversary. Most recently, ISOTT members met in Socorro, NM, USA, and the upcoming 2025 meeting will take place in Thessaloniki, Greece.*

***Duane established the Melvin H. Knisely Award,** ISOTT's highest honor for a young investigator with evidence of sustained interest in oxygen transport to tissue and great potential to make a lasting contribution. A recipient is selected annually by a committee consisting of three international scientists or engineers picked by the president of ISOTT. The award was first given at the 1983 annual meeting of ISOTT. An endowment has now been established for the continued presentation of the award at the annual banquet. To honor Duane, **the Society established the Duane F. Bruley Travel Awards** in 2003, providing travel support for student researchers studying all aspects of oxygen transport to tissue.*

Duane obtained a BS in Chemical Engineering at University Wisconsin, Madison, WI, with honors, and earned an Oak Ridge School of Reactor Technology (ORSORT) degree in Nuclear Engineering with a graduate fellowship. On an Atomic Energy Commission Scholarship, he then attended Stanford University for an MS degree in Mechanical Engineering and then attended the University of Tennessee for a PhD in Chemical Engineering, supported by Texaco and Shell Foundation scholarships.

Duane authored over 180 publications and was instrumental in the development of a Protein C (anticoagulant) research program, in partnership with the American Red Cross, due to a Family hereditary condition of Protein C Deficiency, which was one of the bases of his small-business company, Synthesizer, in Ellicott City, MD, USA. Duane's publications on oxygen transport to tissues date back to the late 1960s. Engineering and mathematical modeling were presented to ISOTT with his work on the use of the B-W-K (Bruley-Williford-Kang) technique, a probabilistic numerical method of mathematical modeling and computer simulation.

Throughout Duane's career, he worked with Cray Research (Seymour Cray) to implement B-W-K on the supercomputer to speed up the simulation processes. Other contributions to ISOTT include: (i) a model of oxygen transport and metabolism in neurons, (ii) interpretation of a model of oxygen transport and metabolism in neurons, (iii) interpretation of oxygen disappearance rates in brain cortex following total ischemia, (iv) measurement prediction based on measurement of pO_2 *and flow at the same locus during seizures and barbiturate suppression of EEG, and (v) simultaneous measurement of* pO_2 *and flow at the same locus during seizures and barbiturate suppression of EEG.*

Duane's distinguished career included prominent leadership positions, including Dean of Engineering at the University of Maryland, Baltimore; Dean of Engineering at California Polytechnic State University, San Luis Obispo; Vice President of Academic Affairs at Rose-Hulman Institute of Technology; Head of Biomedical Engineering and Director of the Rehabilitation Engineering Center at Louisiana Tech University; first Head of the Bioengineering and Environmental Systems Section and Director for the Bioprocess Engineering Program at the National Science Foundation; and Head of the Chemical Engineering Department at Tulane University.

Duane's research was conveyed to the technical community through over 400 speaking engagements, mostly many invited keynote presentations. He was selected to make presentations on Bioengineering as a Distinguished Lecturer by both the American Institute of Chemical Engineers and the American Society of Mechanical Engineers. His research evoked many awards, including international recognition from the Japanese Society for the Promotion of Science and the National Academy of Sciences of the USSR. His educational activities brought him many noteworthy awards, such as the Kerr Award, the Ralph E. Teeter Award (1973), and The Alumni Distinguished Service Award from the University of Wisconsin - Eau Claire (1992).

Duane was elected to Fellow Grade in the American Institute of Chemical Engineers (AIChE) and the American Society of Mechanical Engineers (ASME) and was elected Founding Fellow of the American Institute of Medical and Biological Engineering (AIMBE). His election as a fellow in three diverse and major engineering societies exemplifies Duane's breath and overall contributions to the engineering profession and primarily to Bioengineering.

Duane had a reputation as a versatile individual, with a wide range of interests including an impressive list of accomplishments as a tennis player, umpire, and coach. He served as Head Varsity Tennis Coach at the University of Tennessee (Knoxville), Clemson University, and Tulane University, and also as Assistant Coach at Rose-Hulman Institute of Technology. In 1969, his team had an undefeated season and brought the first Atlantic Coast Conference (ACC) Championship to Clemson University. He was an Honorary Life Member of the United States Tennis Association, a member of the United States Professional Tennis Association, as well as an avid skier, earning a NASTAR gold medal in downhill racing in Vail, Colorado, in 1987.

Duane is survived by his three sons Scott, Randall, and Mark; by four grandchildren; and by two great-grandchildren. He is also survived by many nieces and nephews and their families. He was predeceased by his wife of 61 years Suzanne (Bigler), his parents, his brother US Navy Cmdr. Kenneth Bruley, and his sister-in-law Carole Coleman Bruley.

Peter Vaupel, MD, PhD, ISOTT Founding Member and ISOTT President 1992, Department of Radiation Oncology, University Medical Center, Freiburg im Breisgau, Germany

Edwin M. Nemoto, PhD, ISOTT Founding Member and ISOTT President 1996 and 2019, Department of Neurology, 1 University of New Mexico, Albuquerque, NM, USA
Lin Z. Li, PhD, ISOTT President 2015, Britton Chance Laboratory of Redox Imaging, Department of Radiology, Perelman School of Medicine, University of Pennsylvania, Philadelphia, PA 19104, USA

Kyung A. Kang, PhD, ISOTT President 2006, Department of Chemical Engineering, University of Louisville, Louisville, KY, USA

Sally C. Pias, PhD, ISOTT President 2024, Department of Chemistry, New Mexico Institute of Mining and Technology (New Mexico Tech), Socorro, NM, USA

ISOTT 2024 Meeting

Organizing Team

Sally Pias, ISOTT President 2024 (New Mexico Tech, USA)
Qi Wang (New Mexico Tech, USA; Presbyterian College, USA)
Sheila O'Sullivan (New Mexico Tech, USA)
Ex officio member: Edwin Nemoto, ISOTT Past-President 1996 and 2019 (University of New Mexico, USA)

Awards Selection Committee

Ann Barry Flood (Geisel School of Medicine at Dartmouth, USA)
Alexander Caicedo-Dorado (Pontificia Universidad Javeriana, Colombia)
Oliver Thews (University of Halle, Germany)
Joseph LaManna (Case Western Reserve University School of Medicine, USA)
Harold Swartz (Geisel School of Medicine at Dartmouth, USA)
Ilias Tachtsidis (University College London, UK)

ISOTT Awards

Melvin H. Knisely Award *This award honors Dr. Knisely's professional accomplishments with respect to the transport of oxygen and other metabolites in the human body, as well as his enthusiasm for assisting and encouraging young scientists and engineers in various disciplines. The award acknowledges outstanding young investigators and was first presented during the banquet of the 1983 annual conference of ISOTT in Ruston, Louisiana, USA.*

1983: Antal G. Hudetz (Hungary)
1984: Andras Eke (Hungary)
1985: Nathan A. Bush (USA)
1986: Karlfried Groebe (Germany)
1987: Isumi Shibuya (Japan)
1988: Kyung A. Kang (Korea/USA)
1989: Sanja Batra (Canada)
1990: Stephen J. Cringle (Australia)
1991: Paul Okunieff (USA)
1992: Hans Degens (Netherlands)
1993: David A. Benaron (USA)
1994: Koen van Rossem (Belgium)
1995: Clare E. Elwell (UK)
1996: Sergei A. Vinogradov (USA)
1997: Chris Cooper (UK)
1998: Martin Wolf (Switzerland)
1999: Huiping Wu (USA)
2000: Valentina Quaresima (Italy)
2001: Fahmeed Hyder (Bangladesh)
2002: Geofrey De Visscher (Belgium)
2003: Mohammad Nadeem Khan (USA)
2004: Fredrick Palm (Sweden)
2005: Nicholas Lintell (Australia)
2007: Ilias Tachtsidis (UK)
2008: Kazuto Masamoto (Japan)
2009: Rossana Occhipinti (USA)

2010: Sebatiano Cicco (Italy)
2011: Mei Zhang (USA)
2012: Takahiro Igarashi (Japan)
2013: Malou Friederich-Persson (Sweden)
2014: David Highton (UK)
2015: Alexander Caicedo-Dorado (Belgium)
2016: Sally Pias (USA)
2017: Felix Scholkmann (Switzerland)
2018: Anne Riemann (Germany)
2019: Ting Li (China)
2020: *Meeting not held, due to COVID-19 pandemic*
2021: An Ghysels (Belgium)
2022: Alexander Kalyanov (Switzerland)
2023: Hamoon Zohdi (Switzerland)
2024: Jingjing Jiang (Switzerland)

Dietrich W. Lübbers Award *This award was established in honor of Professor Dietrich W. Lübbers, to recognize his outstanding contributions to the problems of oxygen transport to tissue, as well as his long-standing commitment to the Society. The award was first presented in 1994, during the annual ISOTT conference in Istanbul, Turkey.*

1994: Michael Dubina (Russia)
1995: Philip E. James (UK/USA)
1996: Resit Demit (Germany)
1997: Juan Carlos Chavez (Peru)
1998: Nathan A. Davis (UK)
1999: Paola Pichiule (USA)
2000: Ian Balcer (USA)
2001: Theresa M. Busch (USA)
2002: Link K. Korah (USA)
2003: James J. Lee (USA)
2004: Richard Olson (Sweden)
2005: Charlotte Ives (UK)
2006: Bin Hong (China/USA)
2007: Helga Blockx (Belgium)
2008: Joke Vanderhaegen (Belgium)
2009: Matthew Bell (UK)
2010: Alexander Caicedo Dorado (Belgium)
2011: Malou Friedrich (Sweden)
2012: Maria Papademetriou (UK)
2013: Nannan Sun (China)
2014: Felix Scholkmann (Switzerland)
2015: Shun Takagi (Japan)
2016: Gemma Bale (UK)
2017: Ben Jones (UK)
2018: Hamoon Zohdi (Switzerland)
2019: Kazaki Hotta (Japan)

2020: *Meeting not held, due to COVID-19 pandemic*
2021: Masamichi Moriya (Japan)
2022: Frédéric Lange (UK)
2023: Jingjing Jiang (Switzerland)
2024: Djazia Yacheur (Switzerland)

Britton Chance Award *This award was established in honor of Professor Britton Chance, to recognize his exceptional contributions to the science and engineering aspects of oxygen transport to tissue, as well as his long-standing commitment to the Society. The award was first presented in 2004, during the annual ISOTT conference in Bari, Italy.*

2004: Derek Brown (Switzerland)
2005: James Lee (USA)
2006: Hanzhu Jin (China/USA)
2007: Eric Mellon (USA)
2008: Jianting Wang (USA)
2009: Jessica Spires (USA)
2010: Ivo Trajkovic (Switzerland)
2011: Alexander Caicedo Dorado (Belgium)
2012: Felix Scholkmann (Switzerland)
2013: Tharindi Hapuarachchi (UK)
2014: Anne Riemann (Germany)
2015: Wenhao Xie (China)
2016: Linda Ahnen (Switzerland)
2017: Helene Isler (Switzerland)
2018: Jingjing Jiang (Switzerland)
2019: Alexander Kalyanov (Switzerland)
2020: *Meeting not held, due to COVID-19 pandemic*
2021: Maheen Siddiqui (UK)
2022: Samaneh Davoudi (Belgium)
2023: Djazia Yacheur (Switzerland)
2024: Qi Wang (USA)

Duane F. Bruley Travel Awards *These awards honor Dr. Duane F. Bruley's interest in encouraging and supporting young researchers. As a co-founder of ISOTT in 1973, Dr. Bruley appreciated and encouraged cross-disciplinary research among basic scientists, engineers, medical scientists, and clinicians. He made pioneering contributions in mathematical modeling of oxygen and other metabolite transport in the microcirculation. Established in 2003, the awards were first presented at the 2004 annual ISOTT conference in Bari, Italy.*

2004: Helga Blockx (Belgium), Jennifer Caddick (UK), Charlotte Ives (UK), Nicholas Lintell (Australia), Leonardo Mottola (Italy), Samin Rezania (USA/Iran), Ilias Tachtsidis (UK), Liang Tang (USA/China), Iyichi Sonoro (Japan), Antonio Franco (Italy)
2005: Robert Bradley (UK), Harald Oey (Australia), Kathy Hsieh (Australia), Jan Shah (Australia)

2006: Ben Gooch (UK), Ulf Jensen (Germany), Smruta Koppaka (USA), Daya Singh (UK), Martin Tisdall (UK), Bin Wong (USA), Kui Xu (USA)
2007: Dominique De Smet (Belgium), Thomas Ingram (UK), Nicola Lai (USA), Andrew Pinder (UK), Joke Vanderhaegen (Belgium)
2008: Sebastiano Chicco (Italy)
2009: Lei Gao (UK), Jianting Wang (USA), Obinna Ndubuizu (USA), Joke Vanderhaegen (Belgium)
2010: Zareen Bashir (UK), Tracy Moroz (UK), Mark Muthalib (Australia), Catalina Meßmer (USA), Takashi Eriguchi (Japan), Yoshihiro Murata (Japan), Jack Honeysett (UK), Martin Biallas (Switzerland)
2011: Catherine Hesford (UK), Luke S. Holdsworth (UK), Andreas Metz (Switzerland), Maria D. Papademetriou (UK), Patrik Persson (Sweden), Felix Scholkmann (Switzerland), Kouichi Yoshihara (Japan)
2012: Allann Al-Armaghany (UK), Malou Friederich-Persson (Sweden), Tharindi Hapuarachchi (UK), Benjamin Jones (UK), Rebecca Re (Italy), Yuta Sekiguchi (Japan), Ebba Sivertsson (Sweden), André Steimers (Germany)
2013: Allann Al-Armaghany (UK), Gemma Bale (UK), Alexander Caicedo-Dorado (Belgium), Luke Dunne (UK)
2014: Geraldine De Preter (Belgium), Benjamin Jones (UK), Stefan Kleiser (Switzerland), Nassimsadat Nasseri (Switzerland), Marie-Aline Neveu (Belgium), Shinsuke Nirengi (Japan), Takuya Osawa (Japan)
2015: Gemma Bale (UK), Chenyang Gao (China), Chae Jeong Lim (Korea), Phong Phan (UK), Guennadi Saiko (Canada), Kuangyu Shi (Germany), Nannan Sun (China)
2016: Isabel De Roever (UK), Yuya Enokida (Japan), Jingjing Jiang (Switzerland), Frédéric Lange (UK), Andreas J Metz (Switzerland), Shun Takagi (Japan)
2017: Luca Giannoni (UK), Jingjing Jiang (Switzerland), Alexander Kalyanov (Switzerland), Zuzana Kovacsova (UK), Nassim Nasseri (Switzerland), Maheen Faisal Siddiqui (UK), Shun Takagi (Japan), Yusuke Tsuruno (Japan), Lina Xu (Germany), Hamoon Zohdi (Switzerland)
2018: Andrei Belten (Canada), Aldo Di Costanzo Mata (Switzerland), Yuki Komuro (Japan), Joshua Russell-Buckland (UK), Dries Hendrikx (Belgium), Shun Takagi (Japan)
2019: Gary Angles (USA), Min Feng (USA), Thea Hüsing (Germany), Bhabuk Koirala (USA), Sho Kojima (Japan), Jonathan Nguyen (USA), Labiblais Rahman (Japan), Mandy Rauschner (Germany), Aarti Sethuraman (USA), Eileen Thiessen (USA)
2020: *Meeting not held, due to COVID-19 pandemic*
2021: *Online meeting, due to COVID-19 pandemic*
2022: Xianyin Hu (Japan), Djazia Yacheur (Switzerland), Victor Ochoa-Gutierrez (UK, Mexico), Tarcisi Cantieni (Switzerland), Tong Li (Switzerland)
2023: Hikari Otsuka (Japan)
2024: Masamichi Moriya (Japan), Misa Oba (Japan), Musa Talati (UK)

ISCBFM Bursaries *These bursaries were awarded to defray the cost of registration for the ISOTT 2024 meeting, held in Socorro, NM, USA. They drew on funds provided by the International Society for Cerebral Blood Flow and Metabolism (ISCBFM), to encourage young researchers pursuing interests related to the mission of the ISCBFM and ISOTT.*

2024: Letizia Lanini (Switzerland), Rin Kataoka (Japan), Victor Ochoa-Gutierrez (UK/USA)

Peter Vaupel Honorary Lecture *This award honors Professor Dr. Peter Vaupel, who has made exceptional contributions in the areas of tumor hypoxia, metabolic reprogramming, and radiotherapy. Established in 2016, the award may be given at the discretion of the annual meeting organizers. It was first presented at the 2018 annual ISOTT conference in Seoul, South Korea.*

2018: Gabriele Multhoff (Germany), *Lecture title:* Hypoxia compromises anti-cancer immune responses

2024: Lin Z. Li (USA), *Lecture title:* Role of mitochondrial redox imaging in modern cancer research

Preface

This volume features peer-reviewed papers based on work presented at the 2024 annual meeting of the International Society on Oxygen Transport to Tissue (ISOTT), held on June 9–14. The meeting took place on the beautiful, rural campus of the New Mexico Institute of Mining and Technology (New Mexico Tech) in Socorro, New Mexico, USA. It brought together scientists, physicians, engineers, and mathematicians from at least 10 countries and 16 US states, to exchange information and ideas in a highly interactive setting.

The event opened with the first-ever Young Researchers Symposium, a student-focused day that brought together trainees from Europe, Asia, and North America to present preliminary research, engage in discussion, and receive feedback in an informal, supportive setting. The final presentation of this symposium also marked the opening of the main meeting and was delivered by the 2023 recipient of the Society's prestigious Melvin H. Knisely Award.

The main program included 50 oral and 25 poster presentations, among them 22 plenary lectures by leading researchers and 10 additional invited talks. The meeting emphasized interaction, through ample scientific discussion as well as shared meals and social events. Special sessions addressed in vivo tissue oxygen measurements; redox and cancer; artificial oxygen carriers; blood flow; and membrane-level transport. In addition, the community engaged in forward-thinking discussion in a session organized by young researchers, titled "Confronting Unsolved Challenges in Research."

ISOTT is a worldwide organization known for fostering scientific dialog and lasting professional relationships among experimentalists, technical innovators, theoreticians, and clinical researchers. Given the central biomedical importance of oxygen and the evolving interests of its members, the Society has embraced innovation in many areas—especially in vivo assessments of oxygenation and metabolic activity in brain and tumor tissues using optical and magnetic resonance techniques. With a loyal and internationally diverse membership, the ISOTT annual meeting serves as a unique forum, promoting scientific and technical innovation through open discussion and collaborative thinking.

ISOTT President 2024 and Lead Organizer — Sally C. Pias
Socorro, New Mexico, USA

Photo credit: Bethany Jessen

Acknowledgments The editors gratefully acknowledge the authors, reviewers, and all those who have contributed to this publication and to the ISOTT 2024 conference. In particular, we thank Bethany Jessen and Sheila O'Sullivan, who greatly assisted with logistical aspects of the meeting, along with Nowka Gutierrez, Gloria Gutierrez-Anaya, James Graham, and their respective teams. Harold Swartz generously assisted with organizing the scientific program; Paul Buehler organized the "Artificial Oxygen Carriers" session; Alexander Kalyanov, Qi Wang, and Hamoon Zohdi organized the "Confronting Unsolved Challenges in Research" session; and Qi Wang led the Young Researchers Symposium. We thank student assistants David Koshkin, Isabella Martinez Pruet, Lauren Hall, Deven Raman, Audrey Wilkening, Nusrat Yeasmin, and Chandler Lyon for their practical help during the meeting. We also acknowledge Joyce Fang for developing the meeting website (isott2024.com); Edwin Nemoto for his major role in securing financial support for the meeting; Peter Keipert for his support with financial management; Ursula Wolf for many helpful suggestions; and Felix Scholkmann for valuable assistance with the manuscript review process.

Sponsors The meeting was supported financially by the International Society for Cerebral Blood Flow and Metabolism (ISCBFM) and the New Mexico Consortium.

Other Partners

New Mexico Institute of Mining and Technology (New Mexico Tech)
ClinEPR, LLC
Shearit
O_2M Technologies
Oxford Optronix

Manuscript Review Panel

Denis Bragin (University of New Mexico School of Medicine and Lovelace Biomedical Research Institute, USA)
Paul Buehler (University of Maryland School of Medicine, USA)
Pedro Cabrales (University of California San Diego, USA)
Chris Cooper (University of Essex, UK)
Allan Doctor (University of Maryland School of Medicine, USA)
Alexander Caicedo (Pontificia Universidad Javeriana, Colombia)
Gareth Eaton (University of Denver, USA)
Ryan Field (Kernel, USA)
Howard Halpern (The University of Chicago, USA)
Kazuki Hotta (Kitasato University Graduate School of Medical Sciences, Japan)
Alexander Kalyanov (University of Zürich, Switzerland)
Valery Khramtsov (West Virginia University Robert C. Byrd Health Sciences Center, USA)
Joseph LaManna (Case Western Reserve University, USA)
Lin Z. Li (University of Pennsylvania Perelman School of Medicine, USA)
Petra Mayr (Oxford Optronix, Ltd., UK)
Paul Min (Mayo Clinic, USA)
Edwin Nemoto (The University of New Mexico, USA)
Andre Palmer (The Ohio State University, USA)
George Perdrizet (University of Connecticut, USA)
Sally C. Pias (New Mexico Institute of Mining and Technology, USA)
Brian Pogue (University of Wisconsin-Madison, USA)
Gennadi Saiko (Toronto Metropolitan University, Canada)
Kaoru Sakatani (The University of Tokyo, Japan)
Felix Scholkmann (University of Zürich, Switzerland)
Harold Swartz (Geisel School of Medicine at Dartmouth College, USA)
Ilias Tachtsidis (University College London, UK)
Oliver Thews (University of Halle, Germany)
Mark Tseytlin (West Virginia University, USA)
Peter Vaupel (University Medical Center Mainz, Germany)

Qi Wang (Presbyterian College, USA)
Pamela Wenzel (University of Texas Health Science Center at Houston, USA)
Martin Wolf (University of Zürich, Switzerland)
Ursula Wolf (University of Bern, Switzerland)
He Nucleus Xu (University of Pennsylvania Perelman School of Medicine, USA)
Hamoon Zohdi (University of Bern, Switzerland)

Contents

Part I

Honorary Lectures

Role of Mitochondrial Optical Redox Imaging in Modern Cancer Research (2024 Peter Vaupel Honorary Lecture)

1

Lin Z. Li

Abstract

The mitochondrial optical redox imaging (ORI), pioneered by the late Britton Chance and his coworkers starting from the 1950s, is mainly based on the intrinsic fluorescence of reduced nicotinamide adenine dinucleotide (NADH) and oxidized flavoproteins (Fp) containing flavin adenine dinucleotide (FAD). These two important metabolic coenzymes regulate many intracellular biochemical reactions including glycolysis, the Krebs cycle, and oxidative phosphorylation, and participate in the intricate interplay between metabolism and signaling. The optical redox ratio (ORR) of these two fluorescence signals is an indicator of mitochondrial metabolic and redox status. Chance and his coworkers not only studied redox status in cells and tissues based on the intrinsic fluorescence but also developed the Chance redox scanner in the 1970s–80s, which images the 3D distribution of NADH, Fp, and ORR ex vivo in snap-frozen tissues. The repertoire of ORI has been further strengthened and enriched with the modern technical advancements including state-of-the-art fluorescence, confocal, two-photon, and lifetime imaging microscopy. ORI has been extensively applied to (pre)cancer cells and tissues including clinical biopsies and patient tumor-derived organoids, to develop metabolic biomarkers for (pre)cancer diagnosis, prognosis, and treatment response. In this chapter, I illustrate the basic principles, main concepts and progresses, and future directions of ORI in cancer research.

Keywords

Mitochondrion · NAD · Glickson-Chance Debate · Biomarkers, Breast cancer

1.1 Introduction

It is a great honor for me to receive this Peter Vaupel Lecture Award at the 2024 Annual Meeting of the International Society on Oxygen Transport to Tissue (ISOTT), organized by Dr. Sally Pias at the New Mexico Institute of Mining and Technology, Socorro, New Mexico, USA. My personal career development has benefited greatly from participating in the ISOTT society. My experience with ISOTT

This manuscript is based on the Peter Vaupel Honorary Lecture delivered by the author at the 2024 Annual Meeting of the International Society on Oxygen Transport to Tissue in Socorro, New Mexico, USA

L. Z. Li (✉)
Britton Chance Laboratory of Redox Imaging, Department of Radiology, Perelman School of Medicine, University of Pennsylvania, Philadelphia, PA, USA
e-mail: linli@pennmedicine.upenn.edu

S. C. Pias et al. (eds.), *Oxygen Transport to Tissue XLVI*, Advances in Experimental Medicine and Biology 1498, https://doi.org/10.1007/978-3-032-10389-5_1

started in the 2006 ISOTT meeting organized by Dr. Kyung Kang at the University of Louisville, Louisville, Kentucky, USA. The late Dr. Britton Chance, a long-time senior member and persistent participant of the ISOTT society since the very first meeting in 1973 (Bruley and Thiessen 2016), encouraged me to attend the Louisville meeting. Since then, I have become an ISOTT member and a regular participant of ISOTT meetings. In the 2007 Meeting in Uppsala, I met personally with Dr. Duane Bruley, cofounder of the society, with impressive vision, generosity, wisdom, and kindness. For years, Duane and Kyung mentored me about the ISOTT history, administrative structures, and functions and guided me through the process of bidding for and organizing the 2015 ISOTT meeting in Wuhan, China, as a co-president. The 2015 ISOTT president was Dr. Qingming Luo, a former postdoc of Dr. Chance and the Director of Britton Chance Center for Biomedical Photonics in Huazhong University of Science and Technology, Wuhan, China. Dr. Luo generously allocated local resources and manpower to organize the first ISOTT conference in China very successfully and fulfilled the wish of Dr. Chance. This conference was also a good learning and growing experience for me, with much thanks to the work of local organizing staffs and the support of ISOTT senior members. Throughout the past two decades, I have benefited tremendously from the mentorship of ISOTT senior members including Drs. Peter Vaupel, Harold Swartz, Joseph LaManna, and the help from many other friends and colleagues. Dr. Vaupel is a long-time senior member of ISOTT, and a renowned cancer biologist pioneered in studying tumor oxygenation and hypoxia. I had numerous personal interactions with him at ISOTT meetings as well as the Britton Chance Commemorative Symposia at the University of Pennsylvania. My research benefited from his lectures on tumor hypoxia and the Warburg effect presented at these events.

My first ISOTT presentation was about mitochondrial optical redox imaging (ORI) of human melanoma xenografts under the mentorship of Drs. Jerry D. Glickson and Britton Chance. ORI or optical metabolic imaging (OMI) as referred by some publications, has been developed by the Chance lab since the 1950s. The first detailed study of cellular NADH by fluorescence spectrophotometry in vitro was conducted by Duysens and Amesz in 1957 (Duysens and Amesz 1957). In the next year, Chance and coworkers conducted the measurement of NADH fluorescence in isolated mitochondria and identified how NADH fluorescence changes under different mitochondrial redox states depending on the supply of substrates and ADP (Chance and Baltscheffsky 1958). Within the next several years, Chance et al. reported ex vivo and in vivo measurements of NADH in tissues including muscle, brain, and kidney (Chance and Jobsis 1959; Chance et al. 1962a, b). In the late 1960s, Chance and coworkers studied and identified the fluorescence of oxidized flavoproteins (Fp containing flavin adenine dinucleotide, i.e., FAD) in the mitochondrial respiratory chain (Hassinen and Chance 1968; Garland et al. 1967; Chance et al. 1967). During the 1970s–80s, Chance's lab built the cryogenic redox scanner, imaging the 3D distribution of NADH, Fp, and the redox ratio in tissue (Quistorff et al. 1985; Barlow et al. 1979). Chance predicted that the fluorescence from mitochondrial NADH and Fp would open "a new field of metabolic research" (Li et al. 2014). Indeed, ORI has become an established, valuable metabolic imaging tool for biomedical research including cancer, neurodegenerative diseases, and other disorders. Due to the limited scope of this manuscript, I will only touch upon some main concepts as well as key applications of this technique in cancer so that beginners can quickly grasp the essence. For more in-depth and detailed information, readers should check for more comprehensive reviews (Li and Sun 2014; Li et al. 2009a; Georgakoudi and Quinn 2023; Gu et al. 2002) and original literatures.

1.2 Basic Principles of Optical Redox Imaging

ORI is mainly based on the optical signals of NADH and FAD (or Fp) which have fluorescence in the range of visible light, whereas NAD^+ and $FADH_2$ do not have. Fluorescence of NADH (excitation ~370 nm, emission ~450 nm) and FAD (excitation ~425 nm, emission ~525 nm) has been found to be mainly coming from the mitochondria within cells other than cytosol and nuclei, with the mitochondrial compartment accounting for more than 90% and 60% of the Fp and NADH signals, respectively, in normal and cancer cells (Yu and Heikal 2009; Heikal 2010). Although the cellular NADPH concentration can be several times less than NADH, NADPH has the same fluorescence property as NADH, and currently NAD(P)H is a more commonly used terminology in the field of ORI. The oxidized flavin signal can be contributed by both free and protein-bound flavins including flavin-adenine mononucleotide (FMN) and FAD. However, it was shown that FAD dominates the oxidized flavin signal in tissue (Bessey et al. 1949; Kunz and Gellerich 1993; Kunz and Kunz 1985). Thus, here FAD and Fp are used interchangeably.

The ratio of the two fluorescence signals, i.e., optical redox ratio (ORR), is an indicator of mitochondrial metabolic or redox status (Chance et al. 1979). Often, the normalized ORR, defined as Fp/(NAD(P)H+Fp), is adopted to have a value between 0 and 1 with an improved statistical, normal distribution. A higher ORR represents a more oxidative state and a lower ORR represents a more reductive state. Such definition is aligned with the common definition of redox potential in the mitochondrial electron transport train, by which a higher redox potential represents a more oxidized redox state and tends to grab electrons from the environment (Mobilian and Craft 2022; Alberts et al. 2002). It has been demonstrated that ORR correlates linearly with the redox potential of NAD(H), i.e., $NAD^+/(NADH+NAD^+)$, as measured by liquid chromatography–mass spectrometry (LC-MS) in normal cells and precancerous tissues (Quinn et al. 2013; Varone et al. 2014). Note that a higher ORR is not necessarily an indicator of oxidative stress or a higher level of reactive oxygen species (ROS). Both ORR increase and decrease can be associated with a higher ROS level as shown by studies of breast cancer cells in vitro (Podsednik et al. 2020, 2024). Thus, the redox status probed by ORI is expected to provide independent information that cannot be simply represented by ROS or oxidative stress.

ORI can be acquired with various optical imaging tools including the cryogenic Chance redox scanning, regular fluorescence microscopy or macroscopy, confocal and multiphoton microscopy, etc. Lifetime imaging microscopy is also developed to measure the lifetimes of NAD(P)H and FAD, which are sensitive to the chemical environment and metabolic status and can separate the measurements of free and bound NAD(P)H or FAD (Datta et al. 2020). Single photon fluorescence imaging with visible light is constrained to the surface imaging due to the limited light penetration depth in tissue. Two-photon imaging utilizes the near-infrared light for NAD(P)H and FAD excitation with a larger light penetration depth. Recently, with 3-photon 1100 nm excitation, NAD(P)H imaging has been demonstrated with good signal contrast in murine skin in vivo (Lee et al. 2019). Nevertheless, despite longer wavelengths used, the multiphoton light penetration depth is still less than 1 mm. It remains difficult to image deep organ tissues.

The cryogenic Chance redox scanner is a unique instrument that can image the 3D distribution of NAD(P)H, Fp, and ORR in snap-frozen tissue blocks under low temperature of liquid nitrogen (Quistorff et al. 1985; Li et al. 2009a). Fresh harvested tissue specimens are snap-frozen ex vivo and kept in liquid nitrogen during the storage and imaging to maintain the metabolic status the same as or as close as possible to the in vivo state. During the imaging, the tissue blocks are shaved by a grinder to expose a flat plane at a certain depth, and then line scanned by a moving coaxial fibro-optic probe that excites tissue and collects emitted light to generate 2D fluorescence images. Such procedure is repeated multiple times at different tissue depths to generate 3D maps of NADH, Fp, and ORR. Chance redox scanner has been applied to image organ tissues including brain, heart, liver, and pancreas ex vivo (Barlow et al. 1979; Haselgrove et al. 1990; Xu et al. 2013b, 2014; Xu and Li 2014). Despite its invasiveness, Chance scanner remains one of the gold standards for studying the 3D distribution and heterogeneity of the mitochondrial metabolic/redox status in tissues.

1.3 Optical Redox Imaging for Cancer Diagnosis and Prognosis

To develop imaging biomarkers for cancer diagnosis and prognosis, a general approach is to compare the differences between normal tissue, precancer, and cancer or between cancers with different aggressiveness (invasive/metastatic potential) so that potential biomarkers can be identified. ORI has been applied to image (pre)-cancer cell cultures and organoids, animal models, and clinical tumor specimens, as having been reviewed and summarized (Li and Sun 2014). Here, I would take breast cancer as a representative example to highlight some key concepts, applications, and results for cancer diagnosis and prognosis. Readers can also refer to other publications for detecting precancer (Xu et al. 2013b; Pouli et al. 2020).

It has been shown that ORR is sensitive to breast cancer cell metabolism (Hou et al. 2016) and can differentiate between breast cancer subtypes (Walsh et al. 2014). A review of literature data indicated that higher ORR and more oxidized redox status appear to correlate with higher breast cancer cell aggressiveness in vitro (Li and Sun 2014). Within the triple-generative breast cancer (TNBC) subtype, ORR (=Fp/(NADH+Fp)) was shown to positively correlate with the invasive potential of four human cell lines in vitro (Feng et al. 2021), whereas the intracellular ROS level does not correlate with the invasiveness (data not published). It was also found that the change of ORR due to reoxygenation after acute hypoxia correlates with the rank order of metastatic potential in four murine isogenic TNBC lines (Alhallak et al. 2016). In addition, ORI of the highly metastatic MDA-MB-231 cells responded to the knockdown of oncogene DEK or PGC1α with a concomitant decrease of both ORR and invasive potential (Wen et al. 2019; Lin et al. in press). All these data support the potential of using ORR as a biomarker for TNBC invasive/metastatic potential.

In breast cancer mouse xenografts, ORI by the cryogenic Chance redox scanner showed significant intratumor redox heterogeneity. The more invasive/metastatic MDA-MB-231 tumors exhibit a distinct bimodal distribution with a more oxidized region and a more reduced region, whereas the more indolent MCF-7 tumors are less heterogeneous (Xu et al. 2010). The ORR in the more oxidized region can differentiate the MDA-MB-231 tumors from the MCF-7 tumors significantly better than the globally averaged ORR of whole tumors. We also obtained data showing that the ORR of the more oxidized region correlates better with the aggressiveness of human breast cancer xenografts than the globally averaged ORR (manuscripts in preparation). These data are consistent with the results from the redox scanning of mouse xenografts of human melanoma cell lines with different invasive/metastatic potential. It was found that the ORR in the oxidized region correlates significantly with the invasive potential of the five melanoma xenografts, whereas the ORR averaged over the whole tumor does not (Li et al. 2009b). Furthermore, the NAD(P)H in the oxidized region is also more effective than the globally averaged NAD(P)H to differentiate melanoma xenografts with different aggressiveness, although its differentiating power is less than the ORR in the oxidized region (Peng et al. 2024). In addition, the ORR in the oxidized tumor region is significantly higher in a p53-null xenograft model of human colon cancer than the corresponding wildtype model, whereas the whole-tumor averaged redox indices do not show significant differences between the two models (Xu et al. 2013a). These results demonstrate the importance of imaging intratumor redox heterogeneity and that ORR of the oxidized region can potentially be a good indicator of tumor aggressiveness. H&E staining, TUNEL assay, and the imaging of glucose analog uptake indicated that there were viable cells in the oxidized regions, and the ORR differences observed were more likely due to differential cell metabolism, rather than variations in cell density (Peng et al. 2024; Xu et al. 2012, 2013, 2013a, 2009).

ORI with the Chance redox scanning also identified intrasample redox heterogeneity in clinical specimens of breast cancer patients, which are more heterogeneous than adjacent normal breast tissues (Xu et al. 2016, 2013c). ORR, NADH, and Fp can differentiate the cancerous and normal breast tissue with sensitivity and specificity in the range of 69–81%. Work is in progress to test the value of ORI of clinical specimens for predicting the prognosis in breast cancer patients.

Regarding the metabolic state of mitochondria in cancer cells within the oxidized region, Drs. Jerry Glickson and Britton Chance had a debate for years. Mitochondrial metabolism can be classified into five states from 1 to 5 depending on the availability of ADP, substrate, and oxygen (Xu et al. 2010, Li et al. 2009b). A general working hypothesis proposed by Dr. Glickson for tumor ORI studies is that the oxidized tumor region with higher ORR indicates a State 2, which has limited blood perfusion and oxygen/nutrient supply and favors the survival of cancer cells with higher metastatic potential (Li et al. 2009b). An alternative hypothesis was proposed by Dr. Chance that the optically oxidized region represents high oxidative metabolism of more aggressive cancer cells with ample supply of ADP and substrate, which is characteristic of a State 3 (Xu et al. 2010). This Glickson-Chance debate remains to be resolved by future investigations.

1.4 Optical Redox Imaging Predicting Treatment Response

ORI indices were identified to predict the response to chemotherapy and targeted therapy in breast cancer cell cultures. For example, the ORR of ER^+ breast cancer cell line MCF-7 was responsive to antiestrogen drugs in vitro, whereas the MCF-7-derived drug-resistant line LCC9 and TNBC line MDA-MB-231 were not responsive with insignificant alterations in ORR (Ostrander et al. 2010). Furthermore, the breast cancer $Her2^+$ BT474 mouse xenograft, more responsive to anti-Her2 trastuzumab than a HR6 xenograft with acquired resistance to trastuzumab, was associated with a significant change of ORR as early as day 2 post treatment compared to the control, whereas no significant changes were found in the resistant HR6 xenograft (Walsh et al. 2013). In contrast, FDG-PET detected no changes in both responsive and non-responsive xenografts until day 12. Association between ORR change and trastuzumab sensitivity was also observed for the same cell models in vitro in both 2D cell cultures (Walsh et al. 2013, 2012) and organoids derived from mouse models (Walsh et al. 2014).

The sensitivity of ORI to treatment response was also demonstrated in patient-derived organoids of human breast cancer (Walsh et al. 2014; Sharick et al. 2020) and pancreatic cancer (Sharick et al. 2020). Taking the data from an ER^+ organoid as reported in reference Walsh et al. (2014), the optical redox index is shown to correlate significantly with the cell death index when the organoids were under tamoxifen (a hormone therapy drug), paclitaxel (a standard chemotherapy drug), or in combination ($R^2 = 0.97$, $p = 0.01$) (Fig. 1.1). In contrast, the more aggressive TNBC organoids exhibited no significant response in ORI indices under anti-estrogen or anti-Her2 therapy or combination (Walsh et al. 2014). Under the settings of standard chemotherapy (adriamycin/cyclophosphamide+paclitaxel), the organoids from two TNBC patients with less residual disease burden exhibited larger alterations of ORI index in response to the chemotherapy in vitro compared to the organoids from a TNBC patient with higher disease burden (Sharick et al. 2020). Furthermore, in seven pancreatic cancer patients, a larger effect size on the ORI index induced by in vitro chemotherapy in patient-derived organoids was associated with the responsiveness to adjuvant chemotherapy and a longer relapse-free survival (Sharick et al. 2020).

Xu et al. also compared the treatment response of two TNBC cell cultures, i.e., more invasive/metastatic MDA-MB-231 cells versus less invasive/metastatic HCC1806 cells, to paclitaxel. The more responsiveness with higher proliferation

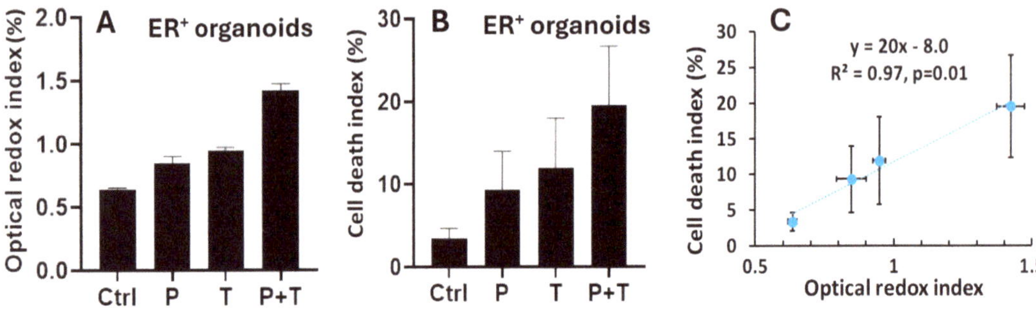

Fig. 1.1 Optical redox imaging indices correlate with therapeutic responses in organoids derived from an ER^+ human breast tumor. (**a**) Optical redox index; (**b**) cell death index (%); and (**c**) their correlation under different 72-hour treatments, i.e., P (paclitaxel), T (tamoxifen), and combination therapy (P+T plus trastuzumab which showed no effect as a single agent). Optical redox index is the inverse of OMI index and conceptually equivalent to ORR. Cell death index is based on the percent of caspase-3 cleavage. OMI index and caspase-3 data were taken from Fig. 5A-B in reference Walsh et al. (2014)

inhibition in HCC1806 is associated with a more significant alteration in ORR induced by paclitaxel compared to MDA-MB-231 (personal communication with Xu HN). ORI of HCC1806 cells in vitro was found to be sensitive to the treatment of paclitaxel, FK866 (an inhibitor of the salvage pathway of NAD biosynthesis), and their combinations (Podsednik et al. 2021). ORI indices such as ORR and Fp in HCC1806 cells were found to correlate positively with the cell growth inhibition induced by such treatment 2021, which is consistent with Fig. 1.1c conceptually. Similarly, the positive correlations of ORR with the growth inhibition were demonstrated in a murine breast cancer line under the FK866 treatment in vitro (Kollmar et al. 2024).

In summary, in response to drug treatment, a common observation is that a higher sensitivity to a treatment is associated with a larger responsiveness in ORI indices. This is a hypothesis that needs further validation by future studies. ORI indices are likely to reflect treatment responses regardless of cancer (sub)types, which is consistent with NAD(H) metabolome alteration as a general hallmark of cancer (Chiarugi et al. 2012) and mitochondrial redox status as a central player in cell death.

1.5 Conclusion and Future Directions

Mitochondrial optical redox imaging is sensitive to perturbation of metabolism and signaling pathways and can provide potential biomarkers for cancer diagnosis, prognosis, and treatment response. Future directions are numerous in both technical developments and biological/clinical applications. Before the in vivo ORI of deep tissue ever becomes feasible in the future, it is worthwhile to explore the feasibility and value of room-temperature ORI of clinical tissue slides, which are more readily available under clinical settings than frozen tissue blocks. It has been under investigation how fixation may affect the NAD(P)H and Fp fluorescence signals and whether the room-temperature ORI of fresh or fixed but unstained tissue slices can generate any biologically useful information (Xu et al. 2000, 2019; Sánchez-Hernández et al. 2023). Other future directions of interest, to name several, include: (i) in-depth study of the relationship between ORI indices and ROS levels; (ii) ORI of cancer cells and tissues under hypoxia, which is known to drive the Warburg effect and cancer progression; (iii) more application and validation of ORI biomarkers under clinical settings of cancer.

Acknowledgments My heartfelt thanks to all my mentors, colleagues, collaborators, and students. The path for my academic career is like climbing a huge mountain. It is a mixture of adventures filled with serendipity and payoffs from persistent hard work. Regardless of whether I will be fortunate enough to reach the mountain top, I will always owe a debt of earnest gratitude to people who have helped me along the way. Finally, this paper is dedicated to my mentors, Drs. Britton Chance and Jerry D. Glickson, and to ISOTT cofounder Dr. Duane F. Bruley.

References

Alberts B, Johnson A, Lewis J et al (2002) Molecular biology of the cell. Garland Science

Alhallak K, Rebello LG, Muldoon TJ et al (2016) Optical redox ratio identifies metastatic potential-dependent changes in breast cancer cell metabolism. Biomed Opt Express 7:4364–4374

Barlow CH, Harden WR 3rd, Harken AH et al (1979) Fluorescence mapping of mitochrondrial redox changes in heart and brain. Crit Care Med 7:402–406

Bessey OA, Lowry OH, Love RH (1949) The fluorometric measurement of the nucleotides of riboflavin and their concentration in tissues. J Biol Chem 180:755–769

Bruley DF, Thiessen EE (2016) ISOTT from the beginning: a tribute to our deceased members (icons). In: Luo Q, Li LZ, Harrison DK, Shi H, Bruley DF (eds) Oxygen transport to tissue XXXVIII, Advances in Experimental Medicine and Biology, Springer International Publishing, v923, pp 1–12

Chance B, Baltscheffsky H (1958) Respiratory enzymes in oxidative phosphorylation. VII. Binding of intramitochondrial reduced pyridine nucleotide. J Biol Chem 233:736–739

Chance B, Jobsis F (1959) Changes in fluorescence in a frog sartorius muscle following a twitch. Nature 184: 195–196

Chance B, Cohen P, Jobsis F et al (1962a) Intracellular oxidation-reduction states in vivo. Science 137:499–508

Chance B, Cohen P, Jobsis F et al (1962b) Localized fluorometry of oxidation-reduction states of intracellular pyridine nucleotide in brain and kidney cortex of the anesthetized rat. Science 136:325

Chance B, Ernster L, Garland PB et al (1967) Flavoproteins of the mitochondrial respiratory chain. Proc Natl Acad Sci USA 57:1498–1505

Chance B, Schoener B, Oshino R et al (1979) Oxidation-reduction ratio studies of mitochondria in freeze-trapped samples. NADH and flavoprotein fluorescence signals. J Biol Chem 254:4764–4771

Chiarugi A, Dolle C, Felici R et al (2012) The NAD metabolome—a key determinant of cancer cell biology. Nat Rev Cancer 12:741–752

Datta R, Heaster TM, Sharick JT et al (2020) Fluorescence lifetime imaging microscopy: fundamentals and advances in instrumentation, analysis, and applications. J Biomed Opt 25:1–43

Duysens LN, Amesz J (1957) Fluorescence spectrophotometry of reduced phosphopyridine nucleotide in intact cells in the near-ultraviolet and visible region. Biochim Biophys Acta 24:19–26

Feng M, Xu HN, Jiang J et al (2021) Potential biomarker for triple negative breast cancer invasiveness by optical redox imaging. Adv Exp Med Biol 1269:247–251

Garland PB, Chance B, Ernster L et al (1967) Flavoproteins of mitochondrial fatty acid oxidation. Proc Natl Acad Sci USA 58:1696–1702

Georgakoudi I, Quinn KP (2023) Label-free optical metabolic imaging in cells and tissues. Annu Rev Biomed Eng 25:413–443

Gu Y, Qian Z, Chen J et al (2002) High-resolution three-dimensional scanning optical image system for intrinsic and extrinsic contrast agents in tissue. Rev Sci Instrum 73:172–178

Haselgrove JC, Bashford CL, Barlow CH et al (1990) Time resolved 3-dimensional recording of redox ratio during spreading depression in gerbil brain. Brain Res 506:109–114

Hassinen I, Chance B (1968) Oxidation-reduction properties of the mitochondrial flavoprotein chain. Biochem Biophys Res Commun 31:895–900

Heikal AA (2010) Intracellular coenzymes as natural biomarkers for metabolic activities and mitochondrial anomalies. Biomark Med 4:241–263

Hou J, Wright HJ, Chan N et al (2016) Correlating two-photon excited fluorescence imaging of breast cancer cellular redox state with seahorse flux analysis of normalized cellular oxygen consumption. J Biomed Opt 21:060503

Kollmar J, Xu J, Gonzalves D et al (2024) Differential mitochondrial redox responses to the inhibition of NAD(+) salvage pathway of triple negative breast cancer cells. Cancers (Basel) 17:7

Kunz WS, Gellerich FN (1993) Quantification of the content of fluorescent flavoproteins in mitochondria from liver, kidney cortex, skeletal muscle, and brain. Biochem Med Metab Biol 50:103–110

Kunz WS, Kunz W (1985) Contribution of different enzymes to flavoprotein fluorescence of isolated rat liver mitochondria. Biochim Biophys Acta 841:237–246

Lee JH, Rico-Jimenez JJ, Zhang C et al (2019) Simultaneous label-free autofluorescence and multi-harmonic imaging reveals in vivo structural and metabolic changes in murine skin. Biomed Opt Express 10: 5431–5444

Li LZ, Sun N (2014) Autofluorescence perspective of cancer diagnostics. In: Natural biomarkers for cellular

metabolism: biology, techniques, and applications. CRC Press, pp 273–297
Li LZ, Xu HN, Ranji M et al (2009a) Mitochondrial redox imaging for cancer diagnostic and therapeutic studies. J Innov Opt Heal Sci 2:325–341
Li LZ, Zhou R, Xu HN et al (2009b) Quantitative magnetic resonance and optical imaging biomarkers of melanoma metastatic potential. Proc Natl Acad Sci USA 106:6608–6613
Li LZ, Leydesdorff L, Sun N et al (2014) Citation analysis of the scientific publications of Britton Chance in ISI citation indexes. J Innov Opt Health Sci 07:1430003
Lin Z, Wen Y, Xu HN et al (in press) Optical redox imaging of breast cancer NADH redox status associated with PGC1α gene expression. Academic Radiology
Mobilian C, Craft CB (2022) Wetland soils: physical and chemical properties and biogeochemical processes. In: Encyclopedia of inland waters, 2nd edn. Elsevier, pp 157–168
Ostrander JH, McMahon CM, Lem S et al (2010) Optical redox ratio differentiates breast cancer cell lines based on estrogen receptor status. Cancer Res 70:4759–4766
Peng A, Xu HN, Moon L et al (2024) Quantitative optical redox imaging of melanoma xenografts with different metastatic potentials. Cancers (Basel) 16:1669
Podsednik A, Jacob A, Li LZ et al (2020) Relationship between optical redox status and reactive oxygen species in cancer cells. React Oxyg Species 9:95–108
Podsednik A, Xu HN, Li LZ (2024) Passage dependence of NADH redox status and reactive oxygen species level in vitro in triple-negative breast cancer cell lines with different invasiveness. Transl Breast Cancer Res 5:27
Podsednik A, Jiang J, Jacob A et al (2021) Optical redox imaging of treatment responses to Nampt inhibition and combination therapy in triple-negative breast cancer cells. Int J Mol Sci 22:5563
Pouli D, Thieu H-T, Genega EM et al (2020) Label-free, high-resolution optical metabolic imaging of human cervical precancers reveals potential for intraepithelial neoplasia diagnosis. Cell Rep Med 1:100017
Quinn KP, Sridharan GV, Hayden RS et al (2013) Quantitative metabolic imaging using endogenous fluorescence to detect stem cell differentiation. Sci Rep 3: 3432
Quistorff B, Haselgrove JC, Chance B (1985) High resolution readout of 3-D metabolic organ structure: an automated, low-temperature redox ratio-scanning instrument. Anal Biochem 148:389–400
Sánchez-Hernández A, Polleys CM, Georgakoudi I (2023) Formalin fixation and paraffin embedding interfere with the preservation of optical metabolic assessments based on endogenous NAD(P)H and FAD two-photon excited fluorescence. Biomed Opt Express 14:5238–5253
Sharick JT, Walsh CM, Sprackling CM et al (2020) Metabolic heterogeneity in patient tumor-derived organoids by primary site and drug treatment. Front Oncol 10:553
Varone A, Xylas J, Quinn KP et al (2014) Endogenous two-photon fluorescence imaging elucidates metabolic changes related to enhanced glycolysis and glutamine consumption in precancerous epithelial tissues. Cancer Res 74:3067–3075
Walsh A, Cook RS, Rexer B et al (2012) Optical imaging of metabolism in HER2 overexpressing breast cancer cells. Biomed Opt Express 3:75–85
Walsh AJ, Cook RS, Manning HC et al (2013) Optical metabolic imaging identifies glycolytic levels, subtypes, and early-treatment response in breast cancer. Cancer Res 73:6164–6174
Walsh AJ, Cook RS, Sanders ME et al (2014) Quantitative optical imaging of primary tumor organoid metabolism predicts drug response in breast cancer. Cancer Res 74: 5184–5194
Wen Y, Xu HN, Privette Vinnedge L et al (2019) Optical redox imaging detects the effects of DEK oncogene knockdown on the redox state of MDA-MB-231 breast cancer cells. Mol Imaging Biol 21:410–416
Xu HN, Li LZ (2014) Quantitative redox imaging biomarkers for studying tissue metabolic state and its heterogeneity. J Innov Opt Health Sci 7:1430002
Xu MG, Williams ED, Thompson EW et al (2000) Effect of handling and fixation processes on fluorescence spectroscopy of mouse skeletal muscles under two-photon excitation. Appl Opt 39:6312–6317
Xu HN, Zhou R, Nioka S et al (2009) Histological basis of MR/optical imaging of human melanoma mouse xenografts spanning a range of metastatic potentials. Adv Exp Med Biol 645:247–253
Xu HN, Nioka S, Glickson JD et al (2010) Quantitative mitochondrial redox imaging of breast cancer metastatic potential. J Biomed Opt 15:036010
Xu HN, Feng M, Moon L et al (2013a) Redox imaging of the p53-dependent mitochondrial redox state in colon cancer ex vivo. J Innov Opt Health Sci 6:1350016
Xu HN, Nioka S, Li LZ (2013b) Imaging heterogeneity in the mitochondrial redox state of premalignant pancreas in the pancreas-specific PTEN-null transgenic mouse model. Biomark Res 1:6
Xu HN, Tchou J, Li LZ (2013c) Redox imaging of human breast cancer core biopsies: a preliminary investigation. Acad Radiol 20:764–768
Xu HN, Zhou R, Moon L et al (2014) 3D imaging of the mitochondrial redox state of rat hearts under normal and fasting conditions. J Innov Opt Health Sci 7: 1350045
Xu HN, Zheng G, Nioka S, Chance B, Li LZ (2012) High-resolution simultaneous mapping of mitochondrial redox state and glucose uptake in human breast tumor xenografts. Adv Exp Med Biol 737:175–179

Xu HN, Zheng G, Tchou J, Nioka S, Li LZ (2013) Characterizing the metabolic heterogeneity in human breast cancer xenografts by 3D high resolution fluorescence imaging. SpringerPlus 2:73

Xu HN, Tchou J, Feng M et al (2016) Optical redox imaging indices discriminate human breast cancer from normal tissues. J Biomed Opt 21:114003

Xu HN, Zhao H, Chellappa K et al (2019) Optical redox imaging of fixed unstained muscle slides reveals useful biological information. Mol Imaging Biol 21:417–425

Yu Q, Heikal AA (2009) Two-photon autofluorescence dynamics imaging reveals sensitivity of intracellular NADH concentration and conformation to cell physiology at the single-cell level. J Photochem Photobiol B Biol 95:46–57

Mind the Individual: Physiological Reactions to Long-Term Colored Light Exposure Assessed by SPA-fNIRS (Knisely Award Honorary Lecture)

2

Hamoon Zohdi, Felix Scholkmann, and Ursula Wolf

Abstract

Colored light exposure (CLE) is a common occurrence in everyday life. In today's modern society, we are increasingly exposed to various sources of colored light, including advertisements, computer and smartphone displays, energy-saving light bulbs, private settings, and street lighting with LEDs. Therefore, the effects of colored light on human physiology are becoming a subject of growing interest to both the research community and society. To gain a deeper understanding of the complex effects of long-term CLE on the human body, we have conducted a series of pioneering studies (with results published in 16 papers) over the past 8 years using the systemic physiology augmented functional near-infrared spectroscopy (SPA-fNIRS) approach. This approach facilitates concurrent brain activity measurement using fNIRS alongside multiple systemic physiological parameters. Our studies yielded a number of new findings: (i) We discovered the need for both group-level and individual-level analyses to fully understand the effects of CLE on human physiology. (ii) We found that CLE has relatively long-lasting effects on cerebrovascular and systemic physiology in humans. (iii) Yellow and short-wavelength light (i.e., blue and violet) were found to induce higher brain activation in the prefrontal cortex and visual cortex, respectively, compared to other colored lights (e.g., red, orange, and green). (iv) We observed that healthy adults exhibited a range of different types of physiological responses. Machine learning enabled us to classify the response patterns and provide new insights into the common response patterns between subjects. Given the rapid exposure of our society to new and increasing lighting levels, this research's findings are of interest and relevance to the scientific community, medical professionals, and society.

Keywords

Colored light exposure · SPA-fNIRS · Intersubject variability · Individual differences · Yellow, blue and violet light ·

H. Zohdi (✉) · U. Wolf
Institute of Complementary and Integrative Medicine, University of Bern, Bern, Switzerland
e-mail: hamoon.zohdi@unibe.ch

F. Scholkmann
Institute of Complementary and Integrative Medicine, University of Bern, Bern, Switzerland

Neurophotonics and Biosignal Processing Research Group, Biomedical Optics Research Laboratory, Department of Neonatology, University Hospital Zurich, University of Zurich, Zurich, Switzerland

Neuroscience Center Zurich, University of Zurich and ETH Zurich, Zurich, Switzerland

S. C. Pias et al. (eds.), *Oxygen Transport to Tissue XLVI*, Advances in Experimental Medicine and Biology 1498, https://doi.org/10.1007/978-3-032-10389-5_2

Machine learning · Cerebrovascular and systemic physiology

2.1 Introduction

Color and colored light have always attracted human attention and played vital roles in daily human life. How colored light is formed and how human color vision works have been widely discussed in various fields, from physics to medicine and from art to psychology. Colored light exposure (CLE) is an experience of everyday life. It is used in public areas to attract customers' attention or within private settings as functional or decorative lighting. Specifically, in our modern society, we are increasingly exposed to different types of colored light, originating from advertisements to computers and from energy-saving light bulbs to smartphones. Therefore, the effects of colored light on human physiology are of rising interest to both the research community and society. Colored light can affect what we see, our emotions and moods, and even our performance. However, our understanding of the psychological and physiological effects of exposure to colored light is incomplete, and many research questions remain.

To elucidate the intricate effects of CLE in humans, we have conducted several leading studies (with results published in 16 papers) over the past 8 years using the systemic physiology augmented functional near-infrared spectroscopy (SPA-fNIRS) approach (Scholkmann et al. 2017, 2022b, 2018, 2020; Metz et al. 2017a, b; Zohdi et al. 2021a, b, c, 2022a, b, 2023, 2024, 2018, 2020; Nasseri et al. 2018). In this chapter, we aim to summarize the key findings of our previous studies investigating the effects of CLE on human physiology using the SPA-fNIRS approach.

2.2 Subjects and Methods

A total of 220 healthy right-handed adults were recruited to participate in our studies, which involved exposure to 22 different colored light conditions. Using SPA-fNIRS neuroimaging, each subject was measured 2–4 times on different days, resulting in 708 single measurements. The SPA-fNIRS approach (Scholkmann et al. 2022a) allows simultaneous measurement of brain activity using fNIRS (frequency domain NIRS; Imagent, ISS Inc., Champaign, IL, USA) alongside multiple systemic physiological parameters. We simultaneously assessed the effects of CLE in the visual cortex (VC), prefrontal cortex (PFC), and systemic physiology (e.g., end-tidal partial pressure of carbon dioxide, heart rate, respiration rate, mean arterial pressure, arterial oxygen saturation, electrodermal activity, and heart rate variability). Subjects were continuously exposed to CLE (e.g., blue, red, green, and yellow) for 15 min in a randomized crossover protocol. Before (baseline, 8 min) and after (recovery, 20 min), the CLE subjects were in darkness. In some studies, we asked our subjects to perform a cognitive task (e.g., verbal fluency task (VFT) and 2-back task) while they were exposed to the CLE for 10 min.

2.3 Results

Our studies have so far produced the following findings:

(i) We discovered the need for both group-level and individual-level analyses for a deeper understanding of the effects of CLE on human physiology (Zohdi et al. 2021b, c; Scholkmann et al. 2022b). For example, we showed that despite the typical hemodynamic response pattern generally observed at the group level (i.e., an increase in oxygenated hemoglobin [O_2Hb] and a decrease in deoxygenated hemoglobin [HHb]), the subgroup analysis revealed that this pattern was found in only 50% of the cases, and the number of patterns was different between red and blue light exposure (Zohdi et al. 2021c). The group-level analysis, although commonly used, only displays the most prominent tendency between subjects. Therefore, the individual-level or subgroup analysis is needed to understand the physiological effects of the CLE completely.

(ii) We found that CLE has relatively long-lasting effects on human cerebrovascular and systemic physiology (Zohdi et al. 2021a). Color-dependent changes were evident in the recovery phase (duration: 20 min) of several systemic physiological parameters (Zohdi et al. 2024, 2021a).

(iii) Yellow light led to greater changes in [O_2Hb] in the PFC compared to the other colored light (Fig. 2.1: during the CLE; Fig. 2.2: in the recovery phase). This effect was also found for other cerebral and oxygenation parameters. In the VC, we found that, in general, the shorter the wavelength, the higher the brain activation, i.e., violet and blue evoked higher changes in [O_2Hb] and [HHb] during the CLE and recovery phase (data not shown).

(iv) Our research underscores the inherent variability between individual responses to CLE (inter-subject variability), which is partially explained by subject-specific systemic physiological changes (Zohdi et al. 2022a, 2018, 2021c).

(v) Changes in arterial oxygen saturation and mean arterial pressure were positively correlated with changes in [O_2Hb] in the PFC during the CLE-VFT (Zohdi et al. 2021c).

(vi) In one study (Zohdi et al. 2024), the subjects were exposed to the sequence of blue and red light or vice versa in a randomized crossover design. CLE sequence-dependent effects were observed for fNIRS signals at the VC and for electrodermal activity (Zohdi et al. 2024).

(vii) Individual-level analysis revealed different responses among healthy adults (Zohdi et al. 2021b, c, 2022a; Scholkmann et al. 2022b). In particular, CLE with blue light resulted in cerebrovascular hemodynamic response patterns that could be classified into eight types (Zohdi et al. 2021b). Machine learning (ML) enabled us to classify the response patterns and provide new insights into the common response patterns between subjects (Zohdi et al. 2022a). We found that unsupervised ML correctly classified subjects into different groups. Using *k*-means clustering, three and five distinct groups of subjects were found for the blue and red light exposure, respectively (Zohdi et al. 2022a).

(viii) The color-specific changes at the PFC were found to be dependent on the amount of changes in mean arterial blood pressure (MAP) (less MAP change, more PFC activation during blue CLE; more MAP change, more PFC activation during red CLE). Furthermore, the color-specific changes at the VC were found to be dependent on the baseline MAP of the individual (lower baseline MAP: stronger activation during red CLE compared to blue CLE) (Scholkmann et al. 2022b).

2.4 Discussion and Conclusions

In our studies conducted over the past 8 years, the effects of colored light on human physiology were investigated with the SPA-fNIRS approach. Our group pioneered this approach, which enables us to measure changes in brain activity (with fNIRS) along with changes in systemic physiology (Scholkmann et al. 2017; Metz et al. 2017a; Scholkmann et al. 2022a, 2013, 2014; Tachtsidis and Scholkmann 2016; Caldwell et al. 2016; Kirilina et al. 2012). The SPA-fNIRS approach enables the identification, removal, and understanding of the effects of systemic physiological changes on fNIRS signals. This approach enhances the accuracy, reliability, and applicability of brain activity studies, making it a valuable tool for researchers in various fields of neuroscience and clinical research.

This chapter summarizes the key findings of our previous studies using the SPA-fNIRS approach to investigate the effects of CLE on

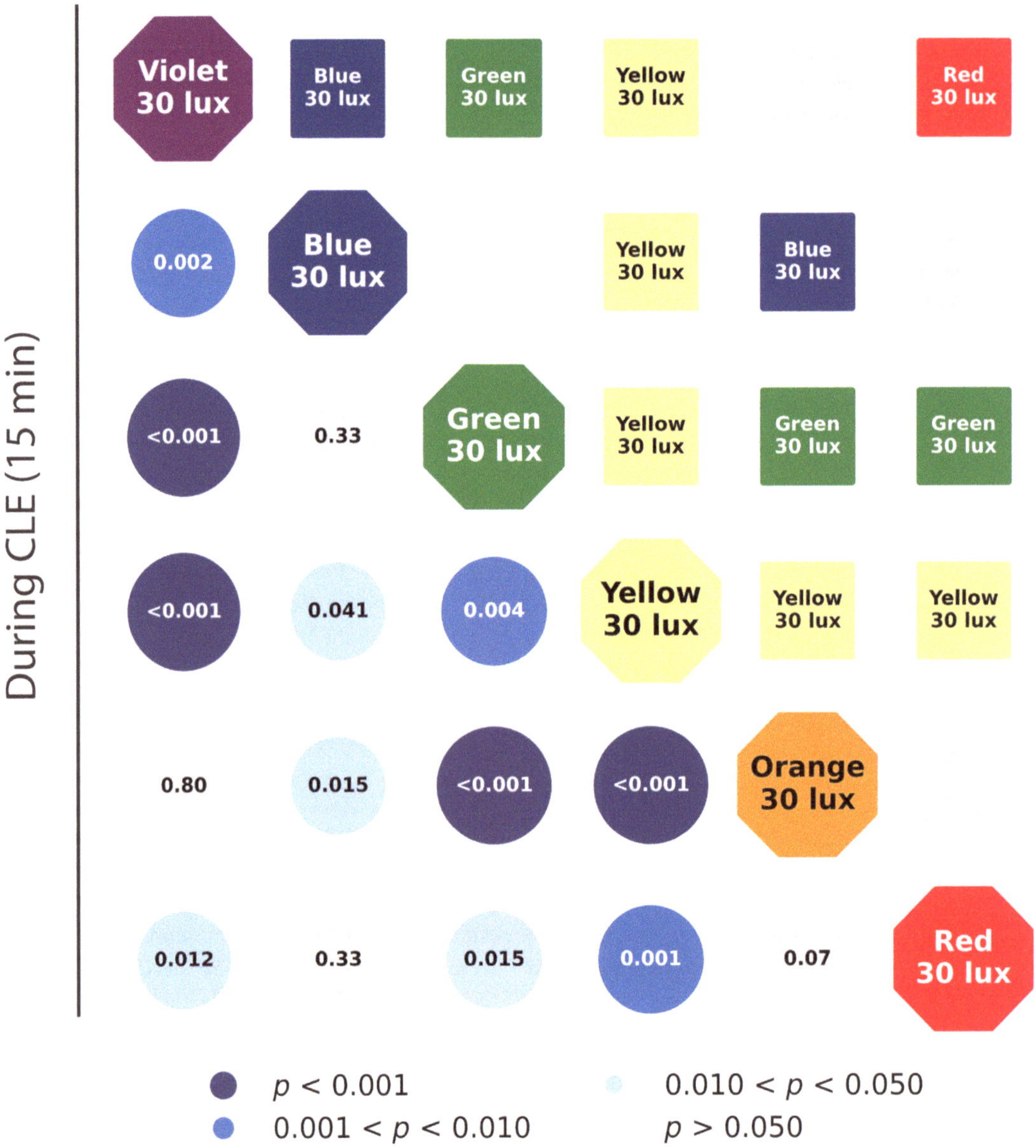

Fig. 2.1 Significance matrix of [O_2Hb] changes in the PFC for CLE with six different colored lights at 30 lux in the CLE phase. Colored light conditions are shown on the diagonal of the matrix. The lower triangular part of the matrix displays p values (Wilcoxon signed-rank test). The upper triangular part of the matrix shows which of the two conditions leads to a higher activation

human physiology. It is essential to conduct both group- and individual-level analyses to understand the impact of long-term CLE on humans (Zohdi et al. 2021a, c). Although group-level analyses offer valuable insights, single-subject analyses reveal crucial nuances in brain structure and function. These approaches allow for the detection of unique activation patterns and functional connectivity that may be obscured when data are averaged across groups (Bolt et al. 2019; Mennes et al. 2010). Furthermore, evidence indicates that both situational factors and

Fig. 2.2 Significance matrix of [O_2Hb] changes in the PFC for CLE with six different colored light at 30 lux in the recovery phase. Colored light conditions are shown on the diagonal of the matrix. The lower triangular part of the matrix displays *p* values (Wilcoxon signed-rank test). The upper triangular part of the matrix shows which of the two conditions leads to a higher activation

individual traits contribute to variations in brain activation during cognitive tasks (Zohdi et al. 2020, 2021c; Kanai and Rees 2011). Moreover, the research studies highlighted the potential of personalized neuroimaging techniques in identifying biomarkers for a range of neurological and psychiatric disorders, as well as in tailoring treatment strategies to individual patients (Woo et al. 2017; Calhoun and Sui 2016). This shift toward personalized neuroimaging has the potential to enhance our understanding of brain function, while also offering the prospect of improving diagnostic accuracy and treatment efficacy in clinical settings (Brammer 2009). Additionally, although a group-level analysis is commonly used in research, it does not adequately account for the inter-individual variation in responses, which is relevant to understanding the impact of CLE on human physiology. Therefore, for a reliable interpretation of the effects of long-term CLE in humans, the use of individual-level (subject-specific) analysis is highly recommended (Zohdi et al. 2021c). The individual-level analysis of our studies revealed that, in most cases, subjects have different reactions to the CLE (Zohdi et al. 2021b). Interestingly, the individual characteristics of the responses to CLE were more noticeable in systemic physiological parameters compared to cerebrovascular variables. Specifically, it was not easy to draw general conclusions about the effects of the CLE of different colors on systemic physiological changes during the CLE at the group level. In contrast, the color-dependent changes were more evident in the recovery phase of the most systemic physiological parameters, confirming the CLE's long-lasting effects on human physiology (Zohdi et al. 2021a). Such long-lasting effects, as well as color-dependent differences during the CLE, are also observed in cerebrovascular variables of the VC and the PFC. Regardless of the impact of a cognitive task, blue light exposure leads to stronger effects in the VC compared to colors associated with longer wavelengths. This effect was proven in both studies with and without a VFT (Zohdi et al. 2021a, b). The finding of this study that blue light has an activating effect on the VC should be taken into consideration when assessing the impact of modern light sources such as screens (e.g., smartphones and computers) and LEDs on the human body. Moreover, we have shown that CLE (namely, yellow) affects the PFC. It is also known that the hemodynamic response in the PFC represents the emotional status and mood (Metz et al. 2017a). Therefore, colored light affects the emotional state. Our findings that CLE leads to PFC activation, i.e., yellow light causes a higher PFC activation, may be applied in complementary medicine, i.e., specifically chromotherapy (color therapy), to restore balance and health on an emotional, mental, or even physical level. Yellow light can be applied as a healing tool to boost positive feelings like hope, courage, happiness, and optimism (Metz et al. 2017a; Al-Ayash et al. 2016). Finally, an analysis conducted at the individual level uncovered varying responses among healthy adults. By applying ML techniques, we were able to categorize these different response patterns, leading to new insights into the commonalities and differences in how subjects responded (Zohdi et al. 2022a). This approach not only highlighted the unique characteristics of each individual's response but also revealed common patterns that might otherwise have gone unnoticed.

We conclude that in a civilization that is increasingly exposed to new and more intense lighting, the findings of this research have significant relevance for the scientific community, medical professionals, and society. Future studies should aim to determine which colors in CLE provide the greatest benefits to specific groups. Utilizing advanced ML techniques, along with distinct and informative features from our recorded SPA-fNIRS signals, will be essential in addressing this question. Moreover, computational physiological modelling (Caldwell et al. 2015; Russell-Buckland et al. 2019) of our SPA-fNIRS data has the potential to yield further insights into the human physiological reactivity to long-term CLE. This research topic will be the subject of further investigation in the near future.

References

Al-Ayash A, Kane RT, Smith D, Green-Armytage P (2016) The influence of color on student emotion, heart rate, and performance in learning environments. Color Res Appl 41:196–205

Bolt T, Nomi JS, Bainter SA et al (2019) The situation or the person? Individual and task-evoked differences in BOLD activity. Hum Brain Mapp 40:2943–2954

Brammer M (2009) The role of neuroimaging in diagnosis and personalized medicine-current position and likely future directions. Dialogues Clin Neurosci 11:389–396

Caldwell M, Hapuarachchi T, Highton D et al (2015) BrainSignals revisited: simplifying a computational model of cerebral physiology. PLoS One 10:e0126695

Caldwell M, Scholkmann F, Wolf U et al (2016) Modelling confounding effects from extracerebral contamination and systemic factors on functional near-infrared spectroscopy. NeuroImage 143:91–105

Calhoun VD, Sui J (2016) Multimodal fusion of brain imaging data: a key to finding the missing link(s) in complex mental illness. Biol Psychiatry Cogn Neurosci Neuroimaging 1:230–244

Kanai R, Rees G (2011) The structural basis of inter-individual differences in human behaviour and cognition. Nat Rev Neurosci 12:231–242

Kirilina E, Jelzow A, Heine A et al (2012) The physiological origin of task-evoked systemic artefacts in functional near infrared spectroscopy. NeuroImage 61:70–81

Mennes M, Kelly C, Zuo XN et al (2010) Inter-individual differences in resting-state functional connectivity predict task-induced BOLD activity. NeuroImage 50:1690–1701

Metz AJ, Klein SD, Scholkmann F, Wolf U (2017a) Continuous coloured light altered human brain haemodynamics and oxygenation assessed by systemic physiology augmented functional near-infrared spectroscopy. Sci Rep 7:10027

Metz AJ, Klein SD, Scholkmann F, Wolf U (2017b) Physiological effects of continuous colored light exposure on Mayer wave activity in cerebral hemodynamics: a functional near-infrared spectroscopy (fNIRS) study. Adv Exp Med Biol 977:277–283

Nasseri N, Caicedo A, Scholkmann F et al (2018) Impact of changes in systemic physiology on fNIRS/NIRS signals: analysis based on oblique subspace projections decomposition. Adv Exp Med Biol 1072:119–125

Russell-Buckland J, Barnes CP, Tachtsidis I (2019) A Bayesian framework for the analysis of systems biology models of the brain. PLoS Comput Biol 15:e1006631

Scholkmann F, Gerber U, Wolf M, Wolf U (2013) End-tidal CO2: an important parameter for a correct interpretation in functional brain studies using speech tasks. NeuroImage 66:71–79

Scholkmann F, Klein SD, Gerber U et al (2014) Cerebral hemodynamic and oxygenation changes induced by inner and heard speech: a study combining functional near-infrared spectroscopy and capnography. J Biomed Opt 19:017002

Scholkmann F, Hafner T, Metz AJ et al (2017) Effect of short-term colored-light exposure on cerebral hemodynamics and oxygenation, and systemic physiological activity. Neurophotonics 4:045005

Scholkmann F, Zohdi H, Nasseri N, Wolf U (2018) Absolute values of optical properties (μa , μ′s , μeff and DPF) of human head tissue: dependence on head region and individual. Adv Exp Med Biol 1072:325–330

Scholkmann F, Zohdi H, Wolf U (2020) Right-left asymmetry of prefrontal cerebral oxygenation: does it depend on systemic physiological activity, absolute tissue oxygenation or hemoglobin concentration? Adv Exp Med Biol 1232:105–112

Scholkmann F, Tachtsidis I, Wolf M, Wolf U (2022a) Systemic physiology augmented functional near-infrared spectroscopy: a powerful approach to study the embodied human brain. Neurophotonics 9:030801

Scholkmann F, Zohdi H, Wolf U (2022b) The role of systemic physiology in individual hemodynamic responses measured on the head due to long-term stimulation involving colored light exposure and a cognitive task: an SPA-fNIRS study. Brain Sci 12:597

Tachtsidis I, Scholkmann F (2016) False positives and false negatives in functional near-infrared spectroscopy: issues, challenges, and the way forward. Neurophotonics 3:031405

Woo CW, Chang LJ, Lindquist MA, Wager TD (2017) Building better biomarkers: brain models in translational neuroimaging. Nat Neurosci 20:365–377

Zohdi H, Scholkmann F, Nasseri N, Wolf U (2018) Long-term changes in optical properties (μa , μ′s , μeff and DPF) of human head tissue during functional neuroimaging experiments. Adv Exp Med Biol 1072:331–337

Zohdi H, Scholkmann F, Wolf U (2020) Frontal cerebral oxygenation asymmetry: Intersubject variability and dependence on systemic physiology, season, and time of day. Neurophotonics 7:025006

Zohdi H, Egli R, Guthruf D et al (2021a) Color-dependent changes in humans during a verbal fluency task under colored light exposure assessed by SPA-fNIRS. Sci Rep 11:9654

Zohdi H, Scholkmann F, Wolf U (2021b) Long-term blue light exposure changes frontal and occipital cerebral hemodynamics: not all subjects react the same. Adv Exp Med Biol 1269:217–222

Zohdi H, Scholkmann F, Wolf U (2021c) Individual differences in hemodynamic responses measured on the head due to a long-term stimulation involving colored light exposure and a cognitive task: a SPA-fNIRS study. Brain Sci 11:54

Zohdi H, Natale L, Scholkmann F, Wolf U (2022a) Intersubject variability in cerebrovascular hemodynamics and systemic physiology during a verbal fluency task under colored light exposure : clustering of subjects by unsupervised machine learning. Brain Sci 12:1449

Zohdi H, Scholkmann F, Wolf U (2022b) Changes in cerebral oxygenation and systemic physiology during a verbal fluency task: differences between men and women. Adv Exp Med Biol 1395:17–22

Zohdi H, Amez-droz V, Scholkmann F, Wolf U (2023) Differences between good, moderate and poor performers of a verbal fluency task under blue light exposure: an SPA-fNIRS study. Adv Exp Med Biol 1438:69–74

Zohdi H, Märki J, Scholkmann F, Wolf U (2024) Cerebral, systemic physiological and behavioral responses to colored light exposure during a cognitive task: a SPA-fNIRS study. Behav Brain Res 462:114884

Part II

Measuring Tissue Oxygen In Vivo

3 A Novel Approach to In Vivo EPR Spectroscopy for Repeatable Assessments of Oxygenation Levels in Tumors at Any Depth: Preliminary Feasibility Studies Utilizing a Multisite Oxygen Sensor Inside HDR Brachytherapy Needles

Daniela Olivera Velarde, Erik Pearson, Yasmin Hasan, Boris Epel, Subramanian Sundramoorthy, Wilson Schreiber, Bulent Aydogan, Ann Barry Flood, Howard Halpern, and Harold M. Swartz

Abstract

There is widespread consensus that hypoxia limits the effectiveness of cancer therapy. This has led to interventions to increase oxygen (O_2) levels in tumors in patients, but success in clinical trials has been very limited and therefore clinical practice has not incorporated such interventions. The limiting step for successful intervention is the need to identify which tumors are hypoxic, whether they respond to interventions to increase O_2, and the timing of the response. Consequently, many techniques have been advanced to measure O_2 in tumors, but to date, none has been able to measure O_2 directly in the tumor repeatedly under clinically applicable conditions (i.e., without perturbing clinical flow). Initial efforts at Dartmouth demonstrated that in vivo electron paramagnetic resonance (EPR) spectroscopy, using three types of injected or implanted O_2 sensors, could provide the desired data under the desired conditions. Two types, injected paramagnetic India ink and an implanted coated derivative of lithium phthalocyanine, were successfully tested in clinical studies. However, their use is limited to tumors <1 cm of the surface. Consequently, Dartmouth developed a third O_2 sensor, an

Senior authors contributed equally to the new MOS approach. DOV with EP performed the preliminary studies reported here under the direction of BE, SS, and WS and clinical consultation with YH.

D. O. Velarde · E. Pearson · Y. Hasan · B. Epel · S. Sundramoorthy · B. Aydogan · H. Halpern
Department of Radiation and Cellular Oncology, University of Chicago, Chicago, IL, USA

W. Schreiber
Clin-EPR, LLC, Lyme, NH, USA

Thayer School of Engineering, Dartmouth College, Hanover, NH, USA

A. B. Flood
Clin-EPR, LLC, Lyme, NH, USA

Geisel School of Medicine, Dartmouth College, Hanover, NH, USA

H. M. Swartz (✉)
Clin-EPR, LLC, Lyme, NH, USA

Thayer School of Engineering, Dartmouth College, Hanover, NH, USA

Geisel School of Medicine, Dartmouth College, Hanover, NH, USA
e-mail: harold.swartz@dartmouth.edu

S. C. Pias et al. (eds.), *Oxygen Transport to Tissue XLVI*, Advances in Experimental Medicine and Biology 1498, https://doi.org/10.1007/978-3-032-10389-5_3

"implantable resonator" (IR), to allow measuring in tumors at any depth; the IR has been successfully tested in preclinical studies. However, because the IR requires implanting at greater depth than the other types, its invasiveness was considered to be a drawback for clinical studies. Therefore, Clin-EPR and colleagues at the University of Chicago made additional technical improvements to the IR and proposed a new approach, called the multisite oxygen sensor (MOS), that allows its use in clinical studies without adding any invasiveness to therapy the patient is already undergoing. Specifically, the MOS is being designed to use in conjunction with a frequently used therapeutic approach (HDR brachytherapy delivered with an afterloader), applied initially to cervical cancer. HDR brachytherapy requires the invasive placement of multiple needles into the tumor and leaves them in situ for days during the course of treatment. Using these same needles, modified to be permeable to O_2, would allow the MOS, inserted inside each needle, to simultaneously measure O_2 at multiple locations throughout the tumor. This O_2 measurement session could be repeated periodically during the course of therapy. We report preliminary technical studies of the modified MOS and the proposed modified brachytherapy needles, demonstrating in vitro the feasibility of our new approach to provide important information about tumor hypoxia during the course of radiation therapy without needing any additional invasiveness beyond standard of care therapy.

Keywords

Electron paramagnetic resonance · In vivo oxygen measurements · Hypoxia in tumors

3.1 Introduction

Tumor hypoxia is a major adverse factor in cancer and is associated with poor outcomes regardless of treatment modality (Vaupel and Mayer 2007; Lehtiö et al. 2004; Sundfør et al. 1999, 2000; Kaanders et al. 2002; Gilbert et al. 2014). The efficacy of radiation therapy has been shown to be significantly more impacted by the level of oxygen (O_2) in the tumor than by stage, tumor morphology, or tumor size (Vaupel and Mayer 2007; Sundfør et al. 2000). Consequently, the longstanding consensus is that if low O_2 levels in tumors could be increased at the time of irradiation, outcomes could be significantly improved. To date, however, clinical trials to test the efficacy of O_2-enhancing interventions (such as the Avastin (Gilbert et al. 2014) [bevacizumab] and ARCON (Kaanders et al. 2002) [Accelerated Radiotherapy with Carbogen and Nicotinamide] trials and others (Sundfør et al. 1999, 2000)) have not achieved practice-changing results.

Studies conducted by Swartz's lab (Schaner et al. 2021, 2020; Swartz et al. 2016) illustrate why applying hyperoxygenation to all tumors may have obscured its usefulness for an appropriate subset of tumors. Swartz's lab used in vivo EPR oximetry to measure O_2 levels in patients' tumors in conjunction with a simple O_2-enhancing intervention (breathing 100% O_2). Twenty-six measurements were made in tumors prior to receiving any treatment. Of these, only 54% of tumors (14) were both hypoxic *and* responsive to O_2 treatments and thus could be expected to have better outcomes if hyperoxygenation were administered during radiation; 27% (7) did *not* respond to breathing O_2-enriched gas. Also of interest, 19% (5) had high levels of O_2 (so they were already *not* resistant to radiation and unlikely to benefit from receiving additional oxygenation) (Swartz and Flood 2024).

These data illustrate why such measurements of O_2 are needed to identify those tumors whose response could be enhanced by using O_2-enhancing interventions. By extension, by also identifying non-responsive tumors, they could then be targeted to receive a different approach to improving outcomes, e.g., increasing the dose. Therefore, clinical studies aiming to evaluate clinical benefit of improving tumor O_2 must have a practical and robust means to measure O_2 changes in tumors.

In vivo EPR oximetry, described further below, can uniquely meet these needs to optimize radiotherapy by providing spatially distributed, repeatable, and reliable O_2 measurements at any depth. The improvements being tested here add significantly to the original approach to measuring tumors at any depth by offering a strategy to place the O_2 sensors without needing to implant them and by allowing significantly more sites to be measured simultaneously.

3.1.1 Comparing Current Alternative Ways to Measure Tumor Oxygen

Although there are currently several clinically applied methods that provide some information on O_2 levels in vivo (particularly ^{18}F-misonidazole [FMISO]-PET (Lehtiö et al. 2004; Lewis and Welch 2001) and BOLD MRI (Hoskin et al. 2007)), and others that are under development using optical and other MRI techniques, none can provide direct measures of O_2 in tumors and surrounding tissues, nor do any allow repeated measurements of O_2 at the same site under clinically applicable conditions to follow the time-course of hypoxia (Swartz and Clarkson 1998; Swartz et al. 2020a). PET-misonidazole derivatives can provide qualitative indications of the occurrence of hypoxia at the time they are administered, but only if: (1) there is sufficient flow to all of the hypoxic sites; (2) the sites have the activating enzymes to generate the hydroxylamines; (3) the hydroxylamines react and bind; and (4) the bound material is not washed out. These studies also cannot be repeated as frequently as required for guiding clinical intervention, require very sophisticated and expensive procedures, and require repeated exposure to ionizing radiation. BOLD MRI provides information only on the amount of unsaturated hemoglobin in the *vasculature*, and not directly on the O_2 levels in the *tissues*. Therefore, BOLD and PET (and other clinically available approaches as well) can reflect factors that are sometimes related to O_2 levels in tumors, but by themselves are not robust indications of the level of O_2 or of changes in the O_2 levels. Moreover, the clinical boost doses used in the [FMISO]-PET studies may have used a suboptimal boost dose to fully exploit the in vivo oxygen enhancement ratio measured in preclinical models (personal comm to HH). They also are much more costly and require specialized personnel, while the current in vivo EPR device can be operated by existing personnel and can readily be placed within the clinical treatment site without perturbation of other equipment. Furthermore, to date, BOLD MRI and PET have not been validated to determine the conditions under which they reflect the direct O_2 in the tissue. Such validation requires comparing their results with *direct* O_2 measures (Swartz et al. 2014; Hou et al. 2014) such as those made by an EPR oximetry system.

3.1.2 Capabilities and Limitations of the Current Clinical In Vivo EPR Approaches to Oximetry

The capabilities and feasibility of in vivo EPR oximetry have already been demonstrated in preclinical and clinical studies, using particulate O_2-sensitive EPR materials directly embedded in tissue such as via India inks (Swartz et al. 2016, 2014, 2024a; Dinguizli et al. 2006) and using OxyChips (Schaner et al. 2021, 2020; Chen et al. 2021; Hou et al. 2018, 2016) made of paramagnetic particulates encapsulated in an O_2-permeable polymer, polydimethylsiloxane (PDMS). Several laboratories, including Swartz's EPR, have shown in preclinical studies that EPR oximetry is a very effective method for measuring O_2 in tissues directly, repeatedly, minimally invasively, and most importantly, that EPR oximetry could be used to improve outcomes (Sundfør et al. 1999; Swartz et al. 2016, 2013; Schaner et al. 2020; Chen et al. 2021; Jeong et al. 2019; Höckel et al. 1993; Elas et al. 2008, 2013; Khan et al. 2013). Swartz's lab, using an implantable resonator (IR) with multiple sensing sites, also established that multiple independent O_2 sensors could be measured simultaneously in deep tumors (Hou et al. 2014; Chen et al. 2021; Caston et al. 2017).

However, when this approach was applied to human subjects in an NCI-supported PPG (P01 CA190193) awarded to Swartz's lab, in vivo EPR oximetry could not *fully* fulfill its promise. Two of the three projects, i.e., using India ink and OxyChips as the O_2 sensor for in vivo EPR oximetry, advanced to clinical studies but still had several limitations:

- The technique required an invasive step, i.e., EPR sensors needed to be injected (India ink) or implanted (OxyChip) into the tissue.
- If the tumor was ≥1 cm below the surface, EPR signals could not be reliably detected (Schaner et al. 2021, 2020; Swartz et al. 2016).
- The location of the oximetry sensors was difficult to determine in relationship to the tumor because they could not be imaged, e.g., with CT or ultrasound.
- They were only able to report a single O_2 level, and that level was a complex average of O_2 over the entire volume that the set of carbon particles (ink) or an OxyChip was querying (Swartz et al. 2020a, b, 2021, 2024b; Swartz and Flood 2024).

To address these last three issues, a third approach, using an IR, was also being developed in Swartz's lab (Hou et al. 2014; Chen et al. 2021; Caston et al. 2017). The basic design of the IR is to use several independent O_2 sensors connected to each other by a wire structure, which allows transmission of the signal to a coupling loop resonator placed on the surface (see Fig. 3.1). When used in conjunction with an EPR device with a magnetic field gradient, each sensor could be measured simultaneously, i.e., providing independent assessments of O_2 at each location.

Each sensor consists of lithium phthalocyanine (LiPc) or a derivative (e.g., LiNc BuO) crystals that have been coated with PDMS. Both the PDMS and the wire used were Food and Drug Administration (FDA)-approved as implantable materials. However, the LiPc and its derivatives were not; therefore, whenever they are implanted

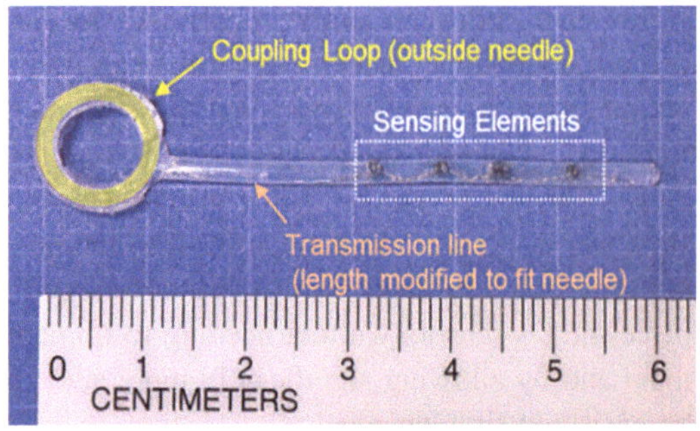

Fig. 3.1 Improved prototype IR. Prototype shows 4 sites (sensing elements); clinical studies will use 8. Modification from IR used in preclinical studies is extension of wire at tail (far right) to improve SNR and use of LiPc in sensing elements (LiPc has a narrower line and can be embedded in PDMS in greater concentration)

as the O_2 sensor for in vivo EPR oximetry in clinical studies, the FDA also needed to approve the research plan for clinical research and approve an Investigational Device Exemption for it.

After consulting with the FDA, the IR was tested in two preclinical studies (Chen et al. 2021; Caston et al. 2017). These studies established the feasibility of using the IR in vivo in tumors to obtain simultaneous, independent measurements of O_2 in different sensors and to repeat them on subsequent days.

Since these studies, researchers at Clin-EPR and the University of Chicago (UChicago) have made two technical improvements to the IR, also portrayed in Fig. 3.1: the transmission wire was extended beyond the last sensor, which resulted in an improved signal-to-noise ratio (SNR), and LiPc was used as the O_2 sensor. LiPc has a narrower line, resulting in improved SNR, and can be used for wider oxygen ranges due to a weaker oxygen concentration dependence. Finally, LiPc crystals can be embedded into the PDMS in greater concentration compared to the derivative previously used, resulting in further SNR improvement.

Nevertheless, the fourth limitation, i.e., the invasiveness of implanting the IR into a deep tumor, was an apparent stumbling block in getting funding and approval for conducting clinical studies. Hence, we embarked on a new approach where EPR oximetry does not *add* any invasive step to the standard therapy being used to treat deep tumors and does not require being implanted in the tissue. Because the O_2 sensor is not implanted in tissue, the process for FDA approval for marketing is somewhat simplified by this approach.

This new sensor, to be used in HDR brachytherapy, is renamed the "multisite oxygen sensor" (MOS). This new approach to in vivo EPR oximetry requires three additional technical developments: modification of the current multisite resonator to fit within the dimensions of the usual brachytherapy needles, modification of the needles to facilitate rapid equilibrium of O_2 in the tissues within the interior of the needles, and a device to load the MOS into the needles precisely to make it an adjunct to a standard type of brachytherapy treatment. These technological improvements and plans and preliminary proofs-of-concept studies are described in the following.

3.1.3 New Non-invasive Approach and Preliminary Results

The new approach is designed to be used in patients over a several-day course of receiving high-dose-rate (HDR) brachytherapy via an interstitial or hybrid interstitial approach (Pötter et al. 2021; Berger et al. 2023; Zeng et al. 2021). HDR brachytherapy is commonly the treatment of choice for gynecologic cancers when the tumor can be anatomically accessible for irradiation with short-range radiation sources by moving them through hollow needles (typically 2–25) that are implanted in the tumor and remain in situ throughout the several-day course of HDR treatment. The therapy uses an "afterloading" system to insert a radioactive source into each needle and, with programmable software, stop the source at several dwelling positions along each needle. Patients receive 2–5 fractions during each hospital admission with a minimum 6-h gap between fractions. The length of time and number of dwelling points are based on a treatment plan that can be changed daily, based on imaging, in order to maximize treatment effectiveness by treating some sites within the tumor with higher doses (Pötter et al. 2021; Berger et al. 2023; Zeng et al. 2021). Currently, the treatment plans are based on imaging of the tumor; we propose to add information about the O_2 in the tumor at multiple points encompassing the dwelling points.

Basics of new approach We propose to modify the needles and the afterloading system for HDR delivery and modify the MOS so that it has 4 to 8[1] independent sensors connected in a linear transmission wire that can be inserted and removed from a needle. The needles will be modified so that they are O_2 permeable but will otherwise be the same length and dimensions of the current HDR needles.

An afterloader and its software will be adapted to be specific for oximetry use, so that it will insert and remove the MOS into the needles. (This does not replace the afterloader used for delivering the HDR brachytherapy.) The 4–8 sensors will provide 4–8 independent measurements of O_2, encompassing the whole region between the dwelling points used for therapy. Since the sensors can be measured simultaneously using gradients, a single positioning of the MOS within each needle will be sufficient for oximetry.

Oximetry measurements will occur at times when the patient is in the therapy suite waiting for daily treatment plans to be finalized (usual wait time is 2–4 h), i.e., while waiting after initial daily imaging has taken place but before daily radiation doses have been programmed. In this way, adding EPR oximetry to HDR brachytherapy can be done without disturbing the clinical workflow, i.e., while the radiation oncologist and medical physicist work on needle segmentation, multi-modality image registration, and treatment planning.

During a daily O_2 measurement session, the MOS can then be withdrawn and re-inserted into as many needles as desired. More specifically, our

[1] The preclinical studies used IR with 2–4 sensors because of the smaller size of the rats and rabbits used in these studies.

planned procedure is to acquire data for 2 min in a given needle. Including the time required to place the resonator (estimated at 2–3 min), measurements in each needle will take about 5 min total and produce 8 independent assessments covering the entire region irradiated by the sources moved through the needle.

It is desirable to make the total daily measurement time as short as possible even though the wait time is several hours. Each session in total is anticipated to be no longer than 30–90 min per day. Note that, if the typical number of needles used is 15 and each needle requires 5 min total, it would require 75 min to measure all 15 needles.

If there is a need to minimize the measuring time, e.g., to 60 min, there are two very feasible approaches. One approach is to measure most but not all needles (e.g., measure 12 needles daily, in the example to limit the session to 60 min), choosing them based on their location within the tumor or on previous measurements of O_2. The second approach would be to use two EPR oximetry systems, so that we could measure two needles simultaneously. The latter would allow up to 24 needles to be measured in 60 min. The choice will depend on the clinical usefulness of the O_2 mapping of the 96 points from the first option (12 × 8) compared to almost 200 points, depending on the actual number of needles used, in the second option.

Measurement sessions could then be repeated daily. Measurements of O_2 will be made ~1–4 h before each treatment, so the initial measurement session will be prior to treatment (i.e., baseline) and subsequent measurements will be made ~20 h after the prior treatment. The results will be immediately available to the treatment planning team.[2]

Modifying the MOS: Preliminary Studies The MOS will be based on modifications of the multisite IR used in preclinical studies in Swartz' lab by Caston et al. (2017) and Chen et al. (2021). Figures 3.1 and 3.2 show three technical improvements made to this prototype by Clin-EPR and UChicago. Figure 3.1 shows the extension of the transmission line beyond the last O_2 sensor. Though not visible, we also changed the sensor to LiPc. Figure 3.2 shows a microcoil, which offers better sensitivity.

Figures 3.3, 3.4 and 3.5 demonstrate that these technical changes resulted in improvements and that this design could be used to simultaneously measure at several sensors placed at clinically relative depths. Figure 3.3 shows that four sensors could provide measurements (using a phantom tissue) at 25 cm). (This overcomes the depth limitations of India ink and the OxyChip as well as their single-site measurement.) Figure 3.4 shows that the SNR was increased 22-fold (from 17 to 375), resulting in faster data acquisition times and shorter measurement times. (This can lead to improved patient compliance and workflow.) Figure 3.5 shows that our new microcoil design maintains robust signal intensity independent of orientation of the resonator within the main magnetic field, unlike the previous design. Axially orienting the sensor loops better preserved orthogonality between the radiofrequency and main magnetic fields. As a result, rotational freedom around the longitudinal axis of the improved IR was achieved without compromising the signal intensity. In the rotational dependence test, signal intensity from the improved IR varied by less than ±7% over the angular range, whereas with the prior design signal intensity dropped by ±73%. (This allows greater flexibility in positioning the patient and still maintaining good signal intensity.)

Future plans To create the MOS, two important changes to the current improved IR are needed: (1) make the MOS suitable for insertion and use inside HDR needles in a clinical setting and (2) enhance the rigidity of the MOS to enable accurate depth control of measurement points inside the HDR needles. To achieve the first goal, the design will be further modified to

[2] At later stages of preclinical and clinical testing, some measurements could be performed while the patient is breathing O_2-enriched air (i.e., 98–100% O_2) through a non-rebreather mask, in order to determine whether hypoxic areas within the tumor are responsive to O_2-enhancing treatment.

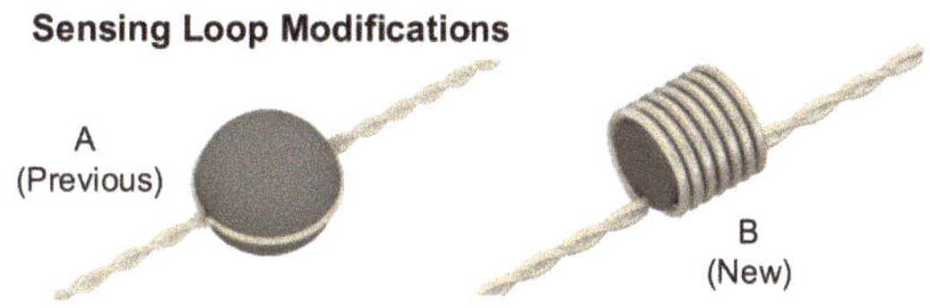

Fig. 3.2 Modification of sensing loops (**a**) Original IR sensor, with transmission wire around perimeter of sensing element (dark gray); (**b**) Improved design with microcoils axial to sensing element and connecting wire

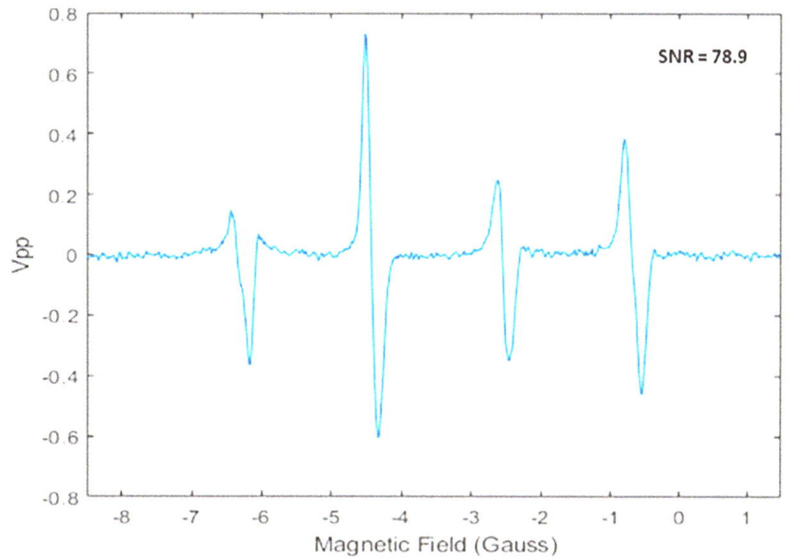

Fig. 3.3 Proof of concept re-depth and simultaneous measurements of multiple sites. EPR spectra, obtained in 1 min using a 23-cm-long transmission line before the four sensing elements on the improved IR in a tissue-mimicking phantom (i.e., gelatin), demonstrate two important improvements: the ability to measure O_2 at clinical depths, i.e., to at least 25 cm deep and the ability to concurrently measure four sensing elements using a magnetic gradient field

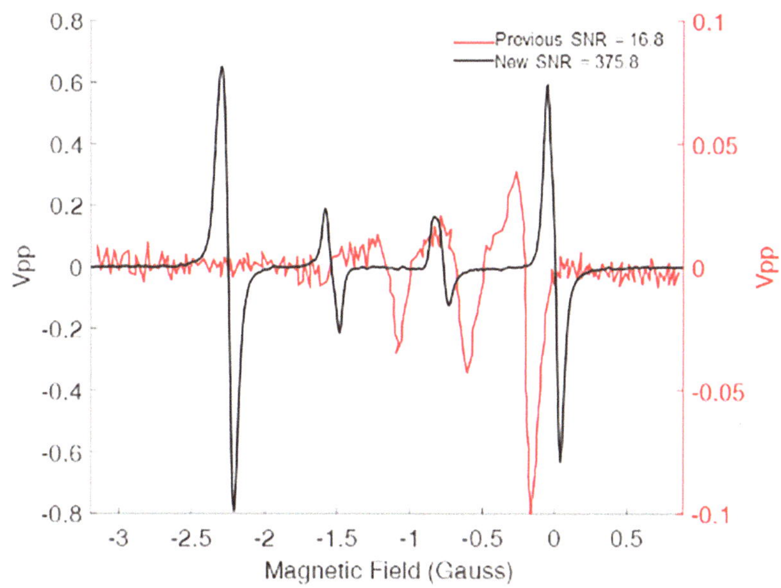

Fig. 3.4 Proof of concept re-improved SNR. Spectra of four sensing elements from original IR (red) and modified IR with microcoil (black) were taken simultaneously using a magnetic field gradient. SNR improved 22-fold (from 17 to 376). Note: drop of signal intensity in central black spectra was due to standing waves [to be eliminated]; note also that linewidth (i.e., data to assess O_2 levels) was unaffected by this artifact

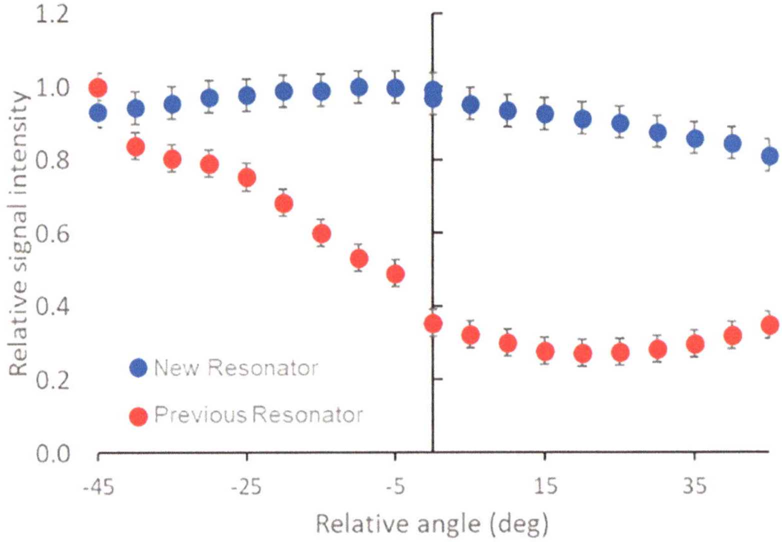

Fig. 3.5 Proof of concept: improved signal intensity and independence of orientation of the sensing elements relative to the main magnetic (B_0) field. One of the motivations for the new design was to reduce the dependency of signal intensity as a function of microcoil angle within the magnet. These data demonstrate that we achieved two improvements: better signal intensity (SNR) in the new design and increased independence of signal amplitude from clinically uncontrollable angular rotations of the device relative to the main magnetic field once it is in the needle/orientation within the patient

match the inner diameter of the HDR needle (~0.9 mm). To achieve the second goal, we will use a robust miniature coaxial transmission line to extend the length without sacrificing signal intensity and to control the depth of insertion.

Due to the length of HDR needles (up to 32 cm), the MOS will need to be placed inside the needle at the depth where the tumor is located. Current plans suggest that the 8 sensors, placed along a 5.6 cm length, would be sufficient to monitor the volume irradiated from each needle. We will achieve these improvements by reducing the diameter of the resonator so that its outer diameter fits cleanly and moves readily within the inner diameter of the HDR needles, replacing the twisted pair transmission line with a miniature coaxial transmission line of the required length and diameter to fit inside the HDR needle, and include a bazooka balun at the feed end to minimize the outer shield currents.

It is expected that reducing the diameter of the microcoil will reduce the signal intensity, given that the volume of LiPc would need to be reduced accordingly. However, the decrease in signal can be offset by replacing the high resistivity of the MP35N wire (103.3 μΩ·cm) used to make the IR with a lower resistivity material such as copper (1.68 μΩ·cm) for the MOS. The MP35N alloy wire had been needed for the long-term (>30 days) implantation in tissue (because it is already FDA approved for this use) but is no longer necessary for the proposed HDR application, i.e., the MOS will not be in contact with tissue.

Signal transmission loss will also be reduced due to the miniature coaxial transmission line (UT-047) with a known characteristic impedance of 50 Ω and attenuation of 0.40 dB/foot, substantially less attenuation per unit length of twisted pair transmission line used in the previous design (~23 dB/foot). This, in combination with the

Table 3.1 O_2 permeability in normalized units for select polymers (McKeen 2016)

Material	(Abbrev./tradename)	Permeability coefficient[a]
Polyetheretherketone	(PEEK/VICTREX)	1.9
Acetyl copolymer	(POM/Celcon)	2.0–2.9
Polytetrafluoroethylene	(PTFE/Teflon)	178–316
Polymethyl Pentene	(PMP/Mitsui TPX)	1800
Polydimethylsiloxane	(PDMS/Elkem Silbione)	39,400

[a]Normalized Units [$(cm^3 \cdot mm)/(m^2 \cdot day \cdot atm)$]; higher coefficient is more O_2 permeable

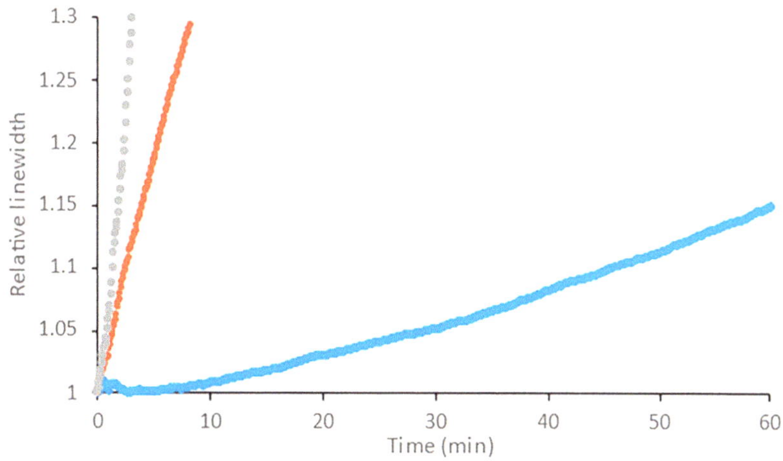

Fig. 3.6 Variation in O_2 permeability of plastics. EPR line width (i.e., used to measure O_2) for new IR with microcoils is measured continuously for up to 60 min, after changing from nitrogen to medical air, in 3 plastic tubes: (1) gray = thin-walled PTFE; (2) orange = regular PTFE; (3) blue = PEEK (see legend for Table 3.1)

bazooka balun end, will minimize outer shield currents, minimizing noise picked up and unwanted transmission line RF radiation.

Modifying the HDR needles We plan to create a minimally modified version of the hollow HDR needles to assure that the O_2 levels within the needle reflect the O_2 levels in the immediately adjacent tumor tissue. This modification will allow for O_2 diffusion and rapid equilibration of the internal volume of the needle with the external surrounding tissue.

Unfortunately, the commercially available plastic HDR needles are made of materials relatively impermeable to O_2 (e.g., PEEK [Varian's product] and Celcon [Best Medical's product]). See Table 3.1 for permeability data (McKeen 2016) for these and initial candidate alternative plastics: polytetrafluoroethylene (PTFE), PDMS, and polymethyl pentene (PMP). Figure 3.6 shows preliminary comparative tests of permeability; the test is the time to equilibrate the sensors in our prototype multisite resonator to the O_2 in the external atmosphere when it changes from nitrogen to medical air. Three conditions are compared: inside a PEEK needle or encased in a regular or thin-walled tube of PTFE.

Both Table 3.1 and Fig. 3.6 suggest that current HDR needles are not sufficiently permeable to O_2 to allow EPR oximetry measurements made within the needles to have adequate rapid equilibrium with O_2 in the tissues. We can systematically pursue several very feasible approaches to achieve a modification of the needles that enables accurate measurements of O_2 in tissue immediately adjacent to the needle, while retaining the full functionality of the needles for their intended use for HDR.

The simplest modification for the needle design would be to substitute a different polymer but maintain all physical dimensions and relevant properties such as rigidity. The best candidate would be PDMS due to its permeability (~20,000 × more O_2-permeable than PEEK, used in current HDR needles) and its high biocompatibility and resistance to biodegradation. (Medical-grade PDMS is widely used in medical devices and implants such as catheters, cardiac pacemakers, cochlear implants, nerve electrodes, microvalves, and micropumps, and is easy to sterilize using the standard autoclave equipment found in oncology departments. It therefore could be readily accepted by the FDA. While medical devices made of PDMS are typically very flexible, the rigidity of PDMS can readily be modified (Seghir and Arscott 2015) to match the rigidity of commercially available HDR needles. (This would ensure our modified HDR needles would retain the high positional precision of the radioactive source required in HDR.) A second alternative is PMP (~ 900 × more O_2-permeable than PEEK).

Modifying the afterloader: Preliminary plans The currently used afterloader for HDR brachytherapy at UChicago is Varian's Bravos Afterloader. We plan to adapt the apparatus for inserting the MOS based on the Bravos Afterloader and have received tentative interest from Varian in helping to design and adapt it for this use. In order to use the Bravos as a predicate device and to ensure that it meets the clinical needs and regulatory requirements, our designs and manufacturing will follow the recommendations for HDR treatment, clinical implementation, and quality assurance as well as standards set by regulatory agencies for devices (Prisciandaro et al. 2022; Food and Drug Administration (FDA) 2016, 1997, 2023). The most relevant standards for afterloaders include: (i) accuracy of geometric localization of all O_2 image measurements compared with gold standard imaging (e.g., CT), (ii) positioning accuracy, and (iii) localization of measurement in anatomic images. Initial preclinical testing for safety and efficacy will use a rabbit animal model. As we have previously done when manufacturing in vivo EPR devices for investigational use, we will use a user-centric design with a modular system for both software and hardware (Food and Drug Administration (FDA) 2016, 1997; Flood et al. 2018; Swartz et al. 2025). Figure 3.7 presents a conceptual prototype, illustrating some

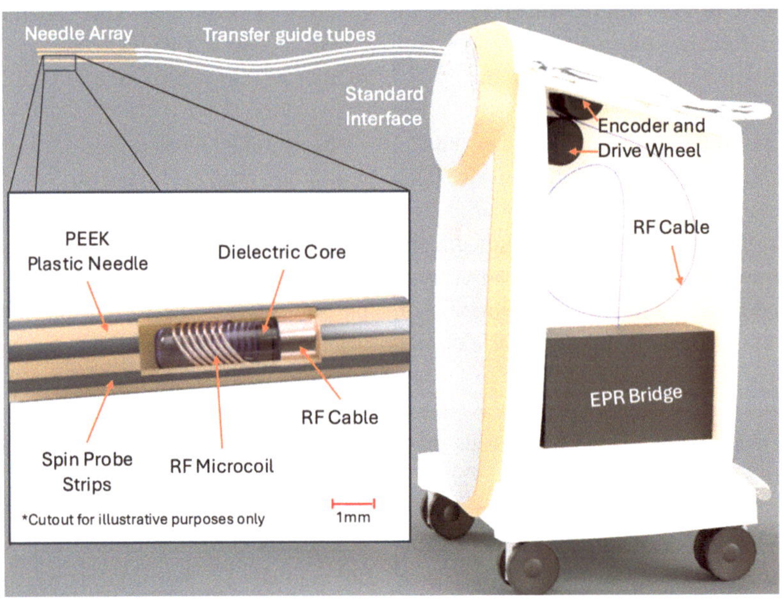

Fig. 3.7 Conceptualization of modified afterloader for O_2 measurements in conjunction with HDR brachytherapy. Modeled after Varian Bravos afterloader, it has the same footprint and interface currently used in HDR at UChicago. Mobile base allows quick removal to transition to using HDR Bravos. Area used in the HDR Bravos to shield the radiation source provides ample space for EPR hardware. Existing hardware for precise radioisotope positioning will be replicated to position the MOS inside O_2-permeable needles, with appropriate modification of the software

of the critical modules. The end result of each measurement session using the EPR oximetry system will be a sparse 3D image superimposing all O_2 measurements (i.e., at ~100–200 geographic locations). Because the location of the MOS within the needle at the time of measurement is readily determined and the location of the tumor with respect to the needle is known, we will be able to anatomically relate our measurements of O_2 to the radiation plan delivered to the tumor by HDR.

3.2 Summary and Conclusions

Our novel approach using the MOS and modifications of the HDR brachytherapy needles and a modified afterloader couples the advantages of multisite oximetry with HDR brachytherapy, thereby eliminating the need to implant the sensor. This approach has the potential of solving all current limitations of in vivo EPR oximetry. It will provide the clinician with both the initial levels of O_2 at many sites within the tumor and the impact of irradiation on the O_2 levels at these sites. Coupling the MOS with HDR brings the further advantage of allowing 12–25 times more measurements of O_2 in the tumor compared to the original IR. These measurement locations, coupled with the images already being captured of the tumor during the course of HDR brachytherapy, allow building a sparse 3D image of O_2 in the tumor. The repeatability of these measurements over the several-day course of HDR treatment allows following treatment effects on the O_2 levels in the tumor.

Although our approach currently is being developed for one specific use of HDR interstitial brachytherapy (cervix), completion of these studies should provide an optimum path to demonstrate the feasibility and utility of our device anywhere HDR interstitial brachytherapy is used for any tumor (prostate seems especially promising). The success of this entire approach also should make it feasible to develop analogous approaches in conjunction with other types of radiotherapy that do not use HDR.

Moreover, the preliminary studies reported here provide proof of concept for the increased sensitivity of the MOS and its ability to detect multiple sites at clinically relevant depths. There is significant work to be done to bring this promising technique into the clinic, but we think we have defined the path forward.

Acknowledgments Clin-EPR, LLC provided funding support, LiPc and multisensory resonator prototypes (AKA the IR), and loaned an in vivo EPR oximetry instrument to the University of Chicago. UChicago Comprehensive Cancer Center also provided funding.

Competing Interests ABF and HMS are co-owners of Clin-EPR, LLC which manufactures clinical EPR instruments for investigational use.

References

Berger D, Van Dyk S, Beaulieu L et al (2023) Modern tools for modern brachytherapy. Clin Oncol 35: e453ee468. https://www.clinicaloncologyonline.net/article/S0936-6555(23)00182-6/pdf (last accessed Sept 5, 2024)

Caston RM, Schreiber W, Hou H et al (2017) Development of the implantable resonator system for clinical EPR oximetry. Cell Biochem Biophys 75(3–4): 275–283. PMCID:PMC5972368

Chen EY, Tse D, Hou H et al (2021) Evaluation of a refined implantable resonator for deep-tissue EPR oximetry in the clinic. Appl Magn Reson 52(10): 1321–1342. PMCID: PMC8570533

Dinguizli M, Jeumont S, Beghein N et al (2006) Development and evaluation of biocompatible films of polytetrafluoroethylene polymers holding lithium phthalocyanine crystals for their use in EPR oximetry. Biosens Bioelectron 21(7):1015–1022

Elas M, Bell R, Hleihel D et al (2008) Electron paramagnetic resonance oxygen image hypoxic fraction plus radiation dose strongly correlates with tumor cure in FSa fibrosarcomas. Int J Radiat Oncol Biol Phys 71(2): 542–549. PMCID: PMC2577780

Elas M, Magwood JM, Butler B et al (2013) EPR oxygen images predict tumor control by a 50% tumor control radiation dose. Cancer Res 73(17):5328–5335

Flood AB, Wood VA, Schreiber W et al (2018) Guidance for academics to transfer 'bench-ready' medical technology into usual clinical practice. Case study: sensors and spectrometer used in EPR oximetry. Adv Exp Med Biol 1072:233–239. Supplement only at 314496_1_En_37_MOESM1_ESM.docx

Food and Drug Administration (FDA) (1997) Design control guidance for medical device manufacturers (related to FDA 21 CFR 820.30 and Sub-clause 4.4 of ISO

9001March 11, 1997). Available at https://www.fda.gov/media/116573/download. Accessed 6 Mar 2025
Food and Drug Administration (FDA) (2016) Applying human factors and usability engineering to medical devices: guidance for industry and Food and Drug Administration staff. Available at https://www.fda.gov/media/80481/download. Accessed 6 Mar 2025
Food and Drug Administration (FDA) (2023) FDA use of international standard ISO10993-1, "Biological evaluation of medical devices - Part 1: Evaluation and testing within a risk management process": guidance for industry and Food and Drug Administration staff. Available at: https://www.fda.gov/media/142959/download. Accessed 13 Mar 2025
Gilbert MR, Dignam JJ, Armstrong TS et al (2014) A randomized trial of bevacizumab for newly diagnosed glioblastoma. N Engl J Med 370(8):699–708
Höckel M, Vorndran B, Schlenger K et al (1993) Tumor oxygenation: a new predictive parameter in locally advanced cancer of the uterine cervix. Gynecol Oncol 51(2):141–149. https://doi.org/10.1006/gyno.1993.1262. PMID: 8276286
Hoskin PJ, Carnell DM, Taylor NJ et al (2007) Hypoxia in prostate cancer: correlation of BOLD-MRI with pimonidazole immunohistochemistry-initial observations. Int J Radiat Oncol Biol Phys 68(4): 1065–1071. PMID:17637389
Hou H, Li H, Dong R et al (2014) Real-time monitoring of ischemic and contralateral brain pO_2 during stroke by variable length multisite resonators. Magn Reson Imaging 32(5):563–569. PMCID: PMC4297608
Hou H, Khan N, Nagane M et al (2016) Skeletal muscle oxygenation measured by EPR oximetry using a highly sensitive polymer-encapsulated paramagnetic sensor. Adv Exp Med Biol 923:351–357. https://doi.org/10.1007/978-3-319-38810-6_46. PMID: 27526163
Hou H, Khan N, Gohain S et al (2018) Pre-clinical evaluation of OxyChip for longterm EPR oximetry. Biomed Microdevices 20(2):29. PMID: 29549438
Jeong JJ, Liu T, Yang X et al (2019) Clinical measurements of oxygen via electron paramagnetic resonance (EPR) during and after breast radiation therapy: Preliminary results of baseline evaluations and response to hyperoxygenation. J Radiol Radiat Ther 7(1):1082. https://doi.org/10.47739/2333-7095/1082
Kaanders JHAM, Pop LAM, Marres HAM et al (2002) ARCON: experience in 215 patients with advanced head-and-neck cancer. Int J Radiat Oncol Biol Phys 52(3):769–778. https://doi.org/10.1016/s0360-3016(01)02678-5. PMID: 11849800
Khan N, Hou H, Chen EY et al (2013) Bench-to-bedside oximetry for real-time monitoring of tumor PO_2: a critical parameter which influences radiotherapeutic outcome. J Radiol Radiat Ther 1(3):1017
Lehtiö K, Eskola O, Viljanen T et al (2004) Imaging perfusion and hypoxia with PET to predict radiotherapy response in head-and-neck cancer. Int J Radiat Oncol Biol Phys 59(4):971–982. PMID: 15234030
Lewis JS, Welch MJ (2001) PET imaging of hypoxia. Q J Nucl Med 45(2):183–188
McKeen LW (2016) Permeability properties of plastics and elastomers, 4th edn. William Andrew Pub, Norwich NY. eBook ISBN: 9780323478885
Pötter R, Tanderup K, Schmid MP, EMBRACE Collaborative Group et al (2021) MRI-guided adaptive brachytherapy in locally advanced cervical cancer (EMBRACE-I): a multicentre prospective cohort study. Lancet Oncol 22(4):538–547. https://doi.org/10.1016/S1470-2045(20)30753-1. PMID: 33794207
Prisciandaro J, Zoberi JE, Cohen G, et al (2022) AAPM task group report 303 endorsed by the ABS: MRI implementation in HDR brachytherapy-Considerations from simulation to treatment. Med Phys 49(8):e983–e1023. https://doi.org/10.1002/mp.15713. Epub 2022 Jun 20. Erratum in: Med Phys 2023 50(2):1258. https://doi.org/10.1002/mp.15991. PMID: 35662032
Schaner PE, Pettus JR, Flood AB et al (2020) OxyChip implantation and subsequent electron paramagnetic resonance oximetry in human tumors is safe and feasible: first experience in 24 patients. Front Oncol. https://doi.org/10.3389/fonc.2020.572060. (last accessed Sept 5, 2024)
Schaner PE, Williams BB, Chen EY et al (2021) First-in-human study in cancer patients establishing the feasibility of oxygen measurements in tumors using electron paramagnetic resonance with the OxyChip. Front Oncologia 11:3878. PMCID: PMC8517507
Seghir R, Arscott S (2015) Extended PDMS stiffness range for flexible systems. Sensors Actuators A230: 33–39
Sundfør K, Tropé C, Suo Z et al (1999) Normobaric oxygen treatment during radiotherapy for carcinoma of the uterine cervix. Results from a prospective controlled randomized trial. Radiother Oncol 50(2): 157–165. PMID: 10368039
Sundfør K, Lyng H, Tropé CG et al (2000) Treatment outcome in advanced squamous cell carcinoma of the uterine cervix: relationships to pretreatment tumor oxygenation and vascularization. Radiother Oncol 54(2):101–107. PMID: 10699471
Swartz HM, Clarkson RB (1998) The measurement of oxygen in vivo using EPR techniques. Phys Med Biol 43(7):1957–1975. PMID: 9703059
Swartz HM, Flood AB (2024) Re-examining what the results of "a measurement of oxygen level in tissues" really mean. Mol Imaging Biol 26:391–402. https://doi.org/10.1007/s11307-023-01887-6
Swartz HM, Williams BB, Jarvis LA et al (2013) Repeated monitoring of tumor oxygen while breathing carbogen to determine the therapeutic potential of hyperoxic therapy. Pract Radiat Oncol 3(2 Suppl 1):S23–S24. https://doi.org/10.1016/j.prro.2013.01.084
Swartz HM, Hou H, Khan N et al (2014) Advances in probes and methods for clinical EPR oximetry. Adv Exp Med Biol 812:73–79. PMCID: PMC4313742
Swartz HM, Williams BB, Hou H et al (2016) Direct and repeated clinical measurements of pO_2 for enhancing cancer therapy and other applications. Adv Exp Med Biol 923:95–104. PMCID: PMC5989722
Swartz HM, Flood AB, Schaner PE et al (2020a) How best to interpret measures of levels of oxygen in tissues to

make them effective clinical tools for care of patients with cancer and other oxygen-dependent pathologies. Physiol Rep 8(15):e14541. PMCID: PMC7422807

Swartz HM, Vaupel P, Williams BB et al (2020b) 'Oxygen level in a tissue' - what do available measurements really report? Adv Exp Med Biol 1232:145–153. https://doi.org/10.1007/978-3-030-34461-0_19

Swartz HM, Flood AB, Williams BB et al (2021) What is the meaning of an oxygen measurement? Analysis of methods purporting to measure oxygen in targeted tissues. Adv Exp Med Biol 1269:301–308. PMID: 33966234. https://doi.org/10.1007/978-3-030-48238-1_48

Swartz HM, Williams BB, Zaki BI et al (2024a) Clinical EPR unique opportunities and some challenges. Acad Radiol 21(2):197–206. PMCID: PMC3921887

Swartz HM, Vaupel P, Flood AB (2024b) Recognizing potential ambiguities in measurements of oxygen in tissues. Adv Exp Med Biol 1463:307–314. https://doi.org/10.1007/978-3-031-67458-7_50

Swartz HM, Flood AB, Schreiber W et al (2025) Lessons learned in transforming in vivo devices from pioneering research into tools used routinely in medical diagnostics. Acad Rad S1076-6332(25):00646-4

Vaupel P, Mayer A (2007) Hypoxia in cancer: significance and impact on clinical outcome. Cancer Metast Rev 26(2):225–239. PMID: 17440684

Zeng J, Liu Z, Jiang S et al (2021) Verification of guiding needle placement by registered ultrasound image during combined intracavitary/interstitial gynecologic brachytherapy. Cancer Manag Res 13:1917–1928. https://doi.org/10.2147/CMAR.S294498

What Is the Information Content of EPR Oximetry?

4

Sandra S. Eaton and Gareth R. Eaton

Abstract

The focus of this overview is on the use of EPR oximetry in biological systems with an emphasis on soluble probes. EPR oximetry measures oxygen concentrations based on changes in electron spin relaxation of probes such as trityl and nitroxide radicals. The experimental relaxation rates depend on the inherent relaxation of the probe, the effects of probe-probe collisions, and the relaxation enhancement caused by collisions of the probe with O_2. To extract the oxygen concentration requires an understanding of all three contributions. The inherent relaxation rates of the probe depend on the rates of tumbling and on microwave frequency. The effects of probe-probe collisions depend on concentration, and for charged trityl radicals depend on ionic strength of the solution. The interactions of O_2 with nitroxide and trityl radicals can be described in terms of exchange interactions and the dynamics of collisions in solution. The spatial volume elements that can typically be observed in EPR imaging experiments are large enough that the volume could contain a range of concentrations of radicals and of O_2, which may result in distributions of linewidths and relaxation times that need to be considered in analyzing data. When using EPR oximetry in living systems, it is also necessary to consider whether the oxygen concentration is changing on the timescale of the experiment.

S. S. Eaton · G. R. Eaton (✉)
Department of Chemistry and Biochemistry, University of Denver, Denver, CO, USA
e-mail: Gareth.Eaton@du.edu

Keywords

Electron spin relaxation · Collisions · Inversion recovery · EPR oximetry · Pulse spin echo decay

4.1 Introduction to the Impact of O_2 on EPR Lineshapes and Relaxation Times

The focus of this chapter is on the soluble trityl (Fig. 4.1) and nitroxide radicals (Fig. 4.2), which provide opportunities for imaging experiments and have spatial resolution that is not available with spatially localized solid probes. EPR spectroscopy is used to monitor local O_2 concentrations based on changes in continuous wave (CW) lineshapes (Fig. 4.3) and relaxation times T_2 (Fig. 4.4) and T_1 (Fig. 4.5).

The CW EPR spectrum of the nitroxide CTPO has three well-separated hyperfine lines due to coupling to the nitrogen spin $I = 1$. The central hyperfine line is shown in Fig. 4.3. In air-saturated solution (Fig. 4.3), there is one broad line with $\Delta B_{pp} = 0.28$ G. The linewidth is

S. C. Pias et al. (eds.), *Oxygen Transport to Tissue XLVI*, Advances in Experimental Medicine and Biology 1498, https://doi.org/10.1007/978-3-032-10389-5_4

a combination of unresolved hyperfine coupling to the protons on the nitroxide ring and 0.24 G line broadening from O_2. In a solution that has been deoxygenated with N_2 gas (Fig. 4.3), the small proton hyperfine couplings are resolved and ΔB_{pp} is reduced to 0.15 G because the impact of collision with paramagnetic O_2 has been removed. The simulated spectrum in N_2 included hyperfine coupling to one proton with $A_H = 0.518$ G and 12 equivalent protons with $A_H = 0.191$ G. As discussed below, the decrease in linewidth is due to an increase in T_2.

T_2 can be measured directly by 2-pulse spin echo using a pulse sequence of $\pi/2$—τ—π—τ—echo. Examples of 2-pulse echo decays for the trityl OX063 radical in water at ambient temperature are shown in Fig. 4.4. As the gas phase in equilibrium with the solution is changed from air to N_2, the time constant for the echo decay (T_2) increases from 0.73 μs to 5.2 μs. The changes in T_2 as a function of O_2 collisions are caused by changes in T_1 which can be measured by 3-pulse inversion recovery and the pulse sequence ($\pi/2$—τ—$\pi/2$—T—$\pi/2$—τ—echo). Examples of T_1 recovery curves for trityl OX063 in water at ambient temperature are shown in Fig. 4.5. As the gas phase in equilibrium with the solution is changed from air to N_2, the time constant for the inversion recovery (T_1) increases from 0.83 μs to 5.2 μs to 14.5 μs. The interpretation of the impact of O_2 on linewidths and relaxation times is discussed below.

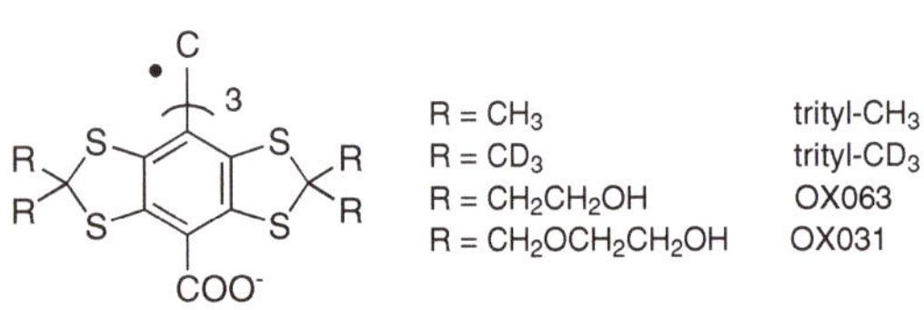

Fig. 4.1 Structures of commonly used trityls

Fig. 4.2 Structures of three widely used nitroxides

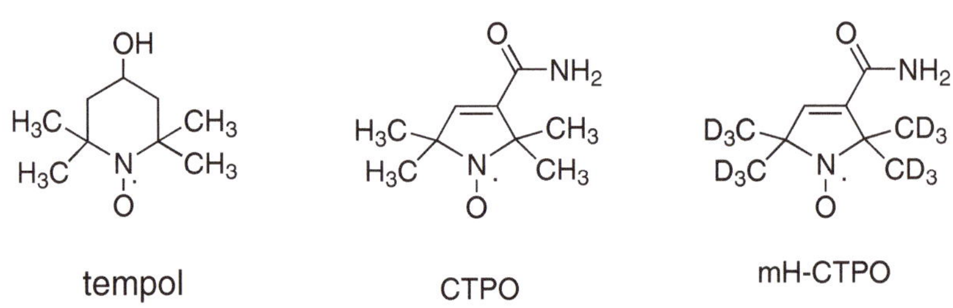

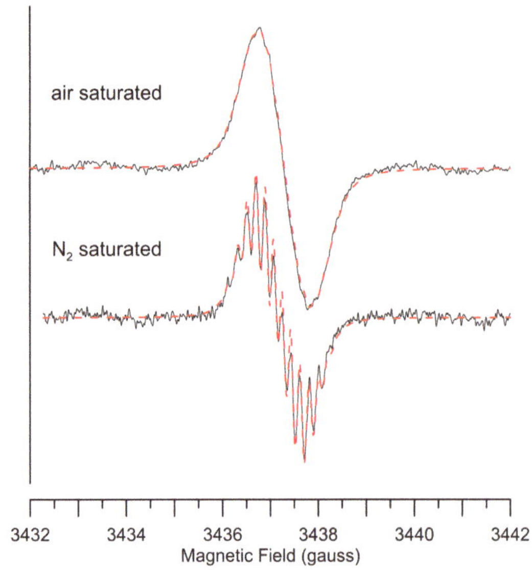

Fig. 4.3 Center line of 0.2 mM CTPO in water at X-band in air-saturated and N_2-saturated solution. Simulations (red) are superimposed on experimental spectra (black)

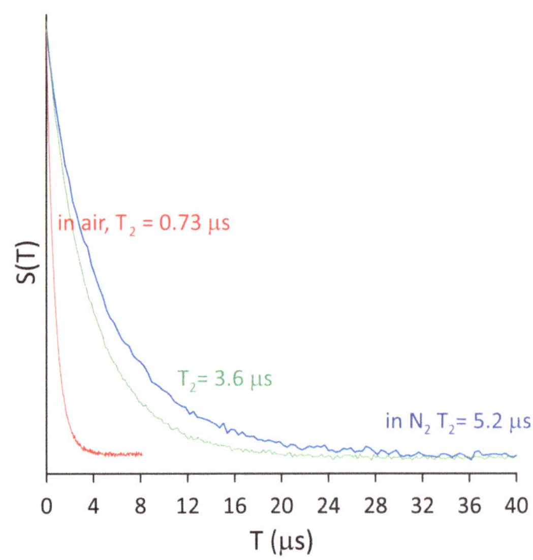

Fig. 4.4 X-band two-pulse echo decays for 0.15 mM OX063 in water for samples equilibrated with air (red), a mixture of air and N_2 (green), and N_2 (blue)

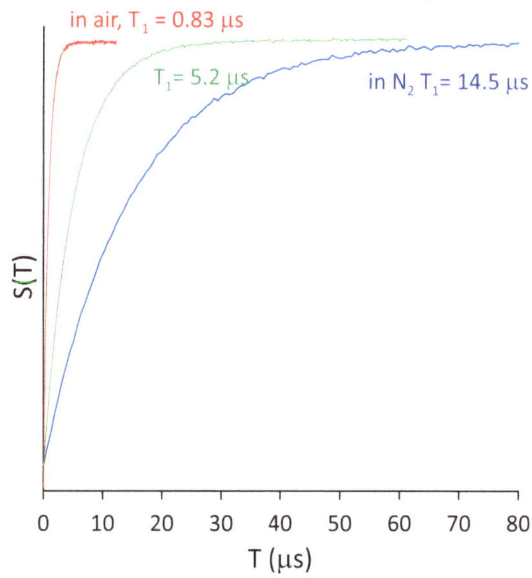

Fig. 4.5 X-band inversion recovery curves for 0.15 mM OX063 in water for samples equilibrated with air (red), a mixture of air and N_2 (green), and N_2 (blue)

4.2 Contributions to Trityl and Nitroxide Relaxation

Users of in vivo oximetry need to be aware of variables that impact the experimental O_2 values, especially in view of the timescale of EPR measurements, the heterogeneity of tissue and the variation in parameters that impact electron spin relaxation within the volumes for which measurements are reported (Swartz et al. 2020, 2021; Swartz and Flood 2024). To address these questions, an understanding of the physical principles that underlie the oximetry experiments is needed. The focus of the discussion is on the soluble trityl (Fig. 4.1) and nitroxide radicals (Fig. 4.2), which provide opportunities for imaging experiments and have spatial resolution that is not available with spatially localized solid probes.

EPR oximetry is based on changes in electron spin relaxation that occur when O_2 collides with paramagnetic probes including trityls, nitroxides, or solids such as LiPc. For relatively rapidly tumbling small molecules in fluid solution, the contributions to relaxation combine as the sum of reciprocals as shown in Eqs. (4.1) and (4.2) (Molin et al. 1980).

$$\frac{1}{T_1^{\text{exp}}} = \frac{1}{T_1^{\text{o}}} + \frac{1}{T_1^{\text{HE}}} + \frac{1}{T_1^{O_2}} \tag{4.1}$$

$$\frac{1}{T_2^{\text{exp}}} = \frac{1}{T_2^{\text{o}}} + \frac{1}{T_2^{\text{HE}}} + \frac{1}{T_2^{O_2}} \tag{4.2}$$

where $\frac{1}{T_1^{\text{o}}}, \frac{1}{T_2^{\text{o}}}$ are the relaxation rates at low concentration in the absence of O_2, $\frac{1}{T_1^{\text{HE}}}, \frac{1}{T_2^{\text{HE}}}$ are the Heisenberg exchange contributions to relaxation from probe-probe collisions, and $\frac{1}{T_1^{O_2}}, \frac{1}{T_2^{O_2}}$ are predicted to be equal and are the contributions to relaxation from collisions with O_2, which are oximetric contributions.

To interpret changes in electron spin relaxation in terms of $[O_2]$, we need to understand the multiple factors that impact each of the terms in Eqs. (4.1) and (4.2). Each of these factors may contribute to variations in relaxation within a sample or between samples and need to be taken into account in designing experiments and interpreting data.

4.3 Impact of Molecular Tumbling and Microwave Frequency on T_1° and T_2°

The processes that dominate electron spin relaxation for organic radicals in fluid solution depend on the molecular tumbling correlation time, τ_R, and on the experimental observation frequency ν (Freed 2005; Eaton and Eaton 2018). Although dominant relaxation mechanisms differ, $1/T_1$ for both trityls and nitroxides is tumbling and frequency dependent in the frequency ranges that commonly are used for EPR oximetry (Owenius et al. 2005, 2004; Biller et al. 2013, 2011, 2012), as shown in Fig. 4.6a (Owenius et al. 2005) and Fig. 4.6b (Owenius et al. 2004). For these relatively rapidly tumbling radicals, $1/T_1{}^{\text{o}}$ contributes to $1/T_2{}^{\text{o}}$, so factors that impact $T_1{}^{\text{o}}$ also impact $T_2{}^{\text{o}}$. Since τ_R is approximately proportional to solution viscosity, τ_R may be different in a membrane than in bulk solution.

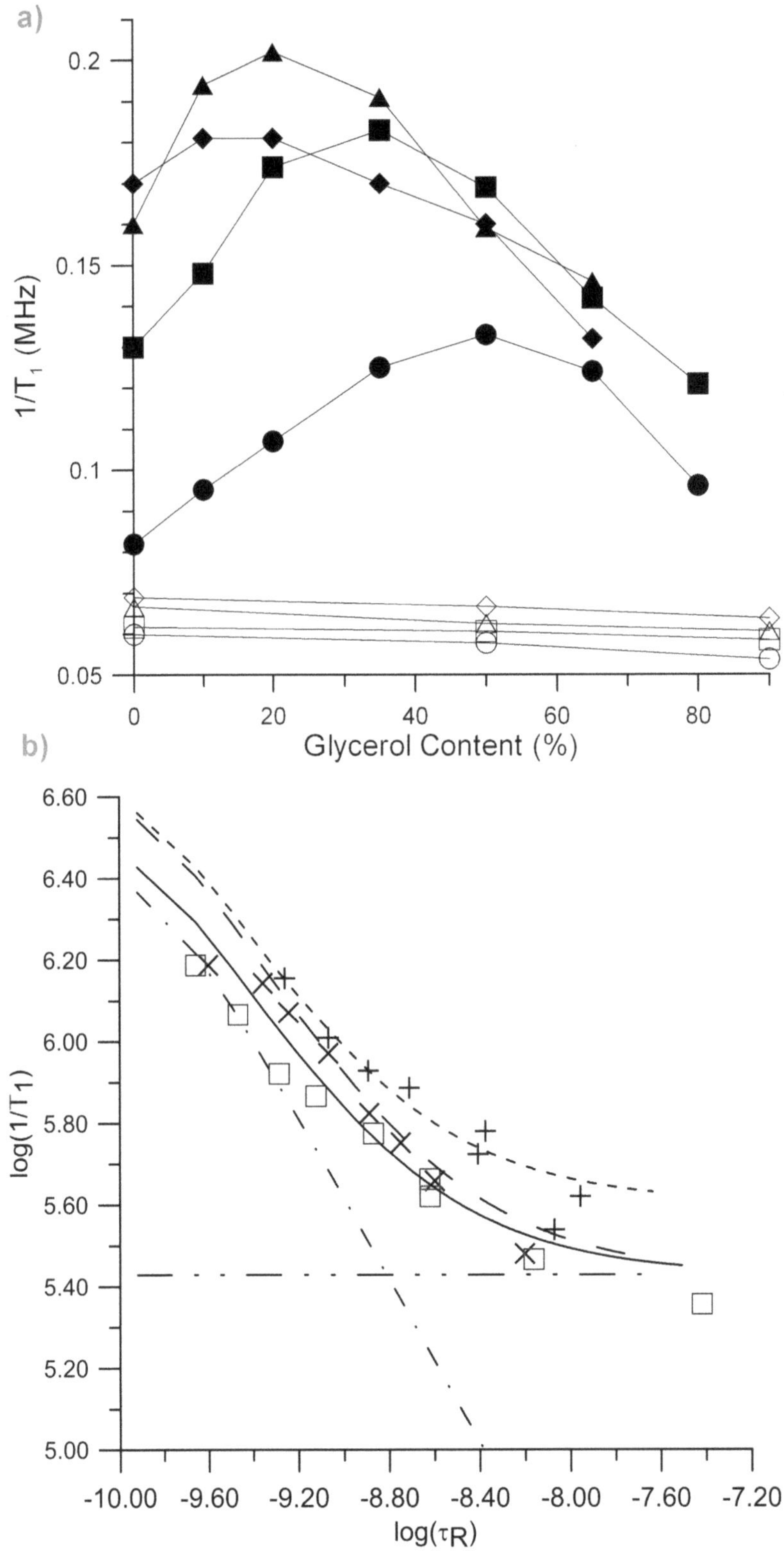

Fig. 4.6 Tumbling dependence of $1/T_1$ in water at 25 °C. (**a**) for trityl-CD_3 (○,●), trityl-CH_3 (□,■), OX063 (Δ,▲), and OX031 (◊,♦) at 250 MHz (closed shapes) and X-band (open shapes) (Owenius et al. 2005). (**b**) for tempol (+) and isotopically substituted tempol (x, □) at 1.0 GHz (Owenius et al. 2004). The fit lines are calculated based on tumbling-dependent models of relaxation

4.4 Heisenberg Exchange Contributions, $\frac{1}{T_1^{HE}}, \frac{1}{T_2^{HE}}$, from Probe-Probe Collisions

Fundamental discussions of Heisenberg exchange in nitroxide-nitroxide collisions are provided by Molin et al. on p. 55 of (Molin et al. 1980) and Salikhov on p. 247 of (Salikhov 2019a). The Heisenberg exchange terms can be expressed as K_eC in Eq. (4.3).

$$\frac{1}{T_2^{\mathrm{exp}}} = \frac{1}{T_2^{\mathrm{o}}} + K_eC \tag{4.3}$$

where K_e is the diffusion rate constant that is probe and solvent dependent, and C is concentration. Molin estimated K_e for a nitroxide in low viscosity solvent at ambient temperature as ~$10^9\ \mathrm{M}^{-1}\ \mathrm{s}^{-1}$ on p. 7 of (Molin et al. 1980). For a nitroxide concentration of 0.2 mM, $K_eC = 2 \times 10^5\ \mathrm{s}^{-1}$. If $T_2^{\mathrm{o}} \sim 0.5 \times 10^{-6}$, then $1/T_2^{\mathrm{o}} = 2.0 \times 10^6\ \mathrm{s}^{-1}$, and the Heisenberg exchange contributions would have a relatively small impact on $1/T_2^{\mathrm{exp}}$. The contributions from Heisenberg exchange are not predicted to vary with microwave frequency.

The initial reports of trityl radicals as DNP (dynamic nuclear polarization) agents showed that concentration dependence was greater for T_2 than for T_1, was solvent dependent, and radical specific (Ardenkjaer-Larsen et al. 1998). Examples of the concentration dependence of T_2 in water are shown in Fig. 4.7a (Halpern 2015) for a trityl at 250 MHz at 37 °C and in Fig. 4.7b for a nitroxide at X-band at 25 °C (Biller et al. 2018). The concentration dependence of T_1 and T_2 depends on the charge on the radical, the type and concentration of counter-ion, the strength of interaction with solvent molecules, including H-bonding ability, and the electron spin density at the periphery of the radical, as well as temperature and viscosity. A larger trityl with succinyl groups appended to the trityl side chains (Shaw et al. 2023) demonstrates that larger size and larger negative charge decrease the effect of collisions between trityl molecules and thus decrease self-relaxation. Concentrations of trityls and nitroxides below about 0.2 mM (Owenius et al. 2005; Ardenkjaer-Larsen et al. 1998) are recommended to minimize the impact of collisions on relaxation. To improve signal intensity for in vivo experiments, it may be necessary to use higher probe concentrations. The concentration dependence of relaxation should be checked for the probe and media of interest. If probe concentration, ionic strength, or viscosity varies through the sample, it is a source of heterogeneity in relaxation. Factors that contribute to the weaker concentration dependence of T_1 than of T_2 are discussed below.

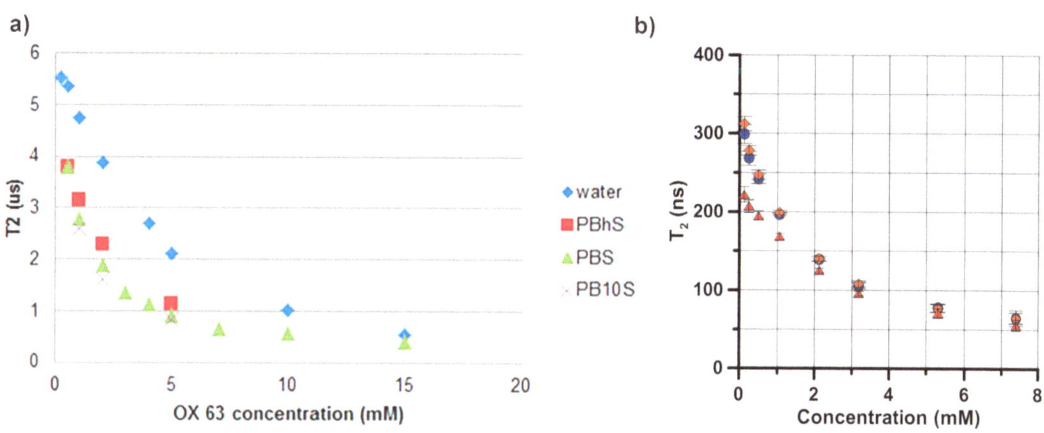

Fig. 4.7 Concentration dependence of T_2 in deoxygenated solutions for (**a**) OX063 at 250 MHz in water, and in phosphate buffer with half-normal saline (PBhS), with normal saline (PBS), and 10× normal saline. (**b**) tempol in deoxygenated water (blue circle, red diamond) or air-saturated water (▲) at X-band

4.4.1 Probe-Probe Collision Effects on T_2

The EPR spectrum of a free radical in solution is a superposition of signals from an ensemble of molecules with all possible combinations of hyperfine couplings, each of which has a different resonance energy. Each collision between radicals with different nuclear spin states results in line broadening and T_2 relaxation. This means that most collisions are between molecules with different nitrogen and/or proton nuclear spin states. For nitroxide, the nitrogen nuclear hyperfine is the dominant hyperfine, with additional contributions from protons. For trityl radicals, about 70% of the signal intensity is in the sharp central line for molecules that include no ^{13}C nuclei. However, about 30% of the intensity is from molecules that contain at least one ^{13}C ($I = ½$), which gives nuclear hyperfine coupling and shifts the resonance. Again, a substantial fraction of the collisions is between molecules with no ^{13}C and one ^{13}C which results in T_2 relaxation.

4.4.2 Probe-Probe Collision Effects on T_1

At room temperature and resonance frequencies between 250 MHz and X-band, the populations of the two electron spin states are almost equal. Thus, half of the spins have $m_s = 1/2$, and half have $m_S = -1/2$. Upon collision of two radicals, Heisenberg exchange results in the molecules diffusing away with either the electron spin with which they came or with the spin of the other radical. If there is no difference in their resonant energies, there is no spin relaxation. If, however, the radicals had different nuclear spin states, then the effect of the spin exchange is to exchange with the thermal energy of the lattice (the solvent) the energy difference between the two nuclear spin states. This is a T_1 relaxation mechanism. Because of the nearly equal spin-up and spin-down populations, relaxation occurs in only half of the encounters. This is in contrast with T_2, where most collisions involve different nuclear spin states and result in T_2 relaxation.

There is another factor that decreases the effect of collisions on T_1. When radicals diffuse close enough to undergo spin exchange, they have multiple close encounters and re-encounters within the solvent shell. There is a roughly 50:50 chance that the radical will emerge from the encounter with the same spin that it had before the encounter, even if the other radical with which it collided had the opposite spin. This re-encounter phenomenon also helps average out any anisotropies in the spin exchange (Molin et al. 1980) p. 71.

4.5 Contributions of O_2 to EPR Relaxation $\frac{1}{T_1^{O_2}}, \frac{1}{T_2^{O_2}}$

For most organic radicals, the overall peak-to-peak linewidth ΔB_{pp} is larger than ΔB_{spkt}, which is the width that the EPR line would have if it were relaxation-determined, and there was no contribution from nuclear hyperfine coupling. The relationship between ΔB_{spkt} and $T_2{}^o$ is given in Eq. (4.4).

$$\Delta B_{spkt}(G) = \frac{2}{\sqrt{3}\gamma T_2} = \frac{6.56 \times 10^{-8}\ \mathrm{G\,s}}{T_2{}^o} \tag{4.4}$$

From the early days of EPR, it has been recognized that collisions of O_2 with organic radicals increase linewidths (Hausser 1960a, b; Deguchi 1960). For radicals tumbling rapidly in fluid solution, linewidth changes (and direct relaxation time changes) are due to interaction of the slowly-relaxing spin with the rapidly relaxing O_2, which has $T_1 \sim 7.5 \times 10^{-12}$ s (Eaton and Eaton 2021). In this motional regime, $\frac{1}{T_1^{O_2}} = \frac{1}{T_2^{O_2}}$ is expected (Eaton and Eaton 1978). For the OX063 radical, it was shown (Epel et al. 2014) that $\frac{1}{T_1^{O_2}} = \frac{1}{T_2^{O_2}} = 8.9 \times 10^3\ s^{-1}$/torr times pO_2, confirming the equivalence of the two effects.

Subczynski and Hyde (1981) report that with strong Heisenberg exchange and negligible dipolar interactions, $1/T_1$ of a nitroxide interacting with O_2 can be written:

$$\frac{1}{T_1^{exp}} = \frac{1}{T_1^{o}} + \frac{2}{3}pZ \qquad (4.5)$$

The factor 2/3 is for the fraction of effective nonsecular transitions for $S = ½$ ^{14}N and $S = 1$ O_2 (Subczynski and Kusumi 2003), and p is a probability to allow for some collisions not being effective (Subczynski and Hyde 1981). The result of this analysis is that $1/T_1$ in the presence of O_2 is linearly dependent on the collision frequency. The extensive ^{13}C hyperfine splitting of common trityl radicals probably makes this 2/3 factor roughly appropriate for trityl as well as for nitroxides. However, given the difficulty of being quantitative about p and Z, it seems reasonable to omit the factor of 2/3. Accepting the approximations used by Lai et al. (1982), collision frequency, Z, between nitroxide or trityl and O_2 can be estimated using the Smoluchowski equation (Subczynski and Hyde 1981; Lai et al. 1982).

$$Z = 4\pi R[O_2](D_{radical} + D_{O_2}) \qquad (4.6)$$

where R is the interaction distance between the radical and O_2, and $D_{radical}$ and D_{O_2} are the diffusion constants for the radical and O_2, respectively. In this expression, $[O_2]$ is in units of molecules cm^{-3}. Stereochemistry and spin density distributions commonly make $p < 1$, but re-encounters can increase p. At 37 °C, $D_{O_2} \sim 2 \times 10^{-5}$ cm^2 s^{-1}. Hyde and Subczynski (1989) comment on the scatter in measurements of the diffusion constant of O_2 in water. A value of ca. 2.8×10^{-5} $cm^2 s^{-1}$ at 37 C seems reasonable for the purposes of this discussion. Lai et al. (1982) assumed $R = 4.5$ Å (4.5×10^{-8} cm), based on the commonly used nitroxide radius of ~2 Å. The O_2 concentration in water is approximately 2×10^{-4} M in equilibrium with air at sea level, which is about 159 torr O_2. For a radical concentration of 0.2 mM (1.2×10^{17} molecules cm^{-3}) and using the values cited, $Z = 1.9 \times 10^6$ s^{-1}. See Hyde and Subczynski for other uncertainties in the calculations (Hyde and Subczynski 1989).

For a trityl radical, the estimated radius is ca. 6 Å (Owenius et al. 2005), so the interaction distance R between the trityl and O_2 is ~7×10^{-8} cm. Using the full radius of the trityl partially compensates for the probability of spin exchange, $p < 1$. With the approximations used by Lai et al. (1982), $D_{O_2} \sim 2.8 \times 10^{-5}$ cm^2 s^{-1} at 37 °C. With trityl OX063 at 37 °C in normal saline (145 mM NaCl), we assume that the trityl diffuses slowly enough that we can ignore its motion relative to that of O_2. The O_2 and radical concentrations are assumed to be 0.2 mM (1.2×10^{17} molecules cm^{-3}). Using these values,

$$Z = 4\pi(7 \times 10^{-8}cm)(1.2 \times 10^{17}cm^{-3})(2.8 \times 10^{-5}cm^2s^{-1}) = 3.0 \times 10^6 s^{-1} \qquad (4.7)$$

This collision frequency for trityl is larger than for nitroxides because of the larger value of R.

Stereochemistry and spin delocalization suggest that the spin density at the location of collision with O_2 will be small which predicts $p < 1$ in Eq. (4.5). Using Epel's results (Epel et al. 2014) of $\frac{1}{T_1^{O_2}} = 8.9 \times 10^3$ s^{-1}/torr times pO_2, the air saturated $pO_2 = 159$ torr, and assuming $p = 1$ Eq. (4.5) gives $\frac{1}{T_1^{O_2}} = 1.4 \times 10^6$ s^{-1}, which is remarkably similar to the value calculated in Eq. (4.7).

$D_{radical}$ usually is smaller than D_{O_2}, approximately by the ratio of the radii. $D_{radical}$ may be further decreased by strong interactions with water. Thus, trityl and nitroxides diffuse more slowly than O_2, and especially so when they H-bond to the water. These strong interactions with water also attenuate the approach of the O_2 to the hydrated radical. Because of the much larger D for O_2 than for nitroxyl or trityls, many papers make the approximation that the motion of the radical is negligible relative to the motion of O_2 and so $(D_{radical} + D_{O_2}) \sim (D_{O_2})$. In biological samples, D_{O_2} may be impacted by differential solubility in compartments and barriers between compartments. Slower motion, increased viscosity, solvation, and decreased spin density at the site of collision also reduce the relaxation effect of collisions with O_2, so the trityl called SOX71 has longer T_1 at higher O_2 concentration and is potentially a better probe at higher concentrations than the trityls more commonly used (Shaw et al. 2023).

4.5.1 Dipolar Spin-Spin Interactions During a Collision

It is generally accepted that dipolar interactions are important only in highly viscous solutions. Nevertheless, it is worth outlining the nature of the dipole interaction to help understand why it makes a negligible contribution to oximetry in fluid solution. Dipolar interaction is a function of distance. The parameter D_{dip} in Table 4.1 is the magnitude of the splitting when the interspin vector is perpendicular to the magnetic field, Eq. (4.8). When the vector is parallel to the magnetic field, the magnitude is –2D (Eaton and Eaton 2000).

$$D_{dip}\,(\text{gauss}) = \frac{2g\beta}{2r^3} = \frac{2.78 \times 10^4}{r^3} \tag{4.8}$$

At the approximate distance of closest approach between a radical and O_2, the dipolar splitting could be of the order of 100 G (Table 4.1). However, the O_2 spin relaxation time of ~7.5×10^{-12} s (Eaton and Eaton 2021) is short relative to the time that the O_2 is close to the radical in low-viscosity solutions, so the O_2 spin dipole is fluctuating rapidly enough to be essentially decoupled from the spin of the radical. Consequently, we will ignore dipolar interactions and focus on exchange interactions.

Table 4.1 Dependence of dipolar couplings on interspin distance[a]

r (Å)	D_{dip} (gauss)
5	222
10	28
20	3.5
40	0.43
60	0.13

[a]The distance r is in Å and g ~2 for nitroxides and trityls

4.5.2 Spin-Spin Exchange Interactions During a Collision

Exchange interactions involve the overlap of molecular orbitals containing the unpaired electrons. This occurs when two molecules get close enough together, the orbitals are similar in energy, and the spins are aligned the same so that they are almost equivalent. This can be envisioned easily when two nitroxides or two trityl radicals collide. When one of the molecules is O_2, modeling of the exchange interaction relies on the fact that the O_2 spectrum, although not observable because of fast relaxation in solution at room temperature, overlaps the radical spectra—that is, it has similar energies. During the exchange interaction, both radical and O_2 are jointly experiencing the same spin relaxation time, provided the exchange interaction J is sufficiently strong. When the O_2 moves away from the radical, the radical spectrum is as broad as the O_2 spectrum. If this were the end of it, the intensity of the radical spectrum would decrease in proportion to the number of collisions with O_2. However, the radicals are also colliding with other radicals, and the exchange interactions among these radicals modulate the relaxation time. Since the EPR observable is an ensemble average, the net reduction in relaxation times results from all of these factors.

In the following, we explore for O_2 effects the use of theory originally derived for interaction between transition metals and organic radicals (Eaton and Eaton 1978; Salikhov et al. 1971; Salikhov 2019b). To select the relevant portions of this theory, we need to restrict the range of observables to dilute fluid solutions in which the frequency, Z, and duration, τ, of collisions is such that $Z\tau << 1$. In condensed phases (e.g., aqueous solution) as opposed to gas phase, it is necessary to consider re-encounters of the same pair of species. If a pair has collided and moved apart one jump length, the most likely next encounter is the same pair (Salikhov 2019b). We also know that the unpaired electron spin density is not uniformly or symmetrically distributed in the nitroxide or trityls. The spin density at the location of closest encounter with O_2 in most cases will be much less than one. The conclusion based on many studies is that with low viscosities and short relaxation times of the faster-relaxing species (O_2 in the present case), the decrease of the radical relaxation time (broadening of the CW linewidth) is determined primarily by exchange interactions (Eaton and Eaton 1978).

Salikhov et al. (1971) describe exchange interactions as weak or strong: the interaction is weak if the translational correlation time or the relaxation time is too short. The case of nitroxides colliding with O_2 in fluid solution is considered to be strong exchange (Subczynski and Hyde 1981; Hyde and Sarna 1978). Hyde and Sarna (1978) use the Salikhov et al. (1971) equation for the exchange frequency. The interaction of a radical with a paramagnetic metal or O_2 with short relaxation time is then described by the equations in Salikhov et al. (1971) and summarized in (Eaton and Eaton 1978; Hyde and Sarna 1978).

$$\Delta\omega_{ex} = Z\frac{2}{3}\frac{J^2 S_2(S_2+1)T_1\tau_c}{1+\frac{2}{3}J^2 S_2(S_2+1)T_1\tau_c} \quad (4.9)$$

$\Delta\omega_{ex}$ is the broadening of the spectrum for the slowly relaxing species, Z is the frequency of collisions of the radical with the O_2, J is the exchange integral, S_2 is the spin of the faster-relaxing species, T_1 is the relaxation time of this faster-relaxing species, and τ_c is the collision time, also called translational correlation time.

This equation contains three important terms that are at best approximations. Since they are multiplied by one another, we have a guess to the fourth power. The collision frequency, Z, is assumed to be as given by the Smoluchowski equation (Eq. 4.5). For O_2 and radicals in water, this is about $2 \times 10^6\ s^{-1}$. J has been estimated to be from 10^{11} to $10^{14}\ s^{-1}$ depending on how close the O_2 is to the radical. Plachy and Kivelson (1967) estimated $J = 6 \times 10^{10}\ s^{-1}$. Edelstein et al. (1964) estimated that for di-tertiary-butyl nitroxide in three hydrocarbon solvents, J ranged from 1.6 to $4.8 \times 10^{11}\ s^{-1}$. Salikhov et al. (1971) state that J should be of the order of the energy of a chemical bond, ca. $10^{16}\ s^{-1}$. At van der Waals distances, the exchange integral is estimated to be of the order of 10^{14} rad/s. Estimates of the exchange integral, the efficiency parameter, the geometric or stereochemistry factors, and the probability, p, are not independent in most of the literature. For our estimate, we use $10^{11}\ s^{-1}$ as a first approximation of J. In 1 cP viscosity solution, $\tau_c \sim 3 \times 10^{-11}$s (Salikhov et al. 1971).

With these assumptions,

$$\Delta\omega_{ex} = (10^6 s^{-1})\frac{2}{3}\frac{(10^{11}s^{-1})^2 1(1+1)(7.5\times10^{-12}s)(3\times10^{-11}s)}{1+\frac{2}{3}(10^{11}s^{-1})^2 1(1+1)(7.5\times10^{-12}s)(3\times10^{-11}s)} = 7.5\times10^5 s^{-1} \quad (4.10)$$

where $\Delta\omega_{ex} = \frac{1}{T_1^{O2}}$. This would be a 0.05 G linewidth increase which is about half of the 0.1 G change between air and N_2 that was observed in model membranes by (Popp and Hyde 1981). Despite the many approximations, this result suggests that an exchange-dominated model of the collision interaction is plausible. With the values used in Eq. (4.10), it is not obvious that $J^2\tau^2 >> 1$ in viscous solutions or membranes, so the collisions may not be in the limit designated as "strong" for some cases. The collision frequency is lower in more viscous regions, so the same relaxation time change in a more viscous region would require a much larger O_2 concentration. Since the terms in the numerator and denominator of Eq. (4.9) nearly cancel when they are significantly larger than 1, we seek to explain most T_1 and T_2 changes in terms of pZ, Eq. (4.5). The more complete model represented by Eq. (4.8) reminds users of the multiple factors that contribute to $\frac{1}{T_1^{O2}}$.

Throughout this discussion, we have restricted the scope to aqueous solutions at room or physiological temperatures and argued, consistent with the literature, that we could assume strong exchange interactions dominate the O_2 effect on the EPR measurements. Extensive studies by Subczynski explore membrane analogs in which the solubilities and Z differ from those in pure water, and the result might be a larger dipole effect and weaker exchange. Application of relaxation time changes to the interpretation of O_2 effects in tissue may be confounded by variations in solubility and diffusion rates in different tissues within the measured voxel.

The fundamentals of spin exchange have intrigued scientists since the early observations of line narrowing and broadening, so it might be enlightening to test some of the quantitative aspects by using cutting-edge instrumentation. For example, the estimates of the exchange energies, J, are of the order of 10^{11} s^{-1}, and a key assumption is that J exceeds the Zeeman interaction gβB, 1 THz EPR would violate this assumption (Pake and Tuttle 1959). What will the effect be on spectra at 1 THz that show exchange broadening at X-band and lower frequencies?

4.6 Imaging Resolution

The common materials used to measure O_2 are nitroxides, trityls, and LiPc and its analogs. LiPc analogs are usually used for localized spectroscopy. Trityl radicals are best imaged with pulsed EPR. Nitroxide T_2 values are too short for practical EPR imaging with pulsed EPR and are best imaged using rapid scan EPR or, with lower signal-to-noise, CW imaging. In pulsed EPR imaging, the FID decay rate is determined by the linewidth (T_2*). The larger the gradient, the broader the line, and the faster the decay. For large gradients, it is necessary to detect spin echoes rather than FIDs. The single-point imaging method (Matsumoto et al. 2003) provides better spatial resolution but reduced spectral resolution.

4.6.1 Ultimate Image Resolution

Resolution limits are discussed in Eaton and Eaton (1991). Many factors contribute to image resolution. If S/N is high enough, the resolution of an image, to first approximation, is determined by the ratio of the linewidth of the EPR signal to the magnetic-field gradient. For example, if the linewidth is 0.1 mT, and the gradient is 10 mT cm^{-1}, the resolution is 0.1 mT/(10 mT cm^{-1}), which is 0.01 cm (100 μm). The resolution can be improved by up to about a factor of 3 by deconvolution of the spectral line shape. It is possible to estimate the inherent broadening (or blurring) of the image by the properties of the imager, and thus define the inherent "point spread function." Deconvolution using the point spread function in a technique called "super-resolution" can improve the spatial resolution of the image (Hirata et al. 2006). However, resolution also is limited by the number of projections, which may be determined by the time available for an in vivo experiment. The shorter the time available for the image, the lower the resolution. For in vivo imaging, the time available may be constrained by the animal physiology. For a large object, the maximum gradient may be limited by the extent to which the gradient broadens the signal. Spatial image resolution increases proportionally to the magnitude of the magnetic field gradient. If the linearity of the gradient is 1%, then the resolution is limited to 1% of the spatial dimension. Microwave/RF B_1 varies over the resonator volume, causing differences in quantitation of the radical spins and the measure of their interaction with O_2.

The ultimate resolution limit is determined by the S/N achievable for the number of spins that can occur in the volume element (voxel) imaged. At X-band, the sensitivity of a modern spectrometer (ca. 10^9 unsaturable spins in a high Q resonator) permits measuring a cube about 10 μm on a side if it contains about 1 mM spins (Eaton and Eaton 1991). At 1 GHz, the sensitivity is much poorer, so the resolution limit will be larger. The frequency dependence in S/N for various cases of sample and resonator scaling has been considered

(Rinard et al. 2004). For example, if the sample size is constant and the size of a loop gap resonator is scaled inversely with frequency, then EPR sensitivity at constant microwave power scales as $\omega^{11/4}$. When X-band and 300 MHz are compared, this is a factor ~10^4 decrease in *S/N*, which implies that the detectable number of spins would be ~10^{13} spins if one could build a high Q 300 MHz resonator that is 32 times as large as an X-band resonator and the spins did not saturate at high incident power. In reality, resonators at 300 MHz to 1 GHz have much lower Q than typical X-band resonators, and resonator efficiencies are much lower for larger size resonators, which increases the minimum detectable spins by several orders of magnitude. Hence, the ultimate spatial resolution for oximetry at lower frequencies will be substantially poorer than at X-band for solutions of trityl or nitroxide radicals. Examples below of current technology suggest that it will be very difficult to achieve image resolution better than about 0.5 mm at 250 MHz to 1 GHz.

The FID decay shape provides a lineshape parameter R_2*. For this type of imaging, narrower trityl lines yield better spatial resolution, which can approach 1 mm for 2 mM OX071 at 300 MHz (Kishimoto et al. 2024). The single point imaging (SPI) method (Subramanian 2023; Subramanian and Krishna 2005) uses spatially dependent phase encoding of a single time point in the FID as a function of gradients, and does not monitor the time-dependence of the response to the microwave pulse. The spatial resolution does not depend on T_2*(linewidth). At 300 MHz, very high resolution has been attained, 0.3 mm in a trityl OX063 phantom. Filtered back-projection imaging of echo-detected projections results in poorer spatial resolution but better accuracy and concentration resolution of O_2.

As in many scientific measurements, in spectroscopy and in imaging, there is a tradeoff between resolution and signal-to-noise. When dealing with animals, this tradeoff is further constrained by physiological limits on the total time of the measurement. As an example, the (Epel et al. 2014) paper that calibrated O_2 effects on trityl relaxation used 80 time steps in nonimaging inversion recovery measurements but 8 time steps in imaging measurements of T_1. In very few cases will this small number of points support multiexponential fits to the experimental data.

4.7 Experimental Considerations

4.7.1 Estimating T_2 from CW Lineshapes

The effect of O_2 on EPR spectra was observed early in the development of EPR (Hausser 1998), and the effect on nitroxide was measured quantitatively by Hyde, Swartz, and Halpern and coworkers (Subczynski and Hyde 1981; Halpern et al. 1990; Subczynski and Swartz 2005). Among nitroxides, the narrowest hyperfine lines are for CTPO (Fig. 4.3). The hyperfine coupling is about 100 mG (Lai et al. 1982). O_2 broadening of CTPO in water at 37 °C was measured by simulation of the highly resolved hyperfine lines (Hyde and Subczynski 1984). Very good simulations were achieved using purely Lorentzian lines, varying only the widths. The 37 °C linewidth increased from 0.14 G in O_2-free solution to 0.24 at 1.76×10^{-4} M O_2. However, for many nitroxides, the proton hyperfine coupling is not well resolved. Analysis of the lineshapes to obtain values of T_2 then requires careful lineshape analysis. The versatility of nitroxides may make them the O_2 probe of choice for many biomedical (and other) applications. However, trityl radicals are much more sensitive to O_2 than are nitroxides and have long enough spin relaxation times that they are amenable to pulse EPR measurement rather than CW EPR. The relaxation times are better reporters of small differences in O_2 concentration, and especially of heterogeneity of concentration. The Halpern and Krishna labs have been leaders in this type of oximetry (Subramanian and Krishna 2005; Williams and Halpern 2005).

4.7.2 Inversion Recovery and Progressive Saturation

Pulsed EPR measurements of trityl T_1 have been calibrated to provide absolute measurements of O_2 concentration. Elas et al. calibrated pulsed EPR of trityl against Oxylite O_2 measurements and found agreement (Elas et al. 2006). The uncertainty of the OX063 measurement was 1 mm and 3 torr. The Oxylite track was measured with 2 mm accuracy. The images reported by Elas et al. also illustrate the large O_2 gradients within a tumor at the resolution limit of the EPR measurements.

Spin echo inversion recovery measurement of trityl T_1 was shown to be superior to T_2 measurements, because of the larger concentration dependence of T_2 than of T_1 (Epel et al. 2014). Calibration of the dependence of trityl T_1 on O_2 concentration was performed by Epel et al. (2014) and by Martin et al. (2023). For OX071 and the nanoencapsulated EPR imaging agent called RADI-Sense, the response of T_1 to O_2 is the same within experimental error, providing confidence in the use of various trityl derivatives (Martin et al. 2023).

Labs that do not have access to direct measurement of relaxation times use CW progressive power saturation curves as a measure of relative relaxation rates. Note that the CW continuous power saturation method of Hyde and Brown (1962), which uses a T_1T_2 product to measure T_1, assumes that you can get T_2 from the width of a fully-resolved Lorentzian hyperfine line. This is rare, and certainly not feasible with any nitroxide other than Fremy's salt or degassed CTPO. Since T_2 and T_1 both are affected by O_2 concentrations, changes in CW saturation behavior should be taken as a measure of overall relaxation and not interpreted as changes in T_1.

4.7.3 Distributions of Relaxation Times

In many oximetry experiments, it is likely that heterogeneity of samples will result in a distribution of relaxation times and inversion recovery or spin echo decay curves will not be single exponentials. We suggest caution about terms like "average concentration," because extremes of concentration may be missed by a measurement technique. The selectivity for fast or slow relaxation could lead to disagreements about O_2 concentration among researchers who use different methods. Data have to be collected for multiple time constants and with enough baseline, to define the time constant(s) well. Narayanan et al. (1995) state, "A common cause of nonexponential spin-lattice relaxation is the presence of several different relaxation environments in the sample." Fitting to two exponentials "is only possible if the S/N is sufficiently large and the time constants are significantly different." This chapter cites many prior papers that have addressed deconvoluting multiple exponentials. Three of these papers are the papers by Provencher that are the basis for the "multifit" computer program used in the Eaton lab (Provencher 1982a). "The accuracy of relaxation time estimates, particularly for two-component relaxation data, is seriously degraded by data nonlinearity." This paper used a $S/N = 100$ "deliberately placing DISCRETE and CONTIN at the limit of their capabilities." These programs were compared by Provencher (1982b).

A stretched exponential is an alternative to fitting to the sum of discrete exponentials. The stretched exponential (also called the Kohlrausch curve) is written as:

$$I(\tau) = \exp\left[-\left(\frac{\tau}{T}\right)^{\beta}\right] \tag{4.11}$$

If $\beta = 1$, this is the normal exponential. Kohlrausch proposed the stretch exponential empirically in 1847. The physical basis for the stretch exponent is considered to be one of the oldest unsolved problems in science (Potuzak et al. 2011). In spin echo experiments, various relaxation mechanisms predict different values of β. Stretched exponential relaxation was shown by Johnston (2006) to arise from a continuous sum of exponential decays. Their study assumes that a system contains many independently relaxing species, each of which decays exponentially.

Discrete and continuous distributions of exponentials and stretched exponentials were compared by Marshall (1989). "The results indicate that typical statistical and information theoretic tests fail to distinguish reliably between a continuous and a discrete model unless the maximum noise in the data is lower than about 0.5% of the maximum signal amplitude, or a signal-to-noise ratio of about 400."

A stretched exponential fit depends on the choice of zero time. In a pulse experiment with finite widths, the zero for echo formation should be approximately the average of the pulse widths plus the dead time of the spectrometer—essentially the d1 parameter in the Bruker Xepr software. However, the Bruker software assumes that the first data point is at time zero. The Denver lab's local software permits you to define zero. The fitting parameters depend on the choices made. This is important for interlaboratory comparisons and reproducibility.

The physical interpretation of the distributions may be specific to each spin system investigated. One could, in some physical systems, imagine a process that is truly not a simple exponential, but in NMR and EPR T_1 measurements, it is plausible to assume that the stretched exponential function is a distribution of exponential decays with different time constants and amplitudes. Stein and Subczynski (2021) interpreted heterogeneity of saturation recovery measures of oxygen transport parameters in the plasma membrane of the eye in terms of stretched exponentials.

Fitting to multiple exponentials is not feasible if the recovery curve consists of only a few data points, which is a common situation when relaxation time is the fourth dimension of an imaging experiment.

4.7.4 Noisy Exponential Data

This section deals with the effect of noise on multi-exponential fits to experimental data. This is a common problem in science, so we take guidance from whatever field we can. The references given do not necessarily deal with spin relaxation.

Moore (1973) concluded that a 1% error in the baseline resulted in a 3% error in the estimated exponential time constant. They also said that sampled points should span at least 2τ. Koenig and Brown (1984) state, "even exponential decays differing by a factor of 2 can be treated as a single exponential decay with a time constant equal within 1% to the average of the two time constants." Rossing and Danford (1968) concluded from prior literature that "curves, generated as the sum of as many as eight exponential terms, may at times be fitted satisfactorily by only two such terms." Analogous conclusions were reached by Whittall et al. (1989): "The inherent non-uniqueness due to the finite number of inaccurate measurements means that a variety of discrete and continuous spectra exist which fit the data equally well." Our UPEN analysis program provides a display of the sum of exponential functions, with their relative contributions, that reproduce the experimental decay. If a single exponential fits the experimental data fairly well, the UPEN program will give a sharp distribution, with maximum value at about the time constant of the single-exponential fit. The width is a measure of the signal-to-noise of the original data—the noisier the data, the wider the distribution. A direct comparison of biexponential fits and UPEN fits to T_1 data of irradiated teeth and synthetic hydroxyapatites was performed by Sato et al. (2007).

4.8 Summary: What Is the Information Content of EPR Oximetry?

Increase in CW linewidth, decrease in T_2, or decrease in T_1 due to interaction with O_2 can be calibrated in terms of the O_2 concentration for specific combinations of probe, solvent, and temperature. For homogeneous low-viscosity fluid solutions of radicals and O_2, such as aqueous solutions of nitroxides and trityls, this calibration can yield absolute O_2 concentrations in molecules per unit volume. In vivo spectroscopy and imaging inherently create problems of interpretation. Where in the biological system is the nitroxide or trityl? Is the concentration uniform, or even what was expected based on the injection? Epel et al. (2014) discussed the extracellular fluid compartment complications in estimating actual trityl concentration in vivo. Swartz has discussed the ambiguities introduced by the distribution of oxygen in tissues and the time needed to acquire the data with both soluble and solid probes (Swartz et al. 2020, 2021; Swartz and Flood 2024). These are ongoing challenges for which the multiple variables discussed in this chapter require consideration.

Competing Interests The authors declare that they have no conflict of interest.

References

Ardenkjaer-Larsen JH, Laursen I, Leunbach I, Ehnholm G, Wistrand L-G, Petersson JS, Golman K (1998) EPR and DNP properties of certain novel single electron contrast agents intended for oximetric imaging. J Magn Reson 133:1–12

Biller JR, Meyer V, Elajaili H, Rosen GM, Kao JPY, Eaton SS, Eaton GR (2011) Relaxation times and line widths of isotopically-substituted nitroxides in aqueous solution at X-band. J Magn Reson 212:370–377

Biller JR, Meyer VM, Elajaili H, Rosen GM, Eaton SS, Eaton GR (2012) Frequency dependence of electron spin relaxation times in aqueous solution for a nitronyl nitroxide radical and per-deuterated-Tempone between 250 MHz and 34 GHz. J Magn Reson 225:52–57

Biller JR, Elajaili H, Meyer V, Rosen GM, Eaton SS, Eaton GR (2013) Electron spin lattice relaxation mechanisms of rapidly-tumbling nitroxide radicals. J Magn Reson 236:47–56

Biller JR, McPeak JE, Eaton SS, Eaton GR (2018) Measurement of T_{1e}, T_{1N}, T_{1HE}, T_{2e}, and T_{2HE} by pulse EPR at X-band for nitroxides at concentrations relevant to solution DNP. Appl Magn Reson 49:1235–1251

Deguchi Y (1960) Proton hyperfine spectra of diphenyl picryl hydrazyl. J Chem Phys 32:1584–1585

Eaton SS, Eaton GR (1978) Metal-nitroxyl interactions. Part 5. Interaction of spin labels with transition metals. Coord Chem Rev 26:207–262

Eaton SS, Eaton GR (1991) Pulsed EPR Imaging. In: Eaton SS et al (eds) EPR imaging and in vivo spectroscopy. CRC Press, Boca Raton, pp 73–78

Eaton SS, Eaton GR (2000) Relaxation times of organic radicals and transition metal ions. Biol Magn Reson 19:29–154

Eaton SS, Eaton GR (2018) Relaxation mechanisms. In: Goldfarb D, Stoll S (eds) EPR spectroscopy: fundamentals and methods. Wiley, Chichester, pp 175–192

Eaton SS, Eaton GR (2021) EPR spectra and electron spin relaxation of O_2. Appl Magn Reson 52:1223–1226

Edelstein N, Kwok A, Maki AH (1964) Effects of hydrostatic pressure and temperature on spin exchange between free radicals in solution. J Chem Phys 41: 3473–3378

Elas M, Ahn KH, Parasca A, Barth ED, Lee D, Haney C, Halpern HJ (2006) Electron paramagnetic resonance oxygen images correlate spatially and quantitatively with OxyLite oxygen measurements. Clin Cancer Res 12:4209–4217

Epel B, Bowman MK, Mailer C, Halpern HJ (2014) Absolute oxygen R1e imaging *in vivo* with pulse electron paramagnetic resonance. Magn Reson Med 72:362–368

Freed JH (2005) ESR and molecular dynamics. Biol Magn Reson 24:239–268

Halpern HJ (2015)

Halpern HJ, Peric M, Nguyen TD, Spencer DP, Teicher BA, Lin YJ, Bowman MK (1990) Selective isotopic labeling of a nitroxide spin label to enhance sensitivity for T_2 oxymetry. J Magn Reson 90:40–51

Hausser KH (1960a) The effect of dissolved oxygen on line broadening of electron spin resonance (EPR) in solution. Naturwissenschaften 47:251

Hausser KH (1960b) Chemical uses of electronresonance. Chimia 14:153–161

Hausser KH (1998) The effect of concentration and oxygen in EPR. In: Eaton GR et al (eds) Foundations of modern EPR. World Scientific, Singapore, pp 469–481

Hirata H, Wakana M, Susaki H (2006) Feasibility study of superresolution continuous-wave electron paramagnetic resonance imaging. Appl Phys Lett 88:254103–254101

Hyde JS, Brown HW (1962) EPR study of tetracene positive ion. J Chem Phys 37:368–378

Hyde JS, Sarna T (1978) Magnetic interactions between nitroxide free radicals and lanthanides or Cu2+ in liquids. J Chem Phys 68:4439–4447

Hyde JS, Subczynski WK (1984) Simulation of ESR spectra of the oxygen-sensitive spin-label probe CTPO. J Magn Reson 56:125–130

Hyde JS, Subczynski WK (1989) Spin-Label oximetry. Biol Magn Reson 8:402–403

Johnston DC (2006) Stretched exponential relaxation arising from a continuous sum of exponential decays. Phys Rev B 74:184430

Kishimoto S, Devasahayam N, Chandramouli GVR, Murugesan R, Otowa Y, Yamashita K, Yamamoto K, Brender JR, Krishna MC (2024) Evaluation of a deuterated triarylmethyl spin probe for in vivo R2*-based EPR oximetric imaging with enhanced dynamic range. Magn Reson Med 91:413–423

Koenig SH, Brown RD (1984) Determinants of proton relaxation in tissue. Magn Reson Med 1:437–449

Lai C-S, Hopwood LE, Hyde JS, Lukiewicz S (1982) ESR studies of O_2 uptake by Chinese hamster ovary cells during the cell cycle. Proc Nat Acad Sci US 79:1166–1170

Marshall DB (1989) Statistical considerations in the analysis of dispersive kinetics data as discrete or continuous distributions of rate constants. Anal Chem 61:660–665

Martin RM, Diaz S, Poncelet M, Driesschaert B, Barth E, Kotecha M, Epel B, Eaton GR, Biller JR (2023) Toward a nanoencapsulated EPR imaging agent for clinical use. Mol Imaging Biol 26:525

Matsumoto K, Chandrika B, Lohman JAB, Mitchell JB, Krishna M, Subramanian S (2003) Application of continuous-wave EPR spectral-spatial image reconstruction techniques for in vivo oximetry: comparison of projection reconstruction and constant-time modalities. Magn Reson Med 50:865–874

Molin YN, Salikhov KM, Zamaraev KI (1980) Spin exchange: principles and applications in chemistry and biology. Springer-Verlag, Berlin

Moore WS (1973) The design, analysis, and performance of resonant and non-resonant microwave transmission devices with theoretically infinite rejection. Rev Sci Instrum 44:158–164

Narayanan A, Hartman JS, Bain AD (1995) Characterizing nonexponential spin-lattice relaxation in solid-state NMR by fitting to the stretched exponential. J Magn Reson A 112:58–65

Owenius R, Terry GE, Williams MJ, Eaton SS, Eaton GR (2004) Frequency dependence of electron spin relaxation of nitroxyl radicals in fluid solution. J Phys Chem B 108:9475–9481

Owenius R, Eaton GR, Eaton SS (2005) Frequency (250 MHz to 9.2 GHz) and viscosity dependence of electron spin relaxation of Triarylmethyl radicals at room temperature. J Magn Reson 172:168–175

Pake GE, Tuttle TR (1959) Anomalous loss of resolution of paramagnetic resonance hyperfine structure in liquids. Phys Rev Lett 3:423–425

Plachy W, Kivelson D (1967) Spin exchange in solutions of di-tertiary-butyl nitroxide. J Chem Phys 47:3312–3318

Popp CA, Hyde JS (1981) Effects of oxygen on EPR spectra of nitroxide spin-label probes of model membranes. J Magn Reson 43:249–258

Potuzak M, Welch RC, Mauro JC (2011) Topological arigin of stretched exponential relaxation in glass. J Chem Phys 135:214502

Provencher SW (1982a) Constrained regularization method for inverting data represented by linear algebraic or integral equations. Comp Phys Commun 27: 213–227

Provencher SW (1982b) CONTIN: a general purpose constrained regularization program for inverting noisy linear algebraic and integral equations. Comp Phys Commun 27:229–242

Rinard GA, Quine RW, Eaton SS, Eaton GR (2004) Frequency dependence of EPR sensitivity. Biol Magn Reson 21:115–154

Rossing RG, Danforth MB (1968) A comparison of continuous distributions of parameters of exponential decay curves. Biometrics 24:117–134

Salikhov KM (2019a) Fundamentals of spin exchange: story of a paramagnetic shift. Springer, Cham

Salikhov KM (2019b) Fundamentals of spin exchange. A paradigm shift. Springer

Salikhov KM, Doctorov AB, Molin YN, Zamaraev KI (1971) Exchange broadening of ESR lines for solutions of free radicals and transition metal complexes. J Magn Reson 5:189–205

Sato H, Filas BA, Eaton SS, Eaton GR, Romanyukha AA, Hayes R, Rossi AM (2007) Electron spin relaxation of radicals in irradiated tooth enamel and synthetic hydroxyapatite. Radiat Meas 42:997–1004

Shaw MA, Poncelet M, Viswakarma N, Vallerini GP, Hameed S, Guth TD, Geldenhuys WJ, Hoblitzell EH, Eubank TD, Epel B, Kotecha M, Driesschaert B (2023) SOX71, a biocompatible succinyl derivative of the triarylmethyl radical OX071 for in vivo quantitative oxygen mapping using electron paramagnetic resonance. Mol Imaging Biol 26(3):542–552

Stein N, Subczynski WK (2021) Oxygen transport parameter in plasma membrane of eye lens fiber cells by saturation recovery EPR. Appl Magn Reson 52:61–80

Subczynski WK, Hyde JS (1981) The diffusion-concentration product of oxygen in lipid bilayers using the spin-label T_1 method. Biochem Biophys Acta 643:283–291

Subczynski WK, Kusumi A (2003) Dynamics of raft molecules in the cell and artificial membranes: approaches by pulse EPR spin labeling and single molecule optical microscopy. Biochim Biophys Acta 1610:231–243

Subczynski WK, Swartz HM (2005) EPR oximetry in biological and model systems. Biol Magn Reson 23: 229–282

Subramanian S (2023) The development of in vivo FT-EPR imaging at 300 MHz: applications in cancer research. Appl Magn Reson 54:807–832

Subramanian S, Krishna MC (2005) Time-domain radiofrequency EPR imaging. Biol Magn Reson 23: 321–383

Swartz HM, Flood AB (2024) Re-examining what the results of "a measurement of oxygen level in tissues" really means. Mol Imaging Biol 26:391–402

Swartz HM, Vaupel P, Williams BB, Schaner PE, Gallez B, Schreiber W, Ali A, Flood AB (2020) 'Oxygen level in a tissue' - what do available measurements really report? Adv Exp Med Biol 1232:145–153

Swartz HM, Flood AB, Williams BB, Pogue BW, Schaner PE, Vaupel P (2021) What is the meaning of an oxygen measurement? analysis of methods purporting to measure oxygen in targeted tissues. Adv Exp Med Biol 1269:301–308

Whittall KP, MacKay AL, Whittall KP, MacKay AL (1989) Quantitative interpretation of NMR relaxation data. J Magn Reson 84:134–152

Williams BB, Halpern HJ (2005) In vivo EPR imaging. Biol Magn Reson 23:283–319

Letting Biology Validate the EPR Imaging of Tumor pO_2

5

Inna Gertsenshteyn, Lucas Berens, Boris Epel, Eugene Barth, Richard C. Miller, Subramanian V. Sundramoorthy, Mihai Giurcanu, John Lukens, Kayla Hall, Viktor M. Tormyshev, Bulent Aydogan, Ralph R. Weichselbaum, Mrignayani Kotecha, Chin-Tu Chen, and Howard J. Halpern

Abstract

Hypoxic resistance of living tissue to cytotoxin, most notably radiation, has been known for over a century but was believed in early trials to be a universal characteristic of all tumors. Oxygen electrode measurements of tumor pO_2 demonstrated that many tumors had insufficient hypoxia to affect tumor resistance. Electron paramagnetic resonance oxygen images have been shown to accurately quantify and locate tumor hypoxia. Using novel radiation delivery for preclinical models with isocentric tumor boosts using 3D printed radiation apertures, we compared tumor control after equivalent integral dose boosts—either to hypoxic tumor or to well-oxygenated tumor—in preclinical models. For the first time, in three preclinical tumor models, we demonstrated a significant advantage for hypoxic boosts relative to well-oxygenated boosts.

Keywords

pO2 Imaging · EPR · Tumor Control · Radiation Boosts

5.1 Introduction

Low molecular oxygen levels, hypoxia, have long been known to be a source of living cell and cancer cell resistance to radiation therapy (Schwarz 1909). After failure during the 1960s and 1970s of clinical trials assuming all tumors were significantly hypoxic (Henk et al. 1977; Henk and Smith 1977), measurements with

I. Gertsenshteyn · L. Berens
Department of Radiation and Cellular Oncology, University of Chicago, Chicago, IL, USA

Center for EPR Imaging In Vivo Physiology, University of Chicago, Chicago, IL, USA

Department of Radiology, University of Chicago, Chicago, IL, USA

B. Epel · E. Barth · R. C. Miller · S. V. Sundramoorthy · J. Lukens · K. Hall · B. Aydogan · R. R. Weichselbaum · H. J. Halpern (✉)
Department of Radiation and Cellular Oncology, University of Chicago, Chicago, IL, USA

Center for EPR Imaging In Vivo Physiology, University of Chicago, Chicago, IL, USA
e-mail: hhalpern@uchicago.edu

M. Giurcanu
Department of Public Health, University of Chicago, Chicago, IL, USA

V. M. Tormyshev
Novosibirsk Institute of Organic Chemistry, Novosibirsk, Russia

M. Kotecha
O2M Technologies, Chicago, IL, USA

C.-T. Chen
Center for EPR Imaging In Vivo Physiology, University of Chicago, Chicago, IL, USA

Department of Radiology, University of Chicago, Chicago, IL, USA

S. C. Pias et al. (eds.), *Oxygen Transport to Tissue XLVI*, Advances in Experimental Medicine and Biology 1498, https://doi.org/10.1007/978-3-032-10389-5_5

quantitative Eppendorf pO_2 electrodes were performed worldwide that tested the assumption (Hockel et al. 1996; Fyles et al. 2002; Brizel et al. 1995, 1997). Means or median values of electrode measurements with a sensitive volume of ~ $(0.25\ mm)^3$ distinguished patient's hypoxic tumors that eventually failed radiation therapy from sensitive well-oxygenated tumors that had significantly enhanced tumor cure. A common distinguishing pO_2 threshold for definition was 10 torr. Technology to define targetable regions of hypoxia defined by retention of nitroimidazole positron emission tomography (PET) radiotracer has to date failed to provide evidence of significant benefit in directing radiation dose boosts in mammalian or patient trials (Welz et al. 2022; Vera et al. 2019). Pulse spin lattice relaxation rate (SLR) images of 3 murine tumor types from iv-injected trityl spin probe provided near absolute extracellular pO_2 images with 1 mm spatial resolution and 1–2 torr pO_2 resolution (Epel et al. 2014). Three murine leg-born tumor types were imaged with registered MRI to define tumor location, CT to direct radiation treatment, and EPR pO_2 images to direct hypoxic or well-oxygenated tumor boosts after whole tumor radiation sufficient to control with 15–20% probability. Well-oxygenated or hypoxic tumor boosts were randomly selected for boost treatment. Mice were followed for 3 or 6 months depending on tumor type and literature indications of follow-up sufficiency. Kaplan-Meier analysis with log-rank test of significance demonstrated $p < 0.001$ significance improvement in the entire animal group (Epel et al. 2019; Gertsenshteyn et al. 2023). This provides the first validation of extra radiation dose as a significant enhancement of radiation treatment with oxygen guidance from pulse spin lattice relaxation spin probe-based images of tumor pO_2.

5.2 Materials and Methods

Pulse spin lattice relaxation rate (SLR) images of 3 murine tumor types from iv-injected trityl spin probe OX071 (GE Healthcare, Little Chalfont, Buckinghamshire, UK) located tumor and leg normal tissue pO_2. As published previously, the rate of relaxation of the unpaired electron on the OX071 spin probe subjected to a 9 mT magnetic field after a 60 ns 80 watt peak power radiofrequency pulse varies from 0.14 μs^{-1} at near 0 torr pO_2 to 4 μs^{-1} at 30 torr. This linear relation between spin probe relaxation rate and the local partial pressure of molecular oxygen is the basis of in vivo pO_2 images. This is primarily due to Heisenberg spin exchange of the spin probe electron with one of the two unpaired electrons in the extracellular oxygen compartment distribution of the spin probe. Thus, this probes the same tissue compartment as was sampled by the Eppendorf electrodes.

Registered MRI defined tumor location, and registered CT obtained with the treatment radiator and EPR pO_2 images directed radiation boosts to hypoxic or well-oxygenated tumor boosts after whole tumor radiation. Animal tumors were initially treated with isocentric radiation therapy delivered using opposed anterior and posterior radiation fields to the entire tumor grown in the gastrocnemius muscle, as shown in Fig. 5.1 (top), with an XRAD225/cx (Precision X-Ray, Madison CT) radiator with a 6 mm margin at 2 Gy/min. A total of 54 mice with FSa fibrosarcomas received 23.5 Gy (a tumor control dose of 15%, TCD_{15}) while 82 mice with SCC VII squamous cell carcinomas and 48 mice with MCa4 mammary adenocarcinomas were treated to a dose of 49 Gy 20% control dose (TCD_{20}). Tumor control probability as a function of tumor dose was determined in a separate set of experiments for each tumor type (Epel et al. 2019; Gertsenshteyn et al. 2023). Opposed oblique fields were designed to treat tumor boost to either all hypoxic tumor subvolumes (Fig. 5.2 left a) or an equal integral volume of well-oxygenated tumor (Fig. 5.2 center b). Tumor pO_2 was located with the EPR SLR images. 3D-printed tungsten-loaded PLA plastic-printed radiation apertures with a 1.2 mm margin selected for hypoxia boosts or 0.6 mm well-oxygenated tumor boosts directed radiation to the proper target (Fig. 5.1 bottom A Central plane projection of tumor hypoxia (magenta contours) bottom B Digital model for the hypoxic boost to 2 hypoxic

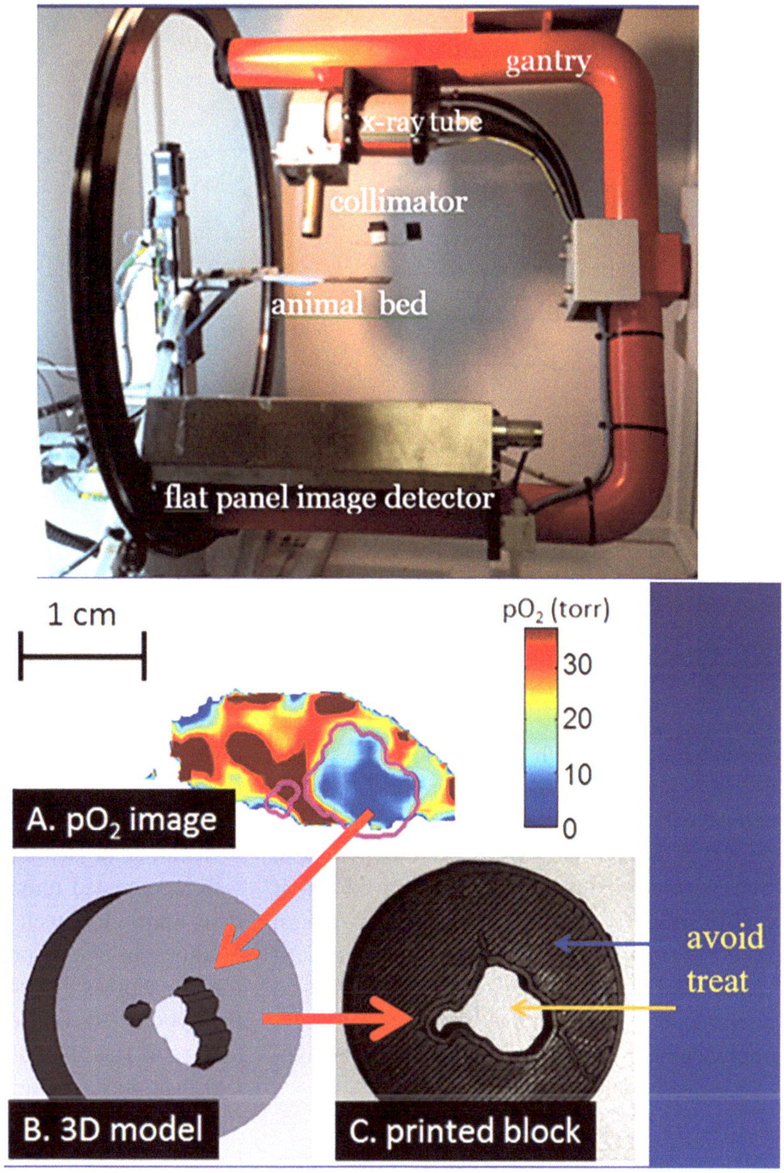

Fig. 5.1 Top: XRAD225cx animal radiator and CT imager. Bottom: (**A**) pO_2 image central pO_2 slice showing two hypoxic volumes (magenta contours) projected onto the central slice for treatment planning. (**B**) 3D model of hypoxia treatment with margin (**C**) Treatment block regions. C Actual tungsten loaded PLA plastic block for one of the opposed oblique treatment fields). Hypoxia was defined as extracellular $pO_2 \leq 10$ torr. This is more extensively described in (Epel et al. 2019; Gertsenshteyn et al. 2023).

5.3 Results

Local tumor control experiments delivered to the full tumor in the separate control versus dose described in (Epel et al. 2019; Gertsenshteyn et al. 2023) provided the dose necessary to control 10% to 20% of tumors and 98% of tumors. The experiments described here used those results to determine a low whole tumor control dose (10–20%) and the dose necessary to control 98% of the three different tumor types. Kaplan-Meier analysis with log-rank test of significance demonstrated $p < 0.00\mathbf{1}$ significance improvement in the tumor control in the entire animal group when tumor boost was directed to hypoxic tumor relative to treatment boosts directed to well-oxygenated tumor. The three tumor types

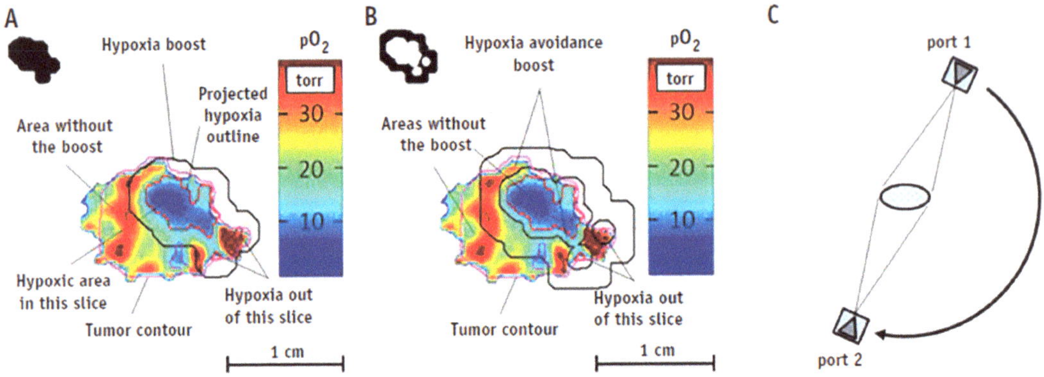

Fig. 5.2 Radiation treatment plans and delivery scheme. (**A**) EPR pO_2 image slice orthogonal to the radiation beam showing hypoxia boost treatment plan. (**B**) The same EPR pO_2 image slice showing hypoxia avoidance boost plan. Magenta contour: MRI-defined tumor margin. Red contour: projection of all in-tumor hypoxic volumes onto the EPR image plane. Black contours: radiation treatment beam shape including additional setup uncertainty margins. The area of the hypoxia-avoiding boost equals the area of the hypoxia boost. Note in both (**A**) and (**B**) the islands of hypoxia out of the plane derived from the DRR of the whole tumor volume as well as the margin about the hypoxia, 1.2 mm for the hypoxia boost and 0.6 mm for the hypoxia avoiding boost. Upper Left corners of (**A**) and (**B**): Black shapes of hypoxic boost (**A**) and well-oxygenated boost (**B**) apertures. (**C**) Illustration of opposed oblique radiation fields for boost treatment with XRAD225Cx gantry-mounted X-ray machine. Each field was treated with half of the total boost dose

which were treated with hypoxic boost are shown in Fig. 5.3 as solid lines. The three tumor types treated with well-oxygenated tumor boosts are shown as dashed lines. Boost dose delivered to well-oxygenated tumor barely improved tumor control relative to the whole tumor control expected from the whole tumor radiation control versus dose experiment by 10–20%. This was an insignificant improvement relative to the improvement by approximately 45% for the hypoxic boost tumor control.

5.4 Discussion

Clearly, the addition of radiation dose to hypoxic tumor alone failed to control tumors at the 98% control probability that might be expected if hypoxia was the only source of radiation treatment resistance. Recently recognized participation of the immune system in recognizing abnormal tumor proteins and other antigens as foreign material to be cleared from normal tissue and its self-protective immune masking provides a source of anti-cancer response other than radiation treatment. Recent work suggests that hypoxia participates in the masking of the immune response (Semenza 2021). This would argue enhanced benefit for hypoxic targeted boosts.

It can be argued that in the boost treatment of well-oxygenated tumor only, shielding the hypoxic tumor left well oxygenated tumor in the shadows of the hypoxic tumor untreated. We tried to address this concern partially by limiting the hypoxic fraction of the MCa4 and SCC VII tumors to less than 30% and the FSa tumors to less than 40%. However, if the well-oxygenated tumor were as resistant or more resistant than the hypoxic tumors, there would be a far greater likelihood of there being no difference or an opposite difference in the control probabilities.

Confounding variation of tumor hypoxia has been recognized by Brown (1979) and Dewhirst (Dewhirst et al. 2008; Braun et al. 1999) in the form of acute hypoxia and temporal changes in hypoxia in some tumor locations. Vaupel et al. estimate based on preclinical models that persistent, chronic hypoxia represents approximately 75% of mammalian hypoxia (Vaupel and Mayer 2014; Maftei et al. 2012). Definition of local, macroscopic regions of clinically relevant hypoxia may be a less burdensome task for the

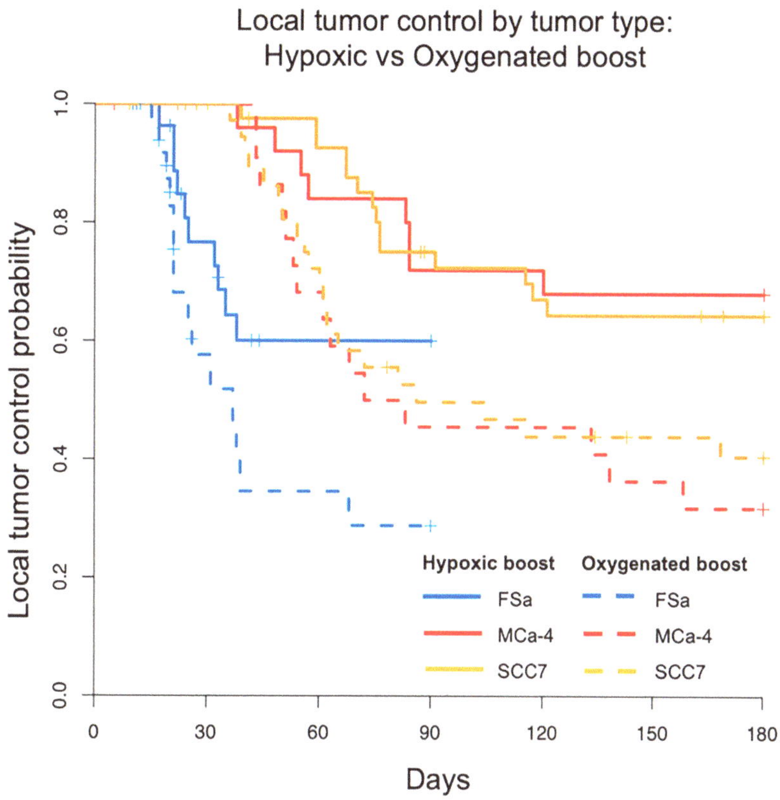

Fig. 5.3 Kaplan–Meier Local tumor control probability (LTCP) vs time comparing hypoxic boost (HB solid line) with oxygenated boost (OB dashed line) for mice with FSa (blue, 29 mice with HB, 25 with OB), MCa-4 (red, 26 HB, 22 with OB), and SCC7 (yellow, 44 HB, 38 with OB) tumor types. For all tumor types combined, LTCP is significantly different with hypoxic boost ($p < 0.001$) compared to oxygenated boost treatments using the log-rank test. Each of the three tumor types individually showed significant improvement in local tumor control ($p < 0.05$)

direction of a localized cytotoxin like radiation than the measurement of the local pO_2 at a particular time and with a particular volume resolution. The ability to direct treatment to a region with a spatial margin of error for the inevitable localization of anatomy inherent in image resolutions and immobilization of living mammalian subjects will further reduce the localization burden.

Oxygen enhancement ratios have been demonstrated in cultured cellular systems exposed to highly controlled oxygen pO_2 environments. These have shown consistent factor of three increases in radiation dose necessary to reduce clonogenic survival under hypoxic conditions relative to ambient oxygenation. Additional preliminary data have been obtained in FSa fibrosarcomas demonstrating an in vivo oxygen enhancement ratio. Whole tumors were radiated to a sequence of radiation doses followed by randomly choosing blocks exposing hypoxic tumor or well oxygenated tumor to doses of radiation sufficient to control the entire tumor with 99% probability.

A preliminary in vivo oxygen enhancement ratio of 1.26 for leg-borne fibrosarcomas is higher than the extra dose that has been used in human trials (Welz et al. 2022) and suggests that hypoxic boosts in human trials need to be increased. A novel PET EPR O_2 imager has obtained near-simultaneous ^{18}F Fluoromisonidazole PET and SLR EPR pO_2 images. Using EPR pO_2 images as ground truth, optimal PET hypoxic thresholds have been estimated in preclinical models (Gertsenshteyn et al. 2022, 2021). Work with dynamic contrast-enhanced MRI and PET is ongoing to develop preclinical corrections with patient available imaging information to sharpen hypoxia location for human hypoxia boosting.

5.5 Conclusion

This work has provided the first validation of extra radiation dose as a significant enhancement of radiation treatment with oxygen guidance from pulse spin lattice relaxation spin probe-based images of tumor pO_2. Conformal radiation therapy was used for a mammalian tumor model in three different syngeneic tumor models. Nonetheless, this provides strong evidence for using spin lattice relation-based pO_2 images in mammalian tumor models to serve as ground truth for further work to correct clinically available hypoxia images for boosting hypoxic human tumors with intensity-modulated radiation therapy.

Acknowledgments The work presented was supported by NIH grants P41 EB002034, R01 CA098575, R01 CA236385, P30CA014599.

Conflict of Interest US patent 8,664,955 has been awarded to H.H. and B.E. for pO_2 imaging methodology; they are also owners of a start-up company, O2M Technologies, LLC (Chicago, IL) that markets the pO_2 imaging technology.

References

Braun RD, Lanzen JL, Dewhirst MW (1999) Fourier analysis of fluctuations of oxygen tension and blood flow in R3230Ac tumors and muscle in rats. Am J Phys 277(2 Pt 2):H551–H568

Brizel DM, Rosner GL, Prosnitz LR, Dewhirst MW (1995) Patterns and variability of tumor oxygenation in human soft tissue sarcomas, cervical carcinomas, and lymph node metastases. Int J Radiat Oncol Biol Phys 32(4):1121–1125

Brizel DM, Sibley GS, Prosnitz LR, Scher RL, Dewhirst MW (1997) Tumor hypoxia adversely affects the prognosis of carcinoma of the head and neck. Int J Radiat Oncol Biol Phys 38(2):285–289

Brown JM (1979) Evidence for acutely hypoxic cells in mouse-tumors, and a possible mechanism of re-oxygenation. Br J Radiol 52(620):650–656

Dewhirst MW, Cao Y, Moeller B (2008) Cycling hypoxia and free radicals regulate angiogenesis and radiotherapy response. Nat Rev Cancer 8(6):425–437

Epel B, Bowman MK, Mailer C, Halpern HJ (2014) Absolute oxygen R-1e imaging in vivo with pulse electron paramagnetic resonance. Magn Reson Med 72(2): 362–368

Epel B, Maggio MC, Barth ED, Miller RC, Pelizzari CA, Krzykawska-Serda M et al (2019) Oxygen-guided radiation therapy. Int J Radiat Oncol Biol Phys 103(4):977–984

Fyles A, Milosevic M, Hedley D, Pintilie M, Levin W, Manchul L et al (2002) Tumor hypoxia has independent predictor impact only in patients with node-negative cervix cancer. J Clin Oncol 20(3):680–687

Gertsenshteyn I, Epel B, Barth E, Leoni L, Markiewicz E, Tsai HM et al (2021) Improving tumor hypoxia location in (18)F-Misonidazole PET with dynamic contrast-enhanced MRI using quantitative electron paramagnetic resonance partial oxygen pressure images. Radiol Imag Cancer 3(2):e200104

Gertsenshteyn I, Epel B, Ahluwalia A, Kim H, Fan X, Barth E et al (2022) The optimal (18)F-fluoromisonidazole PET threshold to define tumor hypoxia in preclinical squamous cell carcinomas using pO2 electron paramagnetic resonance imaging as reference truth. Eur J Nucl Med Mol Imaging 49(12):4014–4024

Gertsenshteyn I, Epel B, Giurcanu M, Barth E, Lukens J, Hall K et al (2023) Absolute oxygen-guided radiation therapy improves tumor control in three preclinical tumor models. Front Med (Lausanne) 10:1269689

Henk JM, Smith CW (1977) Radiotherapy and hyperbaric oxygen in head and neck cancer. Interim report of second clinical trial. Lancet 2(8029):104–105

Henk JM, Kunkler PB, Smith CW (1977) Radiotherapy and hyperbaric oxygen in head and neck cancer. Final report of first controlled clinical trial. Lancet 2(8029): 101–103

Hockel M, Schlenger K, Aral B, Mitze M, Schaffer U, Vaupel P (1996) Association between tumor hypoxia and malignant progression in advanced cancer of the uterine cervix. Cancer Res 56(19):4509–4515

Maftei CA, Bayer C, Shi K, Vaupel P (2012) Intra- and intertumor heterogeneities in total, chronic, and acute hypoxia in xenografted squamous cell carcinomas. Detection and quantification using (immuno-) fluorescence techniques. Strahlenther Onkol 188(7): 606–615

Schwarz G (1909) Uber Desnssibilisierung gegen Roentgen- und Radium-strahlen. Munchener Medizinsche Wochenschriff 24(15 Juni):1218–1219

Semenza GL (2021) Intratumoral hypoxia and mechanisms of immune evasion mediated by hypoxia-inducible factors. Physiology (Bethesda) 36(2):73–83

Vaupel P, Mayer A (2014) Hypoxia in tumors: pathogenesis-related classification, characterization of hypoxia subtypes, and associated biological and clinical implications. Adv Exp Med Biol 812:19–24

Vera P, Mihailescu SD, Lequesne J, Modzelewski R, Bohn P, Hapdey S et al (2019) Radiotherapy boost in patients with hypoxic lesions identified by (18)F-FMISO PET/CT in non-small-cell lung carcinoma: can we expect a better survival outcome without toxicity? [RTEP5 long-term follow-up]. Eur J Nucl Med Mol Imaging 46(7):1448–1456

Welz S, Paulsen F, Pfannenberg C, Reimold M, Reischl G, Nikolaou K et al (2022) Dose escalation to hypoxic subvolumes in head and neck cancer: a randomized phase II study using dynamic [(18)F]FMISO PET/CT. Radiother Oncol 171:30–36

6 Random Acquisition of Spectral Projections in Four-Dimensional Spectral-Spatial EPR Imaging: A Feasibility Study Toward Oxygen Imaging

Misa Oba, Mai Taguchi, Shingo Matsumoto, and Hiroshi Hirata

Abstract

In continuous-wave electron paramagnetic resonance (CW-EPR) imaging, many spectral projections are required to accurately reconstruct four-dimensional (4D) spectral-spatial data, which are used for oxygen mapping in tumors. The ability to obtain fewer spectral projections with less degradation of the resultant image quality is desirable because acquiring many spectral projections leads to a longer time for EPR imaging acquisition. The algebraic reconstruction technique (ART) is one of the most powerful image-reconstruction techniques for 4D spectral-spatial imaging. ART can be applied to incomplete sets of spectral projections. However, data sets of many spectral projections can provide reconstructed images of higher quality. In this study, we aimed to enhance acquisition speed by randomly collecting spectral projections and subsequently synthesizing those that had not been recorded. We applied this strategy to a numerical phantom and a mouse Hs766T xenograft model to confirm the feasibility of our concept. Random acquisition can prevent image degradation in linewidth mapping, which is the foundation for oxygen mapping with CW-EPR.

Keywords

EPR · 4D spectral-spatial imaging · pO_2 mapping · Tumor · Accelerated image acquisition

6.1 Introduction

Most solid tumors have a hypoxic microenvironment, which reduces the efficacy of radiation therapy and chemotherapy (Fyles et al. 1998; Cairns et al. 2006). Therefore, understanding tumor tissue oxygenation is essential for effective treatment. Various imaging modalities have been developed to evaluate tumor hypoxia (Fleming et al. 2014; Xu et al. 2017). Electron paramagnetic resonance (EPR) can quantitatively and noninvasively visualize the partial pressure of oxygen (pO_2), and provide information about the tumor microenvironment (Velan et al. 2000).

Oxygen levels are determined by measuring EPR linewidths using an established calibration function. Mapping the absorption linewidth, which is sensitive to pO_2, requires four-dimensional (4D) spectral-spatial data (three dimensions in space and one dimension in the

M. Oba · M. Taguchi
Graduate School of Information Science and Technology, Hokkaido University, Sapporo, Japan

S. Matsumoto · H. Hirata (✉)
Faculty of Information Science and Technology, Hokkaido University, Sapporo, Japan
e-mail: hhirata@ist.hokudai.ac.jp

S. C. Pias et al. (eds.), *Oxygen Transport to Tissue XLVI*, Advances in Experimental Medicine and Biology 1498, https://doi.org/10.1007/978-3-032-10389-5_6

spectral domain). However, the time required for data acquisition for such an imaging approach is longer because 4D spectral-spatial EPR imaging requires more projection data than 3D pure spatial imaging. In general, the use of fewer spectral projections, to reduce the acquisition time, leads to a degradation of image quality. To address this problem, we recently introduced a concept for preventing a degradation in image quality by computationally synthesizing spectral projections (Oba et al. 2024). However, the difference in linewidth maps with and without our concept was modest in image evaluation in a mouse xenograft model. Thus, we must further investigate the acquisition approach to reduce degradation in the resultant linewidth maps. It is well established that the use of a random sampling approach to acquire less data can prevent the quality degradation of the reconstructed image obtained through compressed sensing (Lustig et al. 2008). This method improves image quality by enabling reconstruction from a minimal number of spectral projections or k-space (Fourier domain) data. Compressed sensing is an image reconstruction technique that uses under-sampled data to recover images using underdetermined linear systems. Although we do not use the compressed sensing approach directly, it provides valuable insights for developing our data acquisition strategy. By adopting a random sampling approach for spectral projections for our CW-EPR imaging, we expect to improve spectral-spatial EPR imaging and help prevent the degradation of image quality. In this report, we demonstrate the random acquisition and computational synthesis of spectral projections for a numerical phantom and an in vivo mouse tumor model to understand how to recover the degradation of linewidth maps with the use of fewer spectral projections.

6.2 Materials and Methods

6.2.1 Concept of Random Spectral Sampling

In our previous report on 4D spectral-spatial EPR imaging, we reduced the number of spectral projections in one direction of Cartesian coordinates and performed a computational spectral projection synthesis for skipped spectral projections (Oba et al. 2024). Spectral projections were recorded with a data acquisition sequence under a systematically changing magnetic field gradient from $+G_{max}$ to $-G_{max}$ in a one-dimensional case for signal encoding. When the number of measured spectral projections was reduced, a skipped spectral projection under a magnetic field gradient G_z was computed from a measured spectral projection under a reversed magnetic field gradient $-G_z$.

This study applied 15-step magnetic field gradients from the positive maximum (first magnetic field gradient) to the negative maximum (15th magnetic field gradient) in each direction. In the data set with sequentially reduced spectral projections, the magnetic field gradients in the X-direction were shifted starting from its positive maximum to the nth magnetic field gradient. In contrast, from the $(n+1)$th to (15th) negative maximum magnetic field gradients, the spectral projections were not generated first and then later obtained from the spectral projections under corresponding opposite magnetic field gradients. The random selection for n spectral projections was made from the 1st through 15th magnetic field gradients in the data set with randomly reduced spectral projections. Image reconstruction of 4D spectral-spatial imaging was conducted using the algebraic reconstruction technique (ART). Ten combinations of spectral projections were randomly selected to avoid overlapping identical combinations.

6.2.2 Numerical Phantom

We first created a numerical model that mimicked three homogeneous solutions in tubes with different EPR absorption linewidths. The Voigt function was used to simulate the EPR signals. The Gaussian linewidth was set to 31 μT for all objects, and the linewidth of the peak-to-peak first-derivative Lorentzian lineshape was set to 33, 39, and 46 μT, respectively. The image matrix was 72^3 voxels for 3D-spatial coordinates and

512 points for the spectral coordinate. Magnetic field gradients were 15 steps with a maximum value of 0.3 mT/cm for each direction. Magnetic field scanning was performed at 1.5 mT to record EPR spectra. White noise was generated with a maximum noise amplitude of 5% of the EPR signal amplitude to investigate the influence of noise on reconstructed images.

6.2.3 In Vivo EPR Imaging of Mouse Xenograft Models

Six-week-old BALB/c-*nu*/*nu* male mice were purchased from Japan SLC (Hamamatsu, Japan). Approximately ten million Hs766T cells were subcutaneously injected into the right hind leg. EPR experiments were performed 9–16 weeks after the implantation of Hs766T cells. Tumor model mice underwent EPR imaging when the tumor volume reached 900 to 1000 mm^3. All animal experiments were performed in compliance with the “Law for the Care and Welfare of Animals in Japan” and approved by the Animal Care and Use Committee of Hokkaido University (approval no. 20-0118).

The nitroxyl radical trans-3,4-dicarboxy-2,2,5,5-tetra($^2H_3$)methylpyrrolidin-(3,4-$^2H_2$)(1^{15}N)-1-oxyl (^{2}H,^{15}N-DCP) was synthesized as described in a previous report and used for 4D spectral-spatial EPR imaging in vivo (Gorodetsky et al. 2016).

6.3 Results

6.3.1 Test of the Random Acquisition Strategy with a Numerical Phantom

We reconstructed EPR linewidths in 3D space using the random acquisition strategy with computational spectral synthesis to test the recovery of image quality. Figure 6.1a shows the Lorentzian linewidth map with the complete set of 3375 spectral projections. We retrospectively reduced the given spectral projections and synthesized those that were skipped with two approaches. In one approach, the data set of spectral projections was reduced sequentially in the x-direction. In the other approach, the data set of spectral projections was reduced randomly in the x-direction. A representative slice-selective linewidth map is shown in Fig. 6.1b. To evaluate the linewidth maps, we obtained the normalized root-mean-square error (NRMSE) and mean absolute error (MAE) of the reconstructed images with various numbers of spectral projections, as shown in Fig. 6.1c. When we computed NRMSE and MAE values, the numerical phantom was used as the ground truth. Thirty ART iterations were performed for image reconstruction. The computation time for a single cycle of ART iterations was 30 min for a complete set of spectral projections.

6.3.2 Experimental Verification of Random Acquisition and Computational Synthesis with a Mouse Xenograft Model

To further test random acquisition in 4D spectral-spatial imaging with a mouse xenograft model, we used a data set for an image of an Hs766T tumor-bearing mouse leg measured with a 750-MHz CW-EPR imager. Figure 6.2a shows a volume-rendered EPR linewidth map of the tumor-bearing leg. Figure 6.2b shows a slice-selective EPR linewidth map of the tumor-bearing right hind leg with a complete set of spectral projections. The data set of spectral projections used was the same as that in our previous study for comparison (Oba et al. 2024). We sequentially and randomly reduced the number of spectral projections. The NRMSE of the linewidth maps is shown in Fig. 6.2c. Ten different data sets of spectral projections that were not of the same combination were selected. We calculated the NRMSE using the linewidth map with a complete set of spectral projections as a reference. Figure 6.2d and e shows the linewidth maps with different numbers of spectral projections reduced sequentially and randomly. Ten ART iterations were performed for image reconstruction.

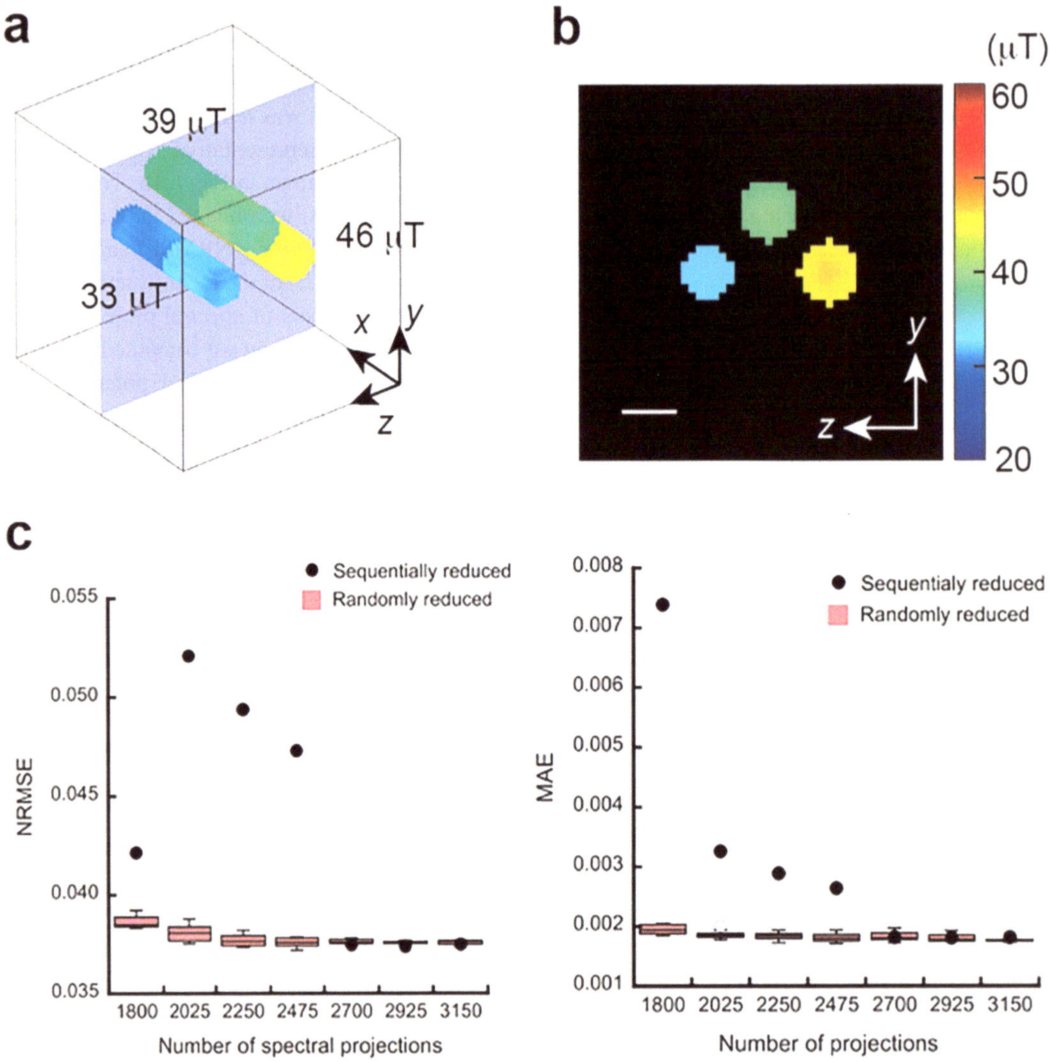

Fig. 6.1 EPR linewidth mapping of a numerical phantom. (**a**) Linewidth map reconstructed with 3375 projections, (**b**) slice-selective map of EPR linewidth with a complete set of spectral projections, (**c**) comparison of image degradation with randomly and sequentially reduced and computationally synthesized spectral projections. The image matrix is 72 × 72 × 72, and the field-of-view (FOV) is 50 × 50 × 50 mm in image reconstruction. Ten data sets of spectral projections were created by randomly reducing the spectral data in the *x*-direction. The scale bar represents 5 mm

6.4 Discussion

We applied our random acquisition approach to the numerical phantom. The NRMSE and MAE values indicated that image degradation was prevented by random acquisition and computational spectral synthesis. This approach helps recover the quality of the linewidth map that has been degraded through the use of fewer spectral projections. However, minor inhomogeneity of the linewidth maps is seen even though the numerical phantom has a uniform linewidth in each tube. In Fig. 6.1c, the NRMSE value at 1800 on the abscissa (the number of spectral projections) was smaller than the NRMSE values at other numbers of spectral projections.

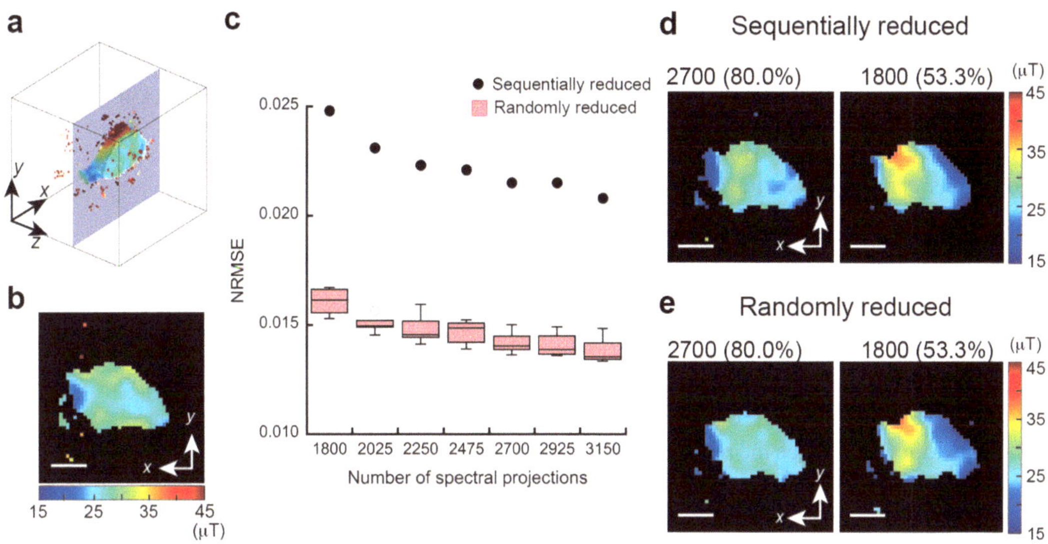

Fig. 6.2 Four-dimensional spectral-spatial imaging of a mouse xenograft model. (**a**) Volume-rendering of the linewidths for a tumor-bearing leg with 3375 spectral projections, (**b**) slice-selective EPR linewidth map reconstructed from 3375 spectral projections, (**c**) comparison of image degradation with randomly and sequentially reduced spectral projections and computational spectral synthesis, (**d**) linewidth maps for sequentially reduced spectral projections, and (**e**) linewidth maps for randomly reduced spectral projections. Ten data sets of randomly obtained projections were created, avoiding duplication. The scale bar represents 5 mm. (Panels **a**, **b**, and **d**: Reproduced with permission from Springer Nature, Oba et al. 2024)

Moreover, there is a significant shift of NRMSE between two conditions at 2475 and 2700 spectral projections. We had to evaluate the reconstructed images with another method (MAE) to understand these results because image evaluation using MAE is more robust for outliers. As a result, there were no significant shifts in the image evaluation using MAE in Fig. 6.1c. The reason for the shifts in NRMSE might be linked to the outliers of the linewidth maps.

We further applied our proposed strategy to the mouse xenograft model. The reconstructed linewidth maps suggested that random spectral acquisition could recover the image quality better than sequential spectral acquisition. The selection and sequence of magnetic field gradients may affect the image quality of mouse tumors because the signal intensity of exogenously infused imaging agents decreases due to the reduction reaction and excretion in the subject mouse in vivo. However, the image quality does not depend on the sequence of magnetic field gradients for the numerical phantom because the signal intensity remains the same over time. For in vivo animal studies, the EPR signal intensity dynamically decreases in time for the bolus injection of EPR imaging agents. If the signal profile in time is known, we can identify the preferred selection and sequence of magnetic field gradients. Such an investigation should give us a better measurement sequence for in vivo mouse tumor experiments. The results in Figs. 6.1 and 6.2 suggested that our random acquisition approach has advantages in noisy measurements, such as in vivo mouse experiments, since there is a large difference between the numerical phantom and the mouse tumor model. Random data acquisition reduces the impact of noise, and ART can average out the noise effect if sufficient spectral projections are provided for image reconstruction. In general, an incomplete data set of spectral projections can be reconstructed by ART. This acquisition approach, which synthesizes spectral projections by estimating the apparent spatial distribution and convoluting the spatial distribution with the EPR spectrum, can be applied to an asymmetrical lineshape. However, additional projections can help to achieve a more accurate representation of details. Since 4D spectral-spatial image reconstruction requires a long time (30 min for a single cycle), 4D image reconstruction should be improved to obtain a precise linewidth map in a limited time. In this work, we reduced and synthesized spectral projections in the X-direction. Then we evaluated the resultant linewidth maps. Since the numerical phantom was uniform in the X-direction, our approach was successful. If a sample geometry is more complex and non-uniform, we must evaluate the reconstructed images with a reduction in the number of spectral projections along another direction. It is important to know the directions in which the spectral projection can be effectively reduced. If a sample contains redundant information in a specific direction, the number of spectral projections in that direction can be reduced with minimal impact on the resultant image.

6.5 Conclusion

This chapter demonstrated the feasibility of recovering image degradation in 4D spectral-spatial imaging of a numerical phantom and a mouse xenograft model. The approach for randomly reducing spectral projections and computationally synthesizing the skipped spectral projections was helpful in CW-EPR-based 4D spectral-spatial imaging.

Acknowledgments The authors are grateful to Dr. Igor A. Kirilyuk, Novosibirsk Institute of Organic Chemistry, Russia, for his generous gift of $^{2}H,^{15}N$DCP probes. This work was supported by JSPS KAKENHI grants JP21K18165 and JP22H00200 (to HH).

References

Cairns RA, Papandreou I, Denko NC (2006) Overcoming physiologic barriers to cancer treatment by molecularly targeting the tumor microenvironment. Mol Cancer Res 4(2):61–70. https://doi.org/10.1158/1541-7786.mcr-06-0002

Fleming IN et al (2014) Imaging tumor hypoxia with positron emission tomography. Br J Cancer 112(2): 238–250. https://doi.org/10.1038/bjc.2014.610

Fyles A et al (1998) Oxygenation predicts radiation response and survival in patients with cervix cancer. Radiother Oncol 48(2):149–156. https://doi.org/10.1016/s0167-8140(98)00044-9

Gorodetsky AA et al (2016) Functional electron paramagnetic resonance imaging of ischemic rat heart: monitoring of tissue oxygenation and pH. Magn Reson Med 76(1):350–358. https://doi.org/10.1002/mrm.25867

Lustig M et al (2008) Compressed sensing MRI. IEEE Signal Process Mag 25(2)72–82. https://doi.org/10.1109/msp.2007.914728

Oba M et al (2024) Partial acquisition of spectral projections accelerates four-dimensional spectral-spatial EPR imaging form mouse xenograft models: a feasibility study. Mol Imaging Biol 26(3):459–472. https://doi.org/10.1007/s11307-024-01924-y

Velan SS et al (2000) Electron paramagnetic resonance oxygen mapping (EPROM): direct visualization of oxygen concentration in tissue. Magn Reson Med 43(6):804–809. https://doi.org/10.1002/1522-2594(200006)43:6

Xu Z et al (2017) 18F-Fluoromisonidazole in tumor hypoxia imaging. Oncotarget 8(55):94969–94979. https://doi.org/10.18632/oncotarget.21662

A Portable Broadband Near-Infrared Spectroscopy Device for In Vivo Oxygenation and Metabolism Measurements

7

Musa Talati, Frédéric Lange, Dimitrios Airantzis, Danial Chitnis, Temisan Ilukwe, Darshana Gopal, Paola Pinti, Niccole Ranaei-Zamani, Olayinka Kowobari, Sarah Johnson, Olivia Newth, Sara Hillman, Dimitrios Siassakos, Anna L. David, Subhabrata Mitra, and Ilias Tachtsidis

Abstract

Broadband NIRS (bNIRS) is an extension of fNIRS that provides the same assessment of oxygenation biomarkers along with a valuable marker for oxygen metabolism at a cellular level, the oxidation state of cytochrome-c-oxidase (oxCCO). bNIRS implements many (100s) wavelengths in the NIR spectrum to address this and provide insight into tissue energetics. To supply these many wavelengths of light, broadband sources are required, and spectrometers are employed to distinguish power per wavelength. Current multi-channel bNIRS instruments are bulky and only semi-portable due to technological limitations. We propose a design for a bNIRS device that has been miniaturised to allow for portable use. This design leverages the innovations in photonic devices that have created a new line of SMD-type micro-spectrometers and broadband NIR high-power LEDs. This first-of-its-kind device, referred to as microCYRIL (after its two predecessors CYRIL and miniCYRIL), has been developed for oxygenation and metabolism measurements with dual channel operation. To validate functionality, concentration changes in oxygenated (HbO_2) and deoxygenated (HHb) haemoglobin and oxCCO were successfully tracked during a cuff-induced brachial arterial occlusion.

Keywords

bNIRS · Haemoglobin · Cytochrome-C-oxidase · Instrumentation · Functional

M. Talati (✉) · F. Lange · D. Chitnis · T. Ilukwe · D. Gopal · I. Tachtsidis
Department of Medical Physics and Biomedical Engineering, University College London, London, UK
e-mail: musa.talati.20@ucl.ac.uk

D. Airantzis
Department of Medical Physics and Biomedical Engineering, University College London, London, UK

School of Computing and Mathematical Sciences, Birkbeck, University of London, London, UK

P. Pinti
Department of Medical Physics and Biomedical Engineering, University College London, London, UK

Department of Psychological Sciences, Birkbeck, University of London, London, UK

N. Ranaei-Zamani · O. Kowobari · S. Johnson · O. Newth · S. Hillman · D. Siassakos · S. Mitra
EGA Institute for Women's Health, University College London, London, UK

A. L. David
EGA Institute for Women's Health, University College London, London, UK

National Institute for Health and Care Research, University College London Hospitals Biomedical Research Centre, London, UK

S. C. Pias et al. (eds.), *Oxygen Transport to Tissue XLVI*, Advances in Experimental Medicine and Biology 1498, https://doi.org/10.1007/978-3-032-10389-5_7

7.1 Introduction

Tissue oxygenation and metabolism measurements are vital for detecting and assessing injuries and diseases. Near-infrared spectroscopy (NIRS) provides a non-invasive method to measure these metrics by quantifying changes in biomarker concentrations, such as oxygenated and deoxygenated haemoglobin (HbO_2 and HHb) and cytochrome-c-oxidase (CCO). NIRS has become a valuable tool in both research and clinical use, with many developments taking functional NIRS (fNIRS) devices from the lab to real-world environments thanks to efforts in portability and wearability. Current commercial implementations only provide oxygenation measurements and have limited ability in measuring the metabolic changes crucial for neurodevelopmental assessments (Bale et al. 2020).

The oxidation state of CCO informs on the metabolic function of the cell, as the enzyme is the terminal electron acceptor in the electron transport chain in the mitochondria, facilitating adenosine triphosphate (ATP) production. The enzyme contains a copper A redox centre which dominates the absorption spectra of the molecule at NIR wavelengths, with a strong peak in the oxidised form around 830–840 nm (Mason et al. 2014). The total CCO concentrations themselves do not change over short periods; therefore, NIRS is used to provide a marker for changes in concentration of oxidised CCO [oxCCO]. However, due to its relatively low concentrations when compared to haemoglobin, NIRS faces challenges when measuring the [oxCCO] changes despite its higher specific extinction spectrum. Attenuation variations due to metabolism can be lost within the larger attenuation changes from the HbO2 and HHb, resulting in a chromophore crosstalk; this is when a change in one element produces a spurious measurement of a change in another. These have been avoided by increasing the number of wavelengths in the NIRS range (Matcher et al. 1995), a concept implemented by broadband NIRS (bNIRS).

bNIRS uses many, at times 100s, of NIR wavelengths of light to resolve these slight changes in the absorption spectrum that CCO is responsible for. Compared to haemoglobin oxygenation changes, this can provide a marker for cell activation, and thereby tissue function, directly via changes in mitochondrial metabolism (Bainbridge et al. 2014). This technique also enables the quantification of absolute haemoglobin concentration and measurement of the optical pathlength of transmitted light.

Current bNIRS instruments are bulky and only semi-portable, requiring significant effort to move from room to room, typically caused by the size of spectrometers and cameras required for high sensitivity measurements and apparatus utilised for mains isolation in clinical environments. There is an urgent need to bring these up to the state of fNIRS systems, in miniaturisation and portability, to provide a tool that can measure tissue oxygen metabolism in vivo in a range of environments. Some “mini” bNIRS devices have been formulated from small-volume spectrometers (Diop et al. 2014; Kaynezhad et al. 2019); these solutions are static systems constrained to laboratories and/or connected to the subject via long, and at times restricting, optical fibres.

Recent advances in microelectronics and photonics, particularly the shift from CCD to CMOS sensors and from Czerny-Turner configurations to integrated spectrometer designs, have led to the development of a novel device called microCYRIL (after its predecessors CYRIL (CYtochrome Research Instrument and appLication) (Bale et al. 2014) and miniCYRIL (Kaynezhad et al. 2019)). This device is designed for oxygenation and metabolism measurements with dual-channel dual-distance operation. To verify functionality, concentration changes in

HbO_2, HHb and oxCCO were tracked during a cuff-induced brachial arterial occlusion.

7.2 Methods

7.2.1 System Design

A summary of the device specifications is given in Table 7.1. Attention has been given to the details that are required for the operation of the device. The design is split into two modules: the control module and the sensing module. This setup allows the larger control electronics to be held away from the monitoring site, while the light-weight sensing module, which includes the source, detector, and additional sensors, can be held directly against the skin. Figure 7.1 shows the complete device with both modules labelled.

The device has two channels, created by alternating power to two individual sources at 3 cm and 6 cm source detector separation (SDS), spaced linearly. The light sources are high-power broadband LEDs which provide a

Table 7.1 Specification table for the microCYRIL device

Specification		Description
Dimensions	Control module (mm)	135 × 104 × 48
	Sensor module (mm)	49 × 135 × 15
	Cable (mm)	1000
Electrical	Supply input	USB-C PD: 15V DC
	Power consumption	45W @ 15V DC
Battery	Power output (W)	100
	Capacity (mAh)	20,000
Single-board computer	Device	LattePanda 3 Delta 864
	Processor	Intel Celeron N5105
	Graphics	Intel UHD Graphics
	Memory	8 GB LPDDR4 2933 MHz
	Storage	64 GB eMMC 5.1 32 GB Micro-SD
	Operating system	Windows 10 IoT 2021
	Video	HDMI 2.0b
	USB	3× USB 3.2 Type-A
Microcontroller	Device	Arduino Leonardo
	Processor	ATmega32u4-MU
	Memory	2.5 kb SRAM, 32 kb Flash, 1 kb EEPROM
	Digital/analogue pins	20/12
	Communication	UART/I2C/SPI
Spectrometer	Device	Hamamatsu C14384MA
	Slit (H × V) (μm)	15 × 300
	Numerical aperture	0.22
	Spectral range (nm)	640 to 1050
	Spectral resolution (FWHM) (nm)	14
Light source	Device	Ushio SMBBIR45A-1100
	Effective spectral range (nm)	400–1000
	Forward current (nA)	150
	Forward voltage (V)	3.1
	Rise/fall time (ns)	90/160
	Optical power (600—1000 nm)	5.4 mW
Additional sensors	Inertial measurement unit (MPU6050)	6 DOF; 3-axis accelerometer, 3-axis gyroscope
	Temperature sensor (LM35)	Typical accuracy: ±0.25 °C @ 25 °C
Environmental conditions	Operating temperature (°C)	+5 to +50
	Storage temperature (°C)	−20 to +70

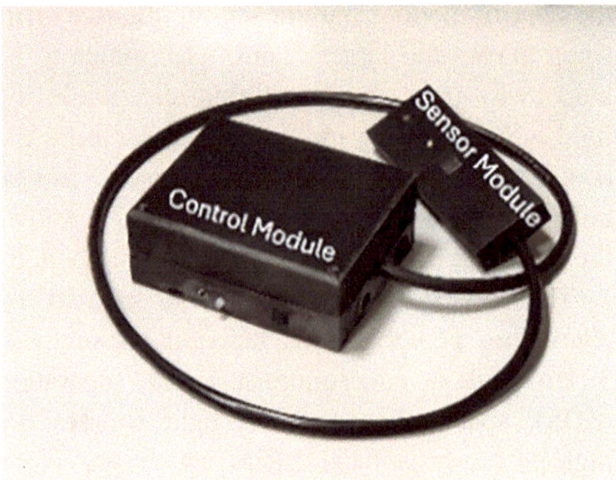

Fig. 7.1 Labelled image of the microCYRIL device

high intensity light output in the visible-NIR region (600–1000 nm) and direct light into the skin via a transparent silicone lens. The exposure to the incoherent LED visible/NIR radiation meets the guidelines as set by the ICNIRP (International Commission on Non-Ionizing Radiation Protection (ICNIRP) 2013). The spectrometer slit is coupled to patient skin via a 90° prism. Additional sensors have been included in the design, a temperature monitoring (LM35) sensor and a 6-axis movement sensor (MPU6050), the LM35 sensor allows for the removal of temperature dependence on measured spectra, and the movement sensor can be used to identify and remove artefacts caused by motion. The sensor module is connected to the control box where the LED power and timing is regulated along with the single-board computer that handles spectrometer acquisitions. A single push button and RGB LED are available for interface, and a removable SD card for data storage.

7.2.2 Software Design

The software for control of the spectrometer acquisitions, the timing sequence of the light sources and the collection of additional sensor data is performed by a Windows application that runs automatically when the device is powered on. As the program steps through each part of the acquisition, an RGB LED on the casing is lit with colours corresponding to device function. At the end of the acquisition, the device can be shut off with a press of the same power button. A flowchart of the high-level software process can be seen in Fig. 7.2. The system has been designed such that after an initial configuration of integration times and monitoring duration, it can be run independently of video output, requiring a single button press, and operation can be halted the same way.

7.2.3 Cuff Occlusion Protocol

To demonstrate the response of the microCYRIL device when measuring in vivo changes in oxygenation and metabolism, a brachial arterial cuff occlusion was performed to induce known changes in chromophore concentration and validate the response in the measured spectra, a protocol approved by the University College London Research Ethics Committee.

The spectrometer was set to an integration time of 1000 ms, and the sensing module was placed on the anterior aspect of the volunteer's bare forearm along with a blood pressure cuff around the upper arm. Initially, the cuff was deflated to monitor baseline changes before being inflated for a period and then deflated once more. Arterial occlusion was achieved by applying inflation pressure of 230 mmHg. Each period lasted 60 seconds for a total acquisition duration of 3 minutes. To convert the measured spectra into values for change in concentrations, the modified Beer-Lambert law-based UCLn algorithm was applied (Bale et al. 2016), for 136 wavelengths from 771 to 906 nm, using the 3 cm SDS and a differential path factor (DPF) of 4.16 for the arm as obtained from previous studies (Duncan et al. 1995).

7.3 Results

Figure 7.3 shows the expected desaturation of oxygen during arterial occlusion, where blood flow into and away from the forearm is halted. There is an increase in $\Delta[HHb]$ by 4.9 μM and a mirrored decrease in $\Delta[HbO_2]$ to −3.9 μM. After

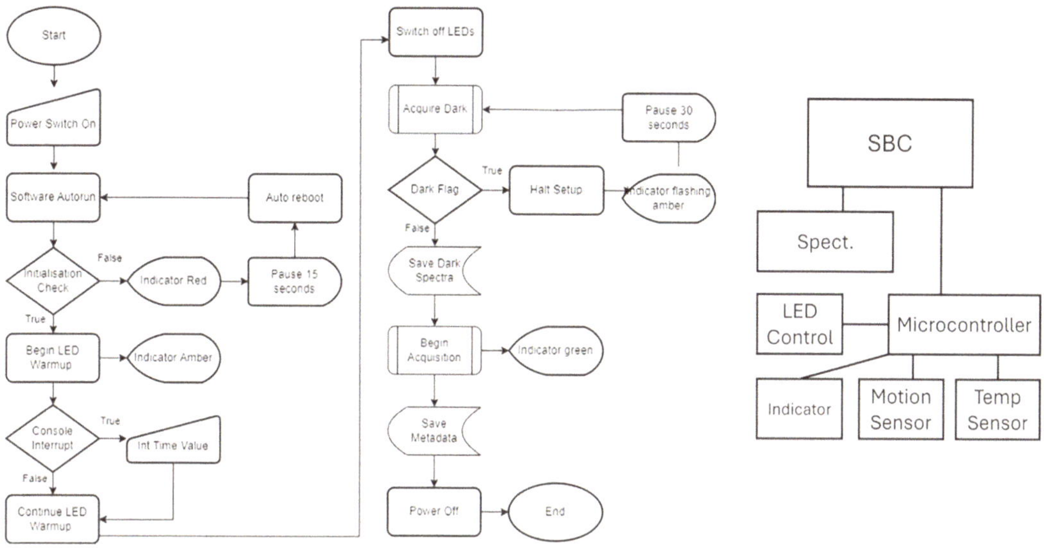

Fig. 7.2 High-level software flowchart for microCYRIL operation and hardware architecture diagram

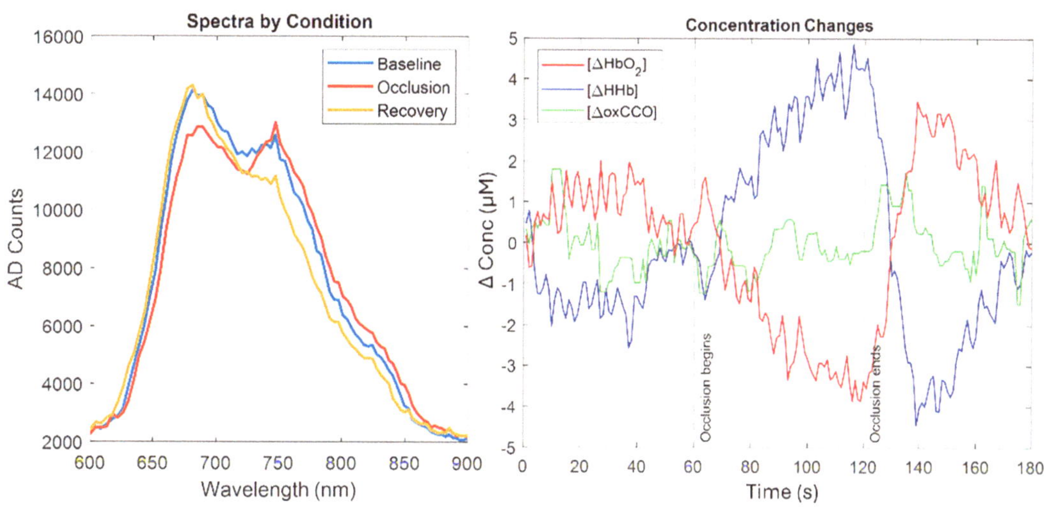

Fig. 7.3 Left: Spectra acquired through arm during baseline (blue), arterial occlusion (red) and recovery (yellow). Right: Concentration changes of chromophores during brachial occlusion

occlusion has ended, $\Delta[HbO_2]$ rises to 3.5 µM and Δ[HHb] falls to −4.4 µM before shortly returning to the baseline measurement. During the arterial occlusion, Δ[oxCCO] demonstrates little overall change as expected, though, during the reactive hyperaemia phase, when pressure is released, we see an increase of +1.7 µM, in a pattern that is consistent with previous studies. Thus, this result validates the measurement of oxygenation with the device and demonstrates the responses of oxCCO, but further testing is required to precisely validate the cytochrome signal.

7.4 Conclusion

By leveraging the innovations in miniaturisation across photonic components, a novel dual-channel bNIRS system that is portable, fibreless and wearable has been designed and conceived and demonstrates promising results in the cuff occlusion validation. This device can be used for oxygenation and metabolism tissue measurements with a capacity for non-restrictive testing, i.e. in real-world environments, without limiting movement.

References

Bainbridge A, Tachtsidis I, Faulkner S, Price D, Zhu T, Baer E et al (2014) Brain mitochondrial oxidative metabolism during and after cerebral hypoxia–ischemia studied by simultaneous phosphorus magnetic-resonance and broadband near-infrared spectroscopy. Neuroimage 102:173–183

Bale G, Mitra S, Meek J, Robertson N, Tachtsidis I (2014) A new broadband near-infrared spectroscopy system for in-vivo measurements of cerebral cytochrome-c-oxidase changes in neonatal brain injury. Biomed Opt Express 5(10):3450

Bale G, Elwell CE, Tachtsidis I (2016) From Jöbsis to the present day: a review of clinical near-infrared spectroscopy measurements of cerebral cytochrome-c-oxidase. J Biomed Opt 21(9):091307

Bale G, Mitra S, Tachtsidis I (2020) Metabolic brain measurements in the newborn: advances in optical technologies. Physiol Rep 8(17):e14548

Diop M, Wright E, Toronov V, Lee TY, St. Lawrence K (2014) Improved light collection and wavelet de-noising enable quantification of cerebral blood flow and oxygen metabolism by a low-cost, off-the-shelf spectrometer. J Biomed Opt 19(5):057007

Duncan A, Meek JH, Clemence M, Elwell CE, Tyszczuk L, Cope M, Delpy D (1995) Optical pathlength measurements on adult head, calf and forearm and the head of the newborn infant using phase resolved optical spectroscopy. Phys Med Biol 40(2): 295–304

International Commission on Non-Ionizing Radiation Protection (ICNIRP) (2013) ICNIRP guidelines on limits of exposure to incoherent visible and infrared radiation. Health Phys 105(1):74–96

Kaynezhad P, Mitra S, Bale G, Bauer C, Lingam I, Meehan C et al (2019) Quantification of the severity of hypoxic-ischemic brain injury in a neonatal preclinical model using measurements of cytochrome-c-oxidase from a miniature broadband-near-infrared spectroscopy system. Neurophotonics 6(04):1

Mason MG, Nicholls P, Cooper CE (2014) Re-evaluation of the near infrared spectra of mitochondrial cytochrome c oxidase: implications for non invasive in vivo monitoring of tissues. Biochimica Et Biophysica Acta (BBA) - Bioenergetics 1837(11): 1882–1891

Matcher S, Elwell C, Cooper C, Cope M, Delpy D (1995) Performance comparison of several published tissue near-infrared spectroscopy algorithms. Anal Biochem 227(1):54–68

Non-contact Oxygenation Assessment of Biological Tissues Based on a Time-of-Flight Method

8

Jingjing Jiang, Letizia Lanini, Djazia Yacheur, Tong Li, Meret Ackermann, Emanuele Russomanno, Aldo Di Costanzo Mata, Martin Wolf, and Alexander Kalyanov

Abstract

Assessment of oxygen saturation for biological tissues is crucial in many medical applications. A non-contact oxygenation measurement is important in clinical scenarios, for example, emergency situations and imaging-guided surgery, because non-contact measurement will speed up the measurement preparation by eliminating the need for cleaning skin and sensor for hygrine and placement of sensor. Near-infrared spectroscopy stands out as a key technology for probing tissue oxygenation. Non-contact measurement requires precise estimation of the distance between the tissue and the probe. However, high precision light detection and ranging in the near-infrared region is challenging to achieve when targeting biological tissues due to the strong light scattering. This challenge limits the design of non-contact instruments for oxygenation assessment. **Aim:** The aim is to utilize time-of-flight (ToF) sensors to accurately determine both oxygenation (optical properties (OP), i.e., absorption and scattering coefficients) of biological tissues and the distance to these tissues. **Methods:** Model-based direct ToF (DToF) models were built with a focus on highly scattering objects. A simulation study was performed to determine the OPs and distances. Two simulation tests were performed for a scattering object placed at 30 cm and 62 cm away. To demonstrate that the ToF measurements are sensitive to the OPs of the scattering object, we placed a tissue-mimicking phantom at 62 cm from the measurement plane, formed by a pulsed laser coupled to a collimator and a pinhole, while a single photon avalanche camera next to the emission point captures the ToF signals at four different wavelengths. **Results**: In both cases, the retrieved OPs and distances were very close to the reference parameters with <0.5% average error, showing the accuracy of the optimization process. The multi-spectral ToF measurements confirmed the simulation. **Conclusion:** We developed a model-based DToF approach for precise oxygenation measurement at a distance. This method holds great potential for non-contact oxygenation measurements, offering valuable insights for various clinical scenarios.

Keywords

Spectroscopy · Oxygen saturation · Light ranging · Diffuse optics · SPAD

J. Jiang (✉) · L. Lanini · D. Yacheur · T. Li · M. Ackermann · E. Russomanno · A. Di Costanzo Mata · M. Wolf · A. Kalyanov
Biomedical Optics Research Laboratory (BORL), Department of Neonatology, University Hospital Zurich and University of Zurich, Zurich, Switzerland
e-mail: jingjing.jiang@usz.ch

S. C. Pias et al. (eds.), *Oxygen Transport to Tissue XLVI*, Advances in Experimental Medicine and Biology 1498, https://doi.org/10.1007/978-3-032-10389-5_8

8.1 Introduction

Oxygen saturation assessment plays an important role in many medical applications. Near-infrared spectroscopy (NIRS) has emerged as a powerful tool for probing tissue oxygenation (Wolf et al. 1997). NIRS relies on the interaction of near-infrared light (700–1000 nm) with biological tissues, including skin, muscles, and bones. By analyzing the absorption and scattering of light, NIRS provides valuable insights into the balance between oxygen supply and consumption. Non-contact measurement approaches have advantages including speed, convenience, and hygiene (Guazzi et al. 2015). For instance, this is advantageous in wound management scenarios. Previous studies have utilized CCD cameras for peripheral blood oxygen saturation measurement (Guazzi et al. 2015) and gated single photon avalanche diode-based NIRS to measure optical properties and oxygenation of the target tissue (Mazurenka et al. 2012; Mazurenka et al. 2013). Our aim is to test the feasibility of time-of-flight (ToF) based non-contact NIRS to determine the optical properties (OP) of the target tissue and the distance to tissue, which might be relevant in imaging-guided surgeries.

8.2 Time-of-Flight Signals from a Scattering Medium at a Distance

Figure 8.1 shows the schematic of the time-of-flight system. A pulsed laser source is placed next to a ToF detector, which measures the ToF signals at a distance s from an object. Assuming an ideal instrument with an instrumental response represented by $g(t)$ as a delta function $\delta(t)$, and the reflected ToF is $f'(t) = g(t)*f(t) = f(t)$, where * stands for convolution. In a non-scattering medium, $f(t)$ acts as a simple shifting operator in time, denoted $t' = t0 + \delta t$. However, in the presence of a scattering medium, $f(t)$ introduces a delay caused by the interaction between light and the medium through scattering μ_s' and absorption μ_a. For simplicity, the instrumental response is not included in the simulation study. We solely focus on the impact of the scattering medium on measuring ToF signals. A Monte Carlo (MC) simulation of light propagation in tissue was conducted using Monte Carlo eXtreme (MCX) (Fang and Boas 2009), executed on a dual-GPU system (GeForce RTX 3090, Nvidia Corp., USA).

As illustrated in Fig. 8.2, the scattering object is a cube of size 60 mm positioned at 200 mm away from the source. Optical properties at 802 nm were used for the tissue model (Table 8.1). The refractive index is 1.4, and 1e8 photons were launched from the source. The source is a Gaussian beam with a waist size of 1 pixel (Fig. 8.2a). Figure 8.2b illustrates the temporal spread functions (TPSFs) detected at the tissue surface and at the source plane 200 mm away from the object, and both are situated 7 mm away from the source emission axis. A temporal shift corresponding to the pathlength of the light traveling forward and backward 200 mm is introduced to the TPSF

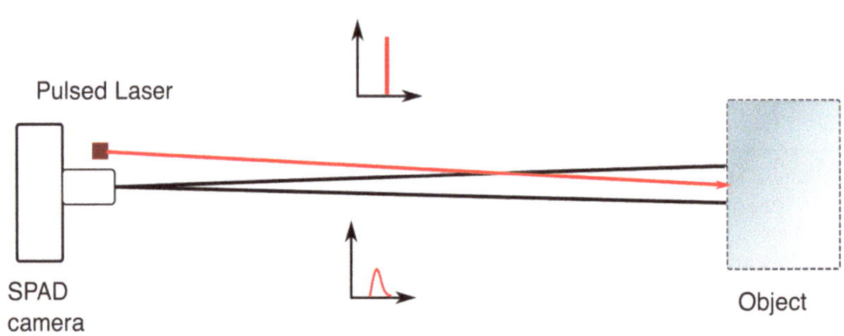

Fig. 8.1 Schematic of the ToF measurement system. The object is placed at a distance from the measurement plane. The laser is connected to a fiber and mounted to a collimator and pinhole, which are fixed on the camera, which detects the light reflected by the object

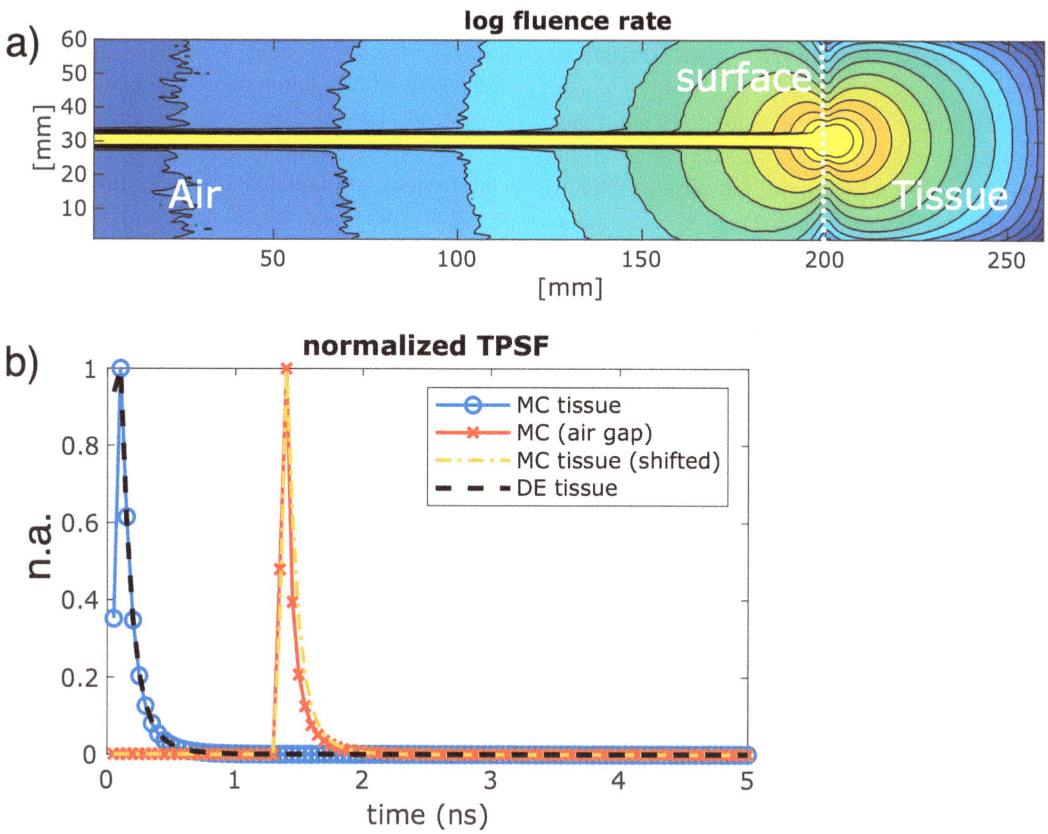

Fig. 8.2 (**a**) Visualization of the impact of a scattering medium on the fluence rate as simulated by MC using the OPs at 802 nm in Table 8.1. (**b**) Normalized TPSFs computed by MC simulation: (i) at the tissue surface (MC tissue), (ii) at a distance of 200 mm from the object (MC (air gap)), (iii) the time-shifted MC tissue with a delay corresponding to the distance of 200 mm (MC tissue (shifted), and (iv) TPSF computed using the DE. The source-detector separation is 1 cm

Table 8.1 Optical properties of the tissue phantom (measured by the Imagent (ISS Inc., USA))

Wavelengths [nm]	689	725	802	840
μ_a [mm^{-1}]	0.031	0.025	0.014	0.0075
μ_s' [mm^{-1}]	0.97	0.98	0.92	0.89

measured at the surface, showing high agreement with the TPSF measured at 200 mm away. Therefore, the modeling of the non-contact approach can be simplified to account for the light-tissue interaction and the additional time delay caused by the light passing through air.

Additionally, we employed a diffusion model to compute the TPSF, showing a high agreement with the MC results. Therefore, in subsequent sections, we use the photon diffusion equation (DE) to facilitate the optical retrieval process. The solution of the DE for a semi-infinite homogeneous medium for a ToF measurement with an impulse point source of light is applied in this study (Kienle and Patterson 1997).

8.3 Inverse Problem-Solving: Retrieval of OPs and Distance

Having investigated the forward problem of measuring ToF signals at a distance of a tissue-like object, we formulate an optimization problem to retrieve the OPs and the distance as stated below (Eq. 8.1).

$$x^* = \arg\min_{\mu} \left(\| f^m - f^{\mathrm{sim}}(x) \|_2^2 + \lambda \Lambda(x) \right), \quad (8.1)$$

where $x = \{\mu_a, \mu_s', s\}$ represents the parameter vector to be optimized and f^m denotes the acquired TPSF at a distance of s from the scattering object with optical properties $\{\mu_a, \mu_s'\}$. The optimization involves two steps. In step 1, a rough distance estimation is performed by utilizing the mean ToF of TPSF, calculated as $E[f(t)] = \int f(t)tdt$. The difference in the distance is then computed as $\Delta s_1 = (E[f^m(t)] - E[f^{\mathrm{sim}}(t, x = x_0)])/c/2$, where $E[f^{\mathrm{sim}}(t, x = x_0)]$ is mean ToF with the initial guess value $x_0 = \{\mu_{a0}, \mu_{s0}', s_0\}$. The estimated distance is then calculated as $s_1 = s_0 + \Delta s_1$. Then the second step is to solve Eq. (8.1) with a non-linear least square solver lsqcurvefit in MATLAB to obtain the optimal x^*..

An example simulation is shown in Fig. 8.3; the reference TPSF was simulated for a distance of 300 mm away from the scattering object of OPs at 802 nm (Table 8.1). The TPSF was simulated with a temporal resolution of 50 ps over a duration of 5 ns, with a vertical source-detector separation of 1 cm. To recover the OPs and distance, we initialized with a random guess x_0={0.01 mm^{-1}, 0.7 mm^{-1}, 100 mm}. The rough distance was computed as s_1= 301.21 mm (error = 0.4%). Through an iterative fitting process, the final optimized values {0.0139 mm^{-1}, 0.9167 mm^{-1}, 300.0382 mm} (error = {0.4%, 0.35%, 0.01%}) (shown in Fig. 8.3a) was achieved for $\{\mu_a, \mu_s', d\}$, reaching the stopping criterion where the minimal step size is smaller than a predefined value of 10^{-12}. Figure 8.3b shows the TPSFs with reference, initial guess, and recovered OP and distance values. To test the robustness of the method, we also performed another test using the reference parameters {0.025 mm^{-1}, 0.98 mm^{-1}, 620 mm} where the object is placed 620 mm away as later in the phantom experiment and measured at 725 nm. We kept the same initial guess and repeated the optimization procedure and obtained the results {0.0250 mm^{-1}, 0.9800 mm^{-1}, 620.0015 mm}. In both cases, there is close agreement in the TPSFs between the reference and the recovered signals, showing the accuracy of the optimization process.

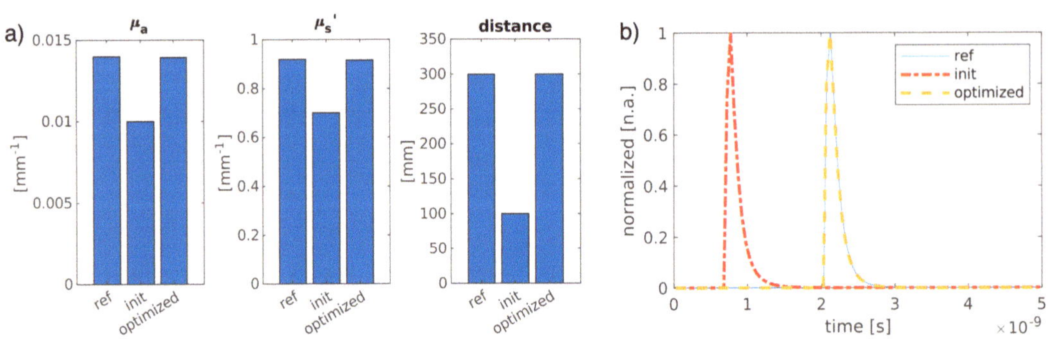

Fig. 8.3 (a) Inverse problem-solving results with the reference value (ref), the initial guess (init), and the final optimized value (optimized). (b) The forward results (normalized) with the reference, the initial guess, and the optimized parameters

8.4 Experiment with a Time-of-Flight Sensor

A pico-second super-continuum laser SuperK Extreme EXR-15 at repetition rate of 80 MHz and an acousto-optic tunable filter (AOTF) (NKT Photonics, Denmark) were used to generate pulsed laser beams at selected wavelengths 689 nm, 725 nm, 802 nm, and 840 nm with <1 mW power. A silicone phantom of the OPs shown in Table 8.1 was placed at a distance of 620 mm away from the detection side as shown in Fig. 8.1. The reflected light was detected by a single-photon avalanche diode (SPAD) sensor (Zhang et al. 2018). The phantom fabrication process is described in (Jiang et al. 2020). A lens 59 lgc 916 (Melles Griot, US) projects the tissue surface to a SPAD array sensor with an f-number 1.4 for this measurement. Measurements at the four wavelengths were conducted under two conditions: (1) with a white paper placed on the phantom and (2) with the phantom alone. The TPSFs were obtained by summing up TPSFs from 9 pixels with the central position at approx. 7.5 mm away from the peak pixel. The acquisition time at 725 nm was ~16.7 s to acquire 26.8 million photons with 13 dark photon counts on average for each time bin.

Figure 8.4 shows the TPSFs of reflected light from a tissue phantom and a white paper placed on the phantom. As anticipated, in the case of the paper, pulsed light was reflected directly to the camera with negligible attenuation by the phantom. In contrast, in the case of phantom measurement, the TPSFs were broadened and delayed due to scattering and absorption of light within the phantom. This shows the limitation of the conventional direct ToF method applied to accurate

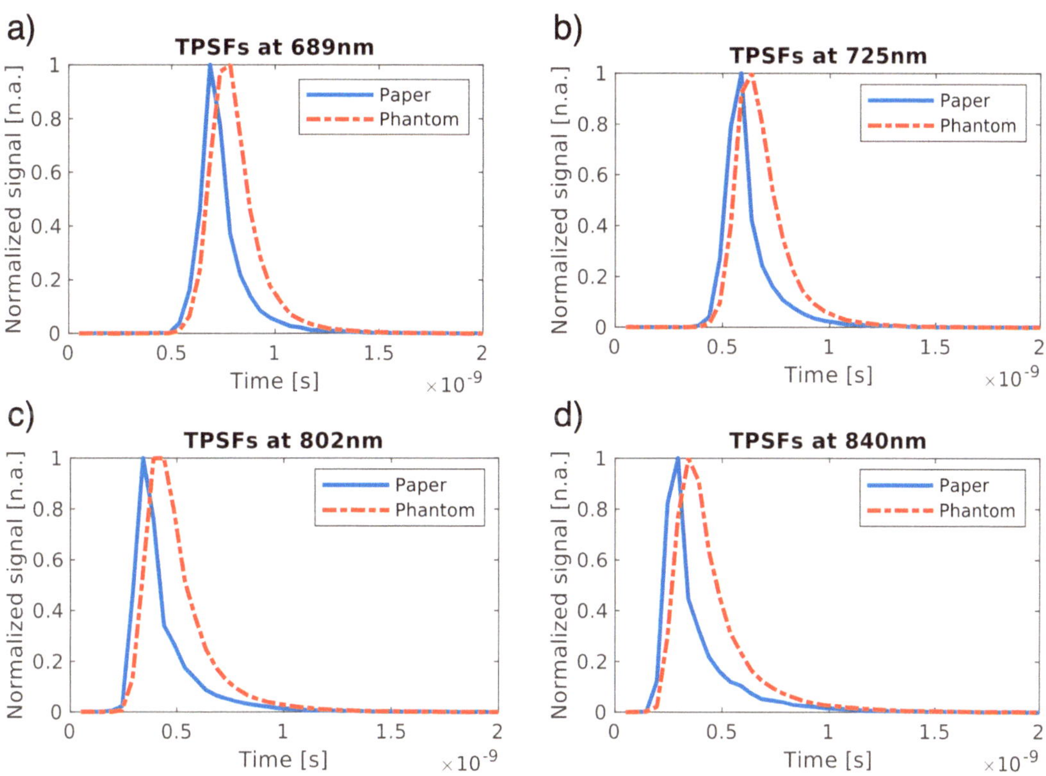

Fig. 8.4 TPSFs of reflected light from both a tissue phantom and a white paper on the phantom positioned at 62 cm from the detection plane. The TPSFs were summed up from 9 pixels with the central position at approx. 7.5 mm away from the peak pixel

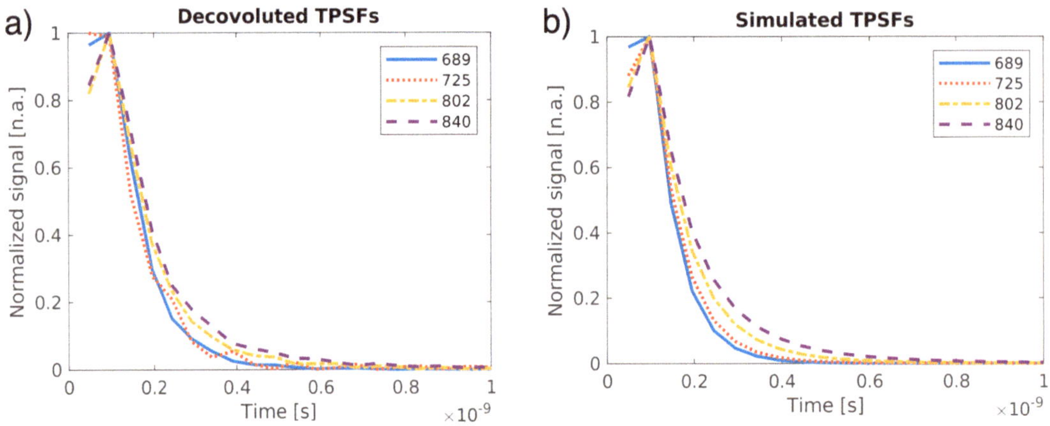

Fig. 8.5 (**a**) Deconvoluted TPSFs at 4 wavelengths. The relation between the TPSFs at 4 wavelengths closely matched the simulated results (**b**). (**b**) Simulated TPSFs at 4 wavelengths with the given OPs (shown in Table 8.1)

distance measurement for scattering objects when it ignores the light-tissue interaction. Deconvolution of the TPSFs was performed to eliminate the direct reflection from the object surface (Fig. 8.5a). To validate the findings, TPSFs were simulated using Eq. (8.1) with the OPs in Table 8.1 and the same source-detector separation (Fig. 8.5b). Although the TPSF at 725 nm was noisier, generally, the widening and delaying patterns observed in the simulations were consistent with those measured at the selected four wavelengths.

8.5 Conclusion

This chapter proposes a new non-contact method for measuring oxygen saturation through multi-spectral ToF measurements. Simulation studies showed the theoretical validation of the concept, i.e., we accurately retrieved the OPs and the distance to a tissue-like object. Multi-spectral experiments have shown the same patterns of TPSFs. This indicates that the results of the simulation can also be achieved experimentally. This study focuses on homogenous scattering media. However, in real clinical scenarios, tissues are heterogeneous, and the depth sensitivity needs to be investigated. Future research will need to address the signal-to-noise challenges in non-contact optical systems to obtain accurate OPs and distance experimentally.

Acknowledgments We gratefully acknowledge funding by Swiss National Science Foundation (159490 and 197079), Innovationspool of University Hospital Zürich Bridge Proof of Concept (40B1-0_221535), SNF Project 197079, Palatin Stiftung, Zurich Neuroscience Center (ZNZ), and Innosuisse.

References

Fang Q, Boas DA (2009) Monte Carlo simulation of photon migration in 3D turbid media accelerated by Graphics Processing Units. Opt Express 17(22): 20178–20190

Guazzi AR, Villarroel M, Jorge J et al (2015) Non-contact measurement of oxygen saturation with an RGB camera. Biomed Opt Express 6(9):3320–3338

Jiang J, Costanzo Mata AD, Lindner S et al (2020) Image reconstruction for novel time domain near infrared optical tomography: towards clinical applications. Biomed Opt Express 11(8):4723–4734

Kienle A, Patterson MS (1997) Improved solutions of the steady-state and the time-resolved diffusion equations for reflectance from a semi-infinite turbid medium. J Opt Soc Am A Opt Image Sci Vis 14(1):246–254. https://doi.org/10.1364/josaa.14.000246

Mazurenka M, Jelzow H, Wabnitz H et al (2012) Non-contact time-resolved diffuse reflectance imaging at null source-detector separation. Opt Express 20(4. Issue 1):283–290

Mazurenka M, Di Sieno L, Boso G et al (2013) Non-contact in vivo diffuse optical imaging using a time-gated scanning system. Biomed Opt Express 4(10):2257–2268

Wolf M, Bucher HU, Dietz V et al (1997) How to evaluate slow oxygenation changes to estimate absolute cerebral haemoglobin concentration by near infrared spectrophotometry in neonates. In: Nemoto EM et al (eds) Oxygen transport to tissue XVIII, Advances in experimental medicine and biology, vol 411. Springer, Boston, MA

Zhang C, Lindner S, Antolovic IM et al (2018) A cmos spad imager with collision detection and 128 dynamically reallocating tdcs for single-photon counting and 3d time-of-flight imaging. Sensors 18(11):4016

Dependence of Reflectance on Optical Properties at Short Distance (Part A: Adaptation of Monte Carlo Simulations)

9

Letizia Lanini, Djazia Yacheur, Tong Li, Alexander Kalyanov, Meret Ackermann, Emanuele Russomanno, Aldo Di Costanzo Mata, Martin Wolf, and Jingjing Jiang

Abstract

Accurate description of light propagation in tissue is a primary requirement in most optical imaging techniques, and enables to quantify for example the oxygen saturation. The Monte Carlo (MC) method, employing Henyey-Greenstein (HG) phase function, is a classical numerical approach to simulate the path of photons in tissue. However, it loses accuracy when describing short light propagation distances (<10 mm), where scattering is anisotropic. The aim of this work was to develop and test different approaches to mitigate this deficiency, employing the modified Henyey-Greenstein (MHG) and Gegenbauer (GB) phase functions. The updated scattering angle probability was implemented in the MC toolbox MCXLAB, written in CUDA and callable in MATLAB. Simulations were performed at source-detector distances from 1.5 to 5 mm, to test the behavior of the new solutions. We observed higher adaptability of the simulated curve due to employing MHG and GB phase functions compared to the conventional HG, due to the presence of the additional γ parameter in the equation that enables to adjust for the anisotropy.

Keywords

Phase function · Backward scattering · Photon transport modeling · Modified Henyey-Greenstein · Gegenbauer

9.1 Introduction

Optical imaging techniques exploit non-ionizing light to investigate biological tissue and retrieve important information from it. Two parameters of interest are the absorption coefficient μ_a and scattering coefficient μ_s, which represent the inverse of the mean path length traveled by a photon before being absorbed or its propagation direction being deviated. Both parameters are determined by the combination of all tissue components and structures, and knowing μ_a allows to retrieve the concentration of molecules. Hereby, for the case of oxygen saturation measurements, employing optical imaging at multiple wavelengths allows to recover the amount of oxygenated and deoxygenated hemoglobin in tissue (Wang and Hsin-I 2009).

Authors Letizia Lanini and Djazia Yacheur have equally contributed to this chapter.

L. Lanini (✉) · D. Yacheur · T. Li · A. Kalyanov · M. Ackermann · E. Russomanno · A. Di Costanzo Mata · M. Wolf · J. Jiang
Biomedical Optics Research Laboratory (BORL), Department of Neonatology, University Hospital Zurich and University of Zurich, Zürich, Switzerland
e-mail: laninil@student.ethz.ch; dyacheur@student.ethz.ch

S. C. Pias et al. (eds.), *Oxygen Transport to Tissue XLVI*, Advances in Experimental Medicine and Biology 1498, https://doi.org/10.1007/978-3-032-10389-5_9

The process employed to retrieve the values of μ_a and μ_s relies on accurate description of the path taken by light in tissue. Well-established models exist to precisely simulate light propagation in materials with specific optical properties. Diffusion equations are an accurate and convenient approximation of the radiative transfer equations when describing light propagation in isotropic media with high scattering. When the latter conditions are not met, numerical alternatives are available, such as Monte Carlo (MC) methods. The latter are computationally more expensive, but have a broader range of applicability. Common MC repositories for optical imaging target turbid media, where high scattering makes light propagation spatially quasi-isotropic. There are however instances that do not fall in this category, such as propagation over short distances (<10 mm), where scattering still is anisotropic (Bevilacqua and Depeursinge 1999). The latter condition is nevertheless relevant in certain clinical situations, for example to investigate peripheral oxygen saturation in preterm neonates where limited space is available for sensor placement. The goal of this work is to investigate the latter case and adapt an MC algorithm to account for the specific requirements of optical imaging at short propagation lengths. Two alternatives proposed in literature are investigated, namely the implementation of the modified Henyey-Greenstein (MHG) and the Gegenbauer (GB) phase functions, which are expected to give a more accurate sampling of the scattering angle.

9.2 Theory and Methods

Classical MC algorithms employ the Henyey-Greenstein (HG) phase function to describe the probability density of the scattering angle θ in tissue. For practical reasons, cosθ, instead of θ itself, is directly sampled in most MC algorithms. The same approach has been considered in this work, and special attention has been paid to consistency of notation regarding the chosen variable in the following declarations. Considering this, the normalization condition for the phase function p reads $\int_{-1}^{1} p(\cos\theta)d\cos\theta = 1$. The resulting HG phase function has the form:

$$p_{HG}(g_{HG}, \cos\theta) = \frac{1}{2}\frac{1-g_{HG}^2}{\left(1+g_{HG}^2-2g_{HG}\cos\theta\right)^{3/2}}, \tag{9.1}$$

and depends on the first-order Legendre moment of the phase function, $g_{HG} = g_1 = \int_{-1}^{1}\cos\theta p(\cos\theta)d\cos\theta$, also called anisotropy factor (Bevilacqua and Depeursinge 1999; Binzoni et al. 2006). Sampling of cosθ for MC simulations is then performed by generating a random number ξ between 0 and 1, and then solving

$$\xi = \int_{-1}^{\cos\theta} p_{HG}(g_{HG}, \cos\theta')d\cos\theta' \tag{9.2}$$

for cosθ (Wang and Hsin-I 2009).

When treating short propagation distances, the scattering angle probability has to be carefully modified to account for the amount of backscattering underestimated by HG. It was proven beneficial for a better description of anisotropic scattering to consider the second-order moment, $g_2 = \int_{-1}^{1}\frac{1}{2}(3\cos^2\theta - 1)\,p(\cos\theta)d\cos\theta$ (Bevilacqua and Depeursinge 1999). Therefore, a second term is present in the MHG phase function:

$$p_{\mathrm{MHG}}(\alpha, g_{HG}, \cos\theta) = \alpha\frac{1}{2}\frac{1-g_{HG}^2}{\left(1+g_{HG}^2-2g_{HG}\cos\theta\right)^{3/2}} + (1-\alpha) \times \frac{3}{2}\cos^2\theta, \tag{9.3}$$

with $\alpha \in [0, 1]$ as weighting factor. In this case, the first- and second-order moments read $g_{1,\mathrm{MHG}} = \alpha g_{HG}$, and $g_{2,\mathrm{MHG}} = \alpha g_{HG}^2 + \frac{2}{5}(1-\alpha)$ (Bevilacqua and Depeursinge 1999; Naglič et al. 2016). A second option that has been adopted for describing scattering in tissue is the GB phase function:

$$p_{GB}(\alpha_{GB}, g_{GB}, \cos\theta) = 2\alpha_{GB} g_{GB} \frac{\left(1 - g_{GB}^2\right)^{2\alpha_{GB}}}{\left[(1 + g_{GB})^{2\alpha_{GB}} - (1 - g_{GB})^{2\alpha_{GB}}\right]\left(1 + g_{GB}^2 - 2g_{GB}\cos\theta\right)^{1+\alpha_{GB}}}, \quad (9.4)$$

where g_{GB} is an anisotropy factor ($g_{GB} \neq g_{1,\ GB}$), and $\alpha_{GB} > -0.5$ a secondary anisotropy term (Naglič et al. 2016; Calabro and Cassarly 2015).

To ease the process of defining the relevant variables in MC simulations, similarity relations are employed, linking different combinations of certain parameters that lead to the same reflectance. The reduced scattering coefficient $\mu_s' = \mu_s(1 - g_1)$ is commonly used instead of the scattering coefficient. When considering second-order moments, a new similarity relation is available and merges first- and second-order moments into a new variable $\gamma \equiv \frac{1-g_2}{1-g_1}$ (Bevilacqua and Depeursinge 1999).

As reference for performing MC simulations of light propagation in tissue, the open-source and GPU-facilitated code package MCXLAB (Fang and Boas 2009) was taken into account. It compiles a CUDA source code, employing the HG phase function, into a MEX file which is directly callable in MATLAB. The source code of MCXLAB was modified to host MHG and GB phase functions, and to enable easy input of γ as an additional variable. The α parameter in MHG was computed inside the algorithm from g and γ, by rearrangement of the equations defining g_1, g_2, γ. The result $\alpha = \left(\gamma - \frac{3}{5}\right)/\left(\gamma g - g^2 + \frac{2}{5}\right)$ leads to a condition for $\gamma < 1 + g$ in MHG. Due to the more complex form of $g_{1,\ GB}$ and $g_{2,\ GB}$, a lookup table was instead built for retrieving the values of α_{GB} and g_{GB} from given γ. Values of γ up to 3 are achievable with GB.

The reflectance from light propagation through short source-detector distances was simulated with 10^6 photons in MATLAB R2022b employing all three presented phase functions. The MC algorithms were run on a GPU workstation Nvidia Gtx Titan Xp (12 GB memory size, 64 GB DDR4 memory card). We considered a $79 \times 51 \times 33$ mm^3 volume with grid unit of 0.5 mm, a detector with diameter of 1 mm in the center of the first surface, and an adjustable amount of pencil sources on the same surface, placed at distances ρ from 1.5 to 5 mm from the detector. The chosen geometry approximates the experimental setup designed to compare the simulations with experimental data, which is described in detail in Part B (Yacheur et al. 2026). Combinations of $\mu_a = 0.04, 0.06, 0.08, 0.1$ mm^{-1} and $\mu_s' = 0.5, 1.0, 1.5, 2.0$ mm^{-1} were considered to simulate neonatal to adult human brain parameters and test the diffusion approximation validity limit, allowing to understand where more attention has to be paid to the phase function employed (Fukui et al. 2003).

9.3 Results and Discussion

The scattering angle distribution obtained from MHG and GB phase functions was investigated by sampling 10^7 random numbers in the interval (0,1). Fig. 9.1a shows the probability density of cosθ sampled using the standard HG phase function, compared to using MHG and GB, for a set of parameters g, α, and γ representative of biological tissue and used throughout this chapter. The discrete distribution of the scattering angle obtained by each phase function is shown in Fig. 9.1.b, where we observe a larger amount of angles $>\pi/2$ being sampled with the new models, in accordance with (Calabro and Bigio 2014). A statistic of angles close to 0° is lost as a consequence of sampling cosθ directly and plotting θ through the nonlinear function arccos. It is possible to counteract this phenomenon with a scaling factor proportional to the derivative of arccos; however, because of the non-direct use of θ in the algorithms and the proven precision of considering cosθ directly, this operation was not performed (Binzoni et al. 2006). Fig. 9.2.a shows cross sections of the fluence rate on the

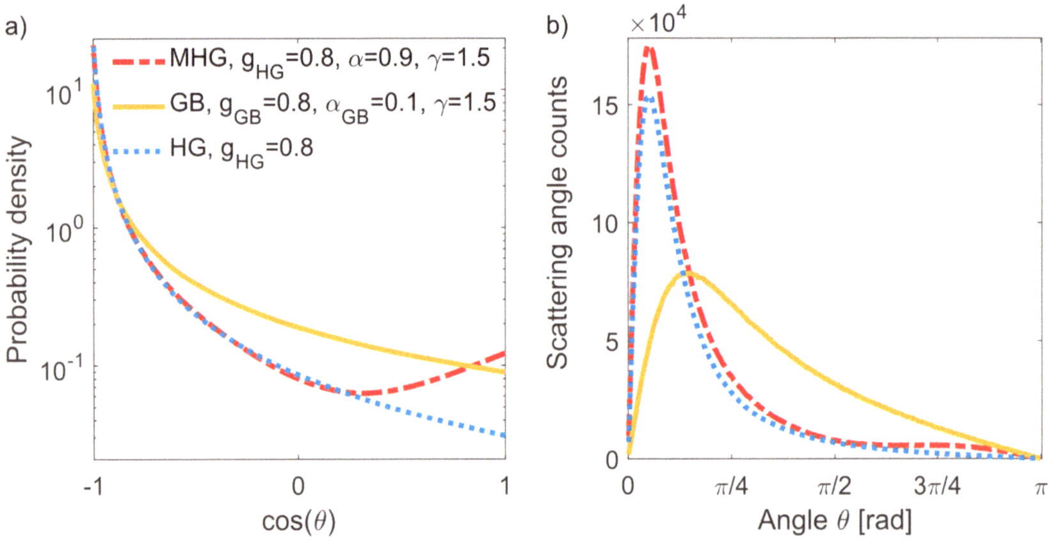

Fig. 9.1 Comparison of (**a**) the probability density of the sampled cosθ and (**b**) the resulting distribution of θ, obtained employing HG, MHG, and GB phase functions with the given parameters

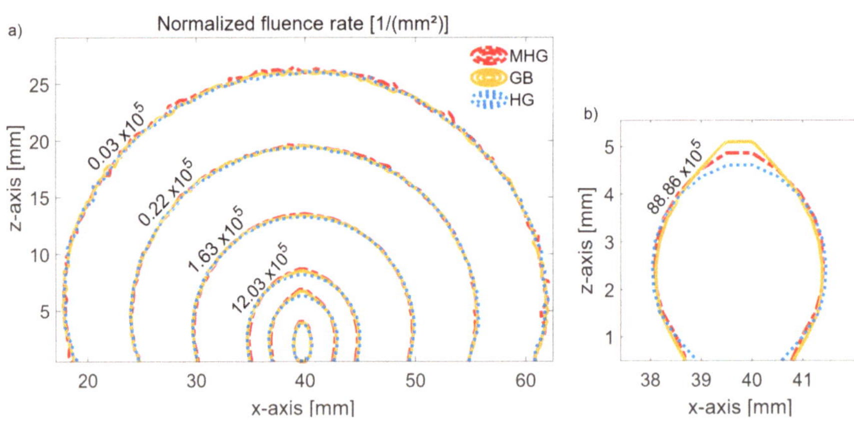

Fig. 9.2 (**a**) Cross section of the fluence rate in the xz-plane, and (**b**) zoom on the inner front, as a comparison between employing HG, MHG, and GB phase functions with the same parameters used in Fig. 9.1 and optical properties of the media $\mu_a = 0.04$ mm^{-1} and $\mu_s' = 0.5$ mm^{-1}

xz-plane, from a light source placed at xyz-coordinates (39.5, 25.5, 0) mm in a media with $\mu_a = 0.04$ mm^{-1} and $\mu_s' = 0.5$ mm^{-1}. We observe comparable fronts at the largest distances from the source, as the amount of scattering events increases and brings light propagation toward homogeneity. Regarding shorter propagation distances, we notice in Fig. 9.2.b a broader front at z = 0 close to the source when using MHG and GB with respect to HG. This is an expected consequence of sampling more angles close to π, which translates into a larger amount of backscattered photons that reemerge earlier.

The reflectance obtained by employing different phase functions is shown in Fig. 9.3. For demonstrative reason, only the reflectance obtained with $\mu_a = 0.04$ mm^{-1} and $\mu_s' = 0.5$ mm^{-1} is shown. Both MHG and GB differ

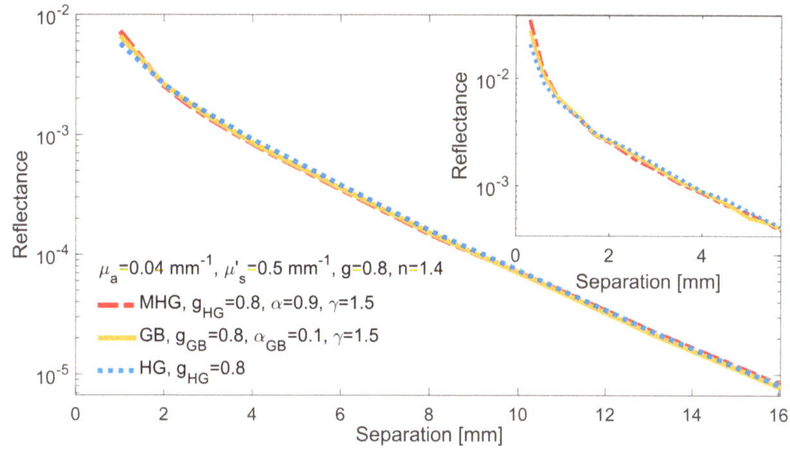

Fig. 9.3 Reflectance over source-detector distance from employing MHG, GB, and HG phase functions. Additional insights into the reflectance at shorter propagation distances are in the insert, which employs a 5× higher spatial resolution of the defined geometry

from standard HG at propagation distances <10 mm, and merge with HG when traveling a longer path. This behavior is again due to a larger amount of backward scattering angles, resulting in more photons reemerging from the surface at very short distance from the source (~2 mm). At propagation distances between 2 mm and 10 mm, a lower reflectance is consequently noticeable, before reaching the diffusion regime where the specific direction of single scattering events cancels out (Bevilacqua and Depeursinge 1999). The reflectance curves from MHG and GB are comparable when the same γ is considered, as expected from the definition of the parameter in relating the amount of backward and forward scattering, and minor differences possibly result from higher order moments (Calabro and Bigio 2014).

Benefits of considering higher-order moments for propagation distances in the range of 1.5–5 mm were assessed in comparison to experimental data and discussed in (Yacheur et al. 2026). Regarding the process of fitting experimental data, the more restricted range of γ applicable in MHG represents a slight limitation of this method, together with a longer computational time needed to perform numerical optimization due to the lack of an analytical solution of eq. (9.2) using p_{MHG}. On the other hand, GB accepts a larger amount of γ values, but necessitates a precise procedure for retrieval of its parameters from the lookup table.

9.4 Conclusions

We successfully incorporated MHG and GB phase functions in MC simulation models, including a new parameter γ whose input enables better tuning of backward scattering. The reflectance simulated with MHG and GB phase functions is consistent with previous literature results. The trend is expected when sampling a larger number of scattering angles >π/2, and this led to accurate fitting of experimental data in Part B (Yacheur et al. 2026) of this study. Employing MHG and GB phase functions at optical distances $\rho\mu'_s$ between 0.75 and 10 showed higher adaptability of the simulations compared to using HG, and was proven beneficial when short geometries are considered and high precision is required.

Acknowledgments We gratefully acknowledge funding by the Swiss National Science Foundation (159490 and 197079), Innovationspool of University Hospital Zürich, Bridge Proof of Concept (40B1-0_221535), Zürich Neuroscience Center (ZNZ), Innosuisse and Vontobel Stiftung.

Competing Interests Martin Wolf declares that he is a co-founder and president of the board of OxyPrem AG and Alexander Kalyanov is working partly for this company.

References

Bevilacqua F, Depeursinge C (1999) Monte Carlo study of diffuse reflectance at source–detector separations close to one transport mean free path. J Opt Soc Am A 16: 2935–2945

Binzoni T, Leung TS, Gandjbakhche AH et al (2006) The use of the Henyey–Greenstein phase function in Monte Carlo simulations in biomedical optics. Phys Med Biol 51:N313

Calabro KW, Bigio IJ (2014) Influence of the phase function in generalized diffuse reflectance models: review of current formalisms and novel observations. J Biomed Opt 19(7):75005

Calabro KW, Cassarly W (2015) Modeling scattering in turbid media using the Gegenbauer phase function. Proc SPIE 9333:93330F

Fang Q, Boas D (2009) Monte Carlo simulation of photon migration in 3D turbid media accelerated by graphics processing units. Opt Express 17(22):20178–20190

Fukui Y, Ajichi Y, Okada E (2003) Monte Carlo prediction of near-infrared light propagation in realistic adult and neonatal head models. Appl Opt 42:2881–2887

Naglič P, Cugmas B, Pernuš F et al (2016) Extraction of optical properties in the sub-diffuse regime by spatially resolved reflectance spectroscopy. In: Proceedings of the SPIE 9706, optical interactions with tissue and cells XXVII, 97061C

Wang LV, Hsin-I W (2009) Monte Carlo modeling of photon transport in biological tissue. In: Wang LV, Wu H-I (eds) Biomedical optics: principles and imaging

Yacheur D, Lanini L, Li T et al (2026) Dependence of reflectance on optical properties at short distance (Part B: experimental validation). Adv Exp Med Biol 1498

Dependence of Reflectance on Optical Properties at Short Distance (Part B: Experimental Validation)

10

Djazia Yacheur, Letizia Lanini, Tong Li, Meret Ackermann, Alexander Kalyanov, Emanuele Russomanno, Aldo Di Costanzo Mata, Martin Wolf, and Jingjing Jiang

Abstract

Background Determining optical properties of biological tissues enables critical clinical insights, e.g., quantifying hemoglobin oxygenation in preterm infants or detecting malignant tissues in cancer diagnostics. In applications such as neonatal monitoring or endoscopic imaging, measurements at short distances (less than 1 cm) are essential due to space constraints.

Aim This study aims to investigate the relationship between the tissue optical properties and the reflectance at short distances using a combination of experimental data and a modified Monte Carlo (MC) simulation.

Materials and Methods Twelve phantoms with different optical properties were created using silicone and validated using the commercial frequency domain near infrared spectroscopy system ISS Imagent. Reflectance measurements were conducted at precise source-detector separations ranging from 1.5 mm to 5 mm. Modified MC simulations incorporating the modified Henyey-Greenstein (MHG) and Gegenbauer (GB) phase functions were employed and simulated and measured data were compared.

Results The reflectance data exhibited a clear dependence on absorption and scattering coefficients. The MHG and GB models provided better fits to experimental data compared to the traditional Henyey-Greenstein (HG) model. The median value of the intraclass correlation coefficient (ICC) among all the investigated separation distances and optical properties of 0.982 for MHG and 0.979 for GB confirms higher agreement with experimental data by using these phase functions compared to HG (ICC = 0.978).

Conclusion The modified MC simulation enabled us to better simulate the experimental data. The MHG and GB models offer improved accuracy over traditional HG models, thus advancing optical imaging and diagnostic applications especially for measuring the peripheral oxygenation for preterm babies.

Keywords

Near-infrared spectroscopy (NIRS) · Short source-detector separation · Optical

Authors Djazia Yacheur and L. Lanini have equally contributed to this chapter.

D. Yacheur (✉) · L. Lanini · T. Li · M. Ackermann · A. Kalyanov · E. Russomanno · A. Di Costanzo Mata · M. Wolf · J. Jiang
Biomedical Optics Research Laboratory (BORL), Department of Neonatology, University Hospital Zurich and University of Zurich, Zurich, Switzerland
e-mail: dyacheur@student.ethz.ch; Laninil@student.ethz.ch

S. C. Pias et al. (eds.), *Oxygen Transport to Tissue XLVI*, Advances in Experimental Medicine and Biology 1498, https://doi.org/10.1007/978-3-032-10389-5_10

properties · Reflectance measurements · Silicone phantoms

10.1 Introduction

In optical spectroscopy, by sending light into biological tissue and measuring the reflectance at different wavelengths, the optical properties of tissue can be determined (Jacques 2013; Bashkatov et al. 2005). They provide important clinical insights in real-time monitoring of peripheral oxygenation in critically ill neonates. Near-infrared spectroscopy (NIRS) is one of the spectroscopic technologies using the non-ionizing light in the near-infrared range, typically between 650 and 1000 nm. When this light penetrates the tissue, two main effects are considered: the absorption, which is mainly an attenuation of the intensity of light due to the absorption coefficient (μ_a), which is defined as the probability of a photon being absorbed per unit path length as it passes through the tissue; and scattering characterized by the reduced scattering coefficient (μ_s') corresponding to the degree of deviation from the propagation direction of photons caused by structural irregularities and interfaces within the tissue. Determining these two parameters enables the quantification of different tissue components. In some clinical applications, measurements at short distances are needed. For example, in measuring the peripheral oxygenation for preterm babies where the measurement space is limited, employing optical imaging with a multi-wavelength approach enables the quantification of both oxygenated and deoxygenated hemoglobin levels in tissue (Pogue and Patterson 2006). The aim of this study is to understand the relationship between tissue optical properties and reflectance at short distances by comparing experimental data to date from a modified Monte Carlo simulation. The primary aim is to experimentally determine the relation between source-detector separation and reflectance at different optical properties.

10.2 Materials and Methods

10.2.1 Phantom Creation

To cover the range of soft tissue optical properties reported in literature (Jacques 2013; Bashkatov et al. 2005), we created 12 phantoms with different optical properties (Table 10.1) by mixing 75 g of each SILPURAN 2420 component silicone A and B (Wacker, Germany) for each phantom. To create tissue-like optical properties, we used silicone-compatible white ink ELASTOSIL RAL 9010 (Wacker, Germany) to obtain specific μ_s'. The μ_a was adjusted by adding carbon black powder (Wacker, Germany). After the mixture was prepared, it was degassed in a vacuum chamber, placed in a mold of 50 mm × 80 mm × 40 mm, then cured at 65 °C in the oven. Target optical

Table 10.1 Phantoms' optical properties at 838 nm

Phantoms	Black powder (μL)	White ink(μL)	μ_a (mm^{-1})	μ_s' (mm^{-1})
Phantom 1	800	83.34	0.04	0.5
Phantom 2	800	166.67	0.04	1
Phantom 3	800	250	0.04	1.5
Phantom 4	800	333.34	0.04	2
Phantom 5	1200	83.34	0.06	0.5
Phantom 6	1200	166.67	0.06	1
Phantom 7	1200	250	0.06	1.5
Phantom 8	1200	333.34	0.06	2
Phantom 9	2000	83.34	0.1	0.5
Phantom 10	2000	166.67	0.1	1
Phantom 11	2000	250	0.1	1.5
Phantom 12	2000	333.34	0.1	2

properties were validated within an error of ~8% using ISS Imagent (measured vs. expected: $R^2 > 0.98$). The rounded μ_a and μ_s' values were determined based on the known concentrations of white ink and the black powder used in phantom preparation, enabling systematic comparison across physiological ranges. These concentrations were carefully calculated using established absorption and scattering coefficients of the materials to achieve the desired optical properties (Kalyanov et al. 2022).

10.2.2 Experimental Setup

A probe was built, where fibers were placed with precise source-detector separations ranging from 1.5 mm to 5 mm with both diagonal and axial configurations (Figs. 10.1 and 10.2). This enables us to investigate the reflectance for different separations.

10.2.2.1 Reflectance Measurements

NIR light at 838 nm emitted by the ISS Imagent device penetrated the phantoms, undergoing scattering and absorption. Reflected light was collected using a 400 μm core diameter optical fiber (NA = 0.39, Thorlabs FT400EMT) at the mentioned source-detector distances. Data were sampled at 2.5 Hz. To reduce the influence of ambient light, measurements were conducted in a dark room, and the probe was additionally covered. We averaged the data for 30 s and repeated measurements five times for each source-detector separation.

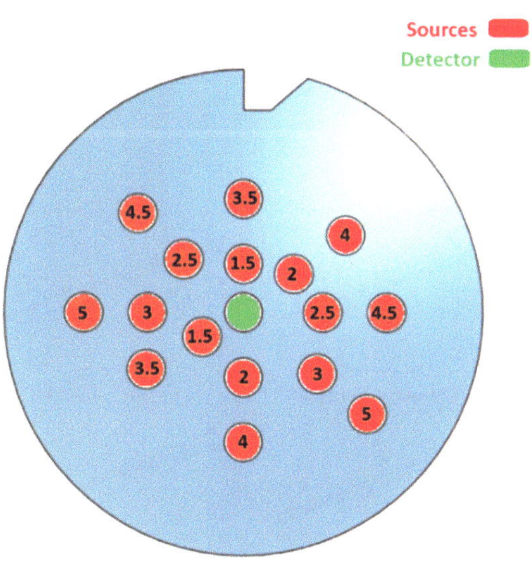

Fig. 10.1 Sources and detector placement. The numbers show the source-detector separation in mm

10.2.3 The Modified Simulation

The HG phase function used in conventional MC simulations underestimates the amount of backward scattering events that occurs in materials with large anisotropy factor (g) such as biological tissue. To address this, given the relevance of this function at short propagation distances outside the diffusive regime, we therefore incorporated the MHG and GB phase functions in our MC simulations. These alternatives improve the modeling accuracy at short distances by including an additional variable that enables us to tune their second-order moment, which improves the model of forward/backward scattering (Lanini et al. 2026). Simulations employed 10^6 photons with a pencil beam point source (838 nm wavelength). The output was scaled to match the experimental source power (3.2 mW, FWHM = 2 nm). Reflectance results were calibrated to experimental measurements using a scaling factor. As light propagation in biological tissue is dominated by scattering, coherence effects were not considered in the radiative transfer models.

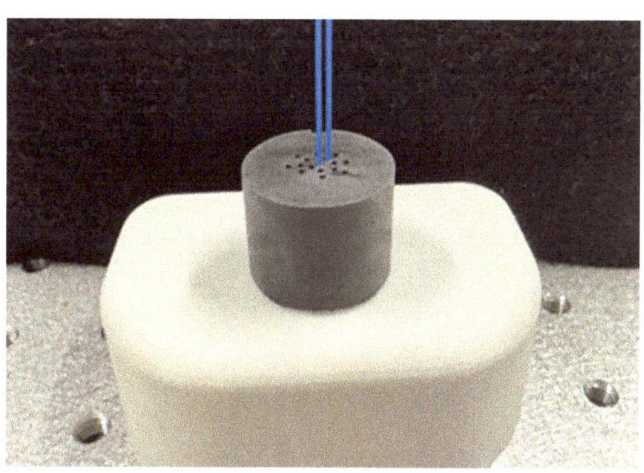

Fig. 10.2 The placement of the probe on the Phantom

10.3 Results and Discussion

Experimental data were collected using the ISS Imagent instrument measured on 12 silicone phantoms with different optical properties (Table 10.1). For axial (Fig. 10.3a) and diagonal (Fig. 10.3b) sources-detector arrangement, the reflectance data at the same distance are perfectly correlated with an average correlation coefficient of 0.9998 with a standard deviation of 0.0003 for a group of 12 data sets for the above 12 phantoms. We selected the axial arrangement of the source detector pairs for the rest of the study, Fig. 10.4, showing the mean and standard deviation of the measurements, the standard deviations are generally very small and most of the curves are linear, showcasing the reproducibility of the measured data. Fig. 10.5a depicts the log of the reflectance as a function of distance for one example of $\mu_a = 0.1$ mm^{-1} and $\mu'_s = 0.5$ mm^{-1} and the corresponding simulated curves with HG, MHG and GB. The HG model consistently deviated from the experimental data particularly at shorter distances, highlighting its limitations in accurately modeling anisotropic scattering in tissues at short distances. In contrast, the GB model

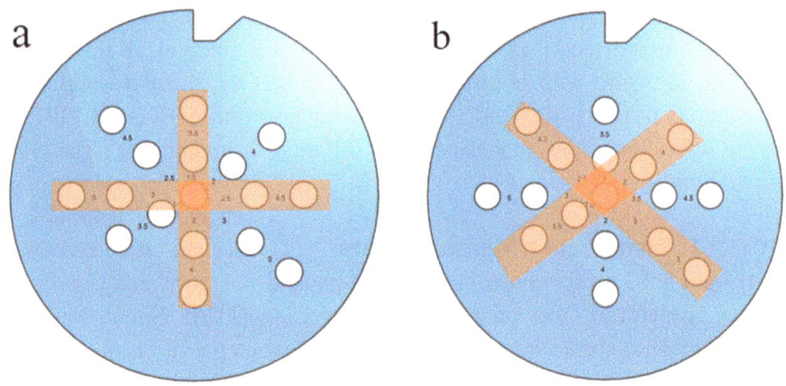

Fig. 10.3 (**a**) The results for the axial configuration. (**b**) The results for the diagonal configuration

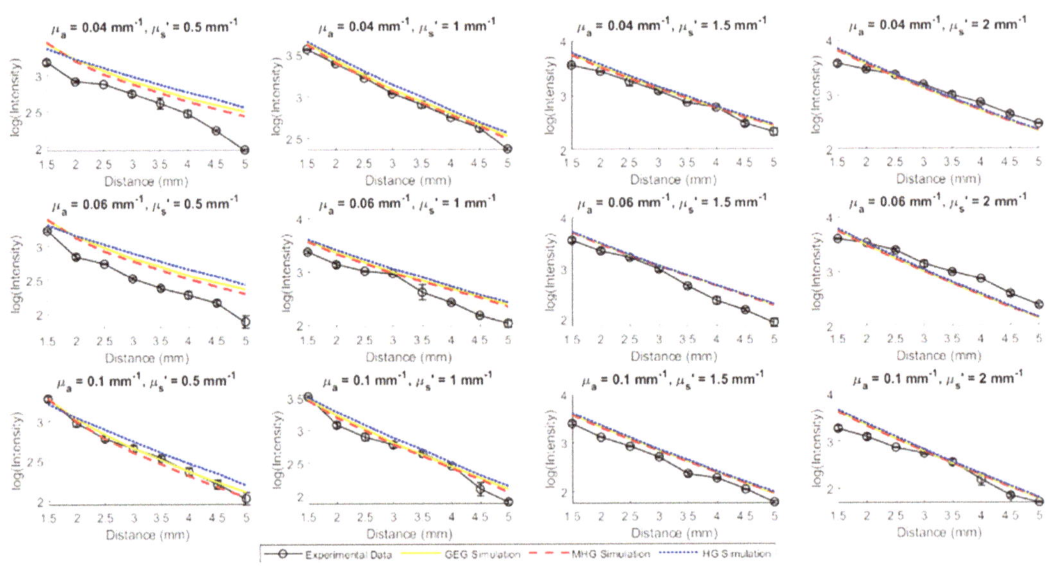

Fig. 10.4 The mean and standard deviation of the axial configuration measurements

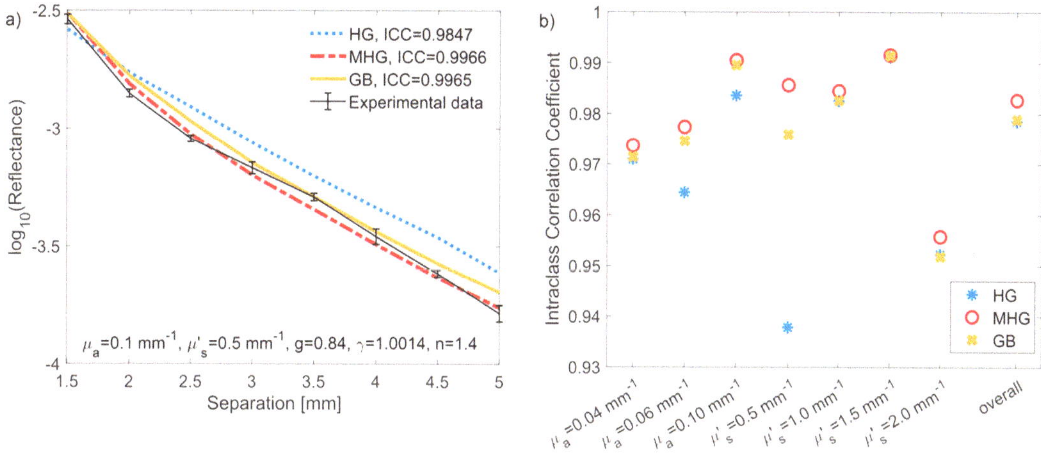

Fig. 10.5 Reflectance over propagation distance simulated with HG, MHG, and GB phase functions using $\mu_a = 0.1\ \text{mm}^{-1}$ and $\mu'_s = 0.5\ \text{mm}^{-1}$. Corresponding experimental data are plotted, and the ICC is added in the legend to display the level of correlation between experimental data and simulations. (**b**) ICC between experimental data and simulations employing HG, MHG, and GB phase functions, computed for each μ_a and μ'_s as median among all phantoms having such μ_a or μ'_s, and lastly as median over all phantoms together

demonstrated considerable improvements over the HG model, offering a better fit to the experimental data across all tested optical properties. Similarly, the MHG model showed an even better match to the experimental data than the GB and the HG models, especially for phantoms with lower μ_s'. Adding a second-order moment term and parameter γ to the HG model improves its ability to accurately describe the scattering phase function.

To assess how well the different phase function models (HG, MHG, and GB) match the experimental data, the intraclass correlation coefficient (ICC) was computed for every phantom. Fig. 10.5b shows the median of the ICC among all phantoms with the same μ_a or μ'_S, and the median over all phantoms in the last entry on the y-axis. We observe higher ICC values for MHG and GB compared to HG in all cases except for $\mu'_S = 2\ \text{mm}^{-1}$. The ICC for MHG and GB is higher particularly for high μ_a and low μ'_S. This relates to the fact that we profit less from MHG and GB when μ_a and μ'_S generate conditions closer to the diffusion approximation (i.e., low μ_a and high μ'_S) and vice versa. As an example, Fig. 10.5a shows a better match of the experimental data from MHG and GB, particularly clear at distances <2.5 mm, thanks to the additional tunable parameter γ. Overall, ICC values of 0.982 for MHG, 0.979 for GB, and 0.978 for HG. To further validate phase function performance, we computed the RMSE between simulations and experiments. MHG achieved the lowest errors 0.160, and outperformed HG 0.205 by 22%, while GB was in between at 0.175. These results are consistent with the ICC trends, and confirm the superior accuracy of MHG/GB in modeling reflectance at short distances.

10.4 Conclusion

An experimental setup was developed and the reflectance was experimentally measured at short source-detector distances ranging from 1.5 to 5 mm for phantoms with different optical properties. We compared these experimental results to the novel MC simulations described in part A (Lanini et al. 2026): We showed that the MHG and GB phase functions provide more accurate simulations compared to the HG phase function in particular for high μ_a and low μ'_s. This higher accuracy is crucial for precise optical measurements in clinical optical imaging and diagnostic applications, where only short source-detector distances are feasible.

Acknowledgments We gratefully acknowledge funding by Swiss National Science Foundation (159490 and 197079), Innovationspool of University Hospital Zürich, Bridge Proof of Concept (40B1-0_221535), Zurich Neuroscience Center (ZNZ), Innosuisse and Vontobel Stiftung.

Competing Interests Martin Wolf declares that he is co-founder and president of the board of OxyPrem AG and Alexander Kalyanov is working partly for this company.

References

Bashkatov AN, Genina EA, Kochubey VI et al (2005) Optical properties of human skin, subcutaneous and mucous tissues in the wavelength range from 400 to 2000 nm. J Phys D Appl Phys 38(15): 2543–2555. https://doi.org/10.1088/0022-3727/38/15/004

Jacques SL (2013) Optical properties of biological tissues: a review. Phys Med Biol 58(11):R37–R61. https://doi.org/10.1088/0031-9155/58/11/R37

Kalyanov A, Jiang J, Russomanno E, et al.(2022). Development and validation of robust and cost-effective liquid heterogeneous phantom for time domain near-infrared optical tomography. Adv Exp Med Biol 2022;1395:199–204. https://doi.org/10.1007/978-3-031-14190-4_33

Lanini L, Yacheur D, Li T et al (2026) Dependence of reflectance on optical properties at short distance (Part A: adaptation of Monte Carlo simulations). Adv Exp Med Biol 1498

Pogue BW, Patterson MS (2006) Review of tissue simulating phantoms for optical spectroscopy, imaging, and dosimetry. J Biomed Opt 11(4):041102

Part III

Membrane-Level Transport

Influence of Breast Tumor-Associated Lipids on Oxygen Transport Across Model Cell Membranes (Part B: Saturated Phospholipids and Cholesterol)

11

Qi Wang and Sally C. Pias

Abstract

Dysregulation of lipid homeostasis is common in cancers, and biophysical studies have suggested that increased incorporation of saturated lipids and cholesterol may reduce cell membrane permeability to oxygen (O_2). Low intracellular oxygen, in turn, may promote tumor aggressiveness, as well as resistance to radiotherapy. A prior lipidomics study found membrane phospholipids associated with de novo fatty acid synthesis to be unusually abundant in breast tumors with poor prognosis, including the saturated lipids 1-**m**yristoyl-2-**p**almitoyl-*sn*-glycero-3-**p**hospho**c**holine ($MPPC_{14,16}$, or PC14:0/16:0) and 1,2-**dip**almitoyl-*sn*-glycero-3-**p**hospho**c**holine ($DPPC_{16,16}$, or PC16:0/16:0). Part A of the current work has found that $MPPC_{14,16}$ and cholesterol together promote reduced oxygen permeability. Here, we have followed up on this finding to determine whether the effect is related to lipid saturation generally or to structural features of $MPPC_{14,16}$ specifically. We have used atomistic molecular dynamics simulations to calculate oxygen permeability coefficients for lipid bilayers composed of either $DPPC_{16,16}$ or 1,2-**dim**yristoyl-*sn*-glycero-3-**p**hospho**c**holine ($DMPC_{14,14}$, or PC14:0/14:0) mixed in a 1:1 ratio with cholesterol. The resulting mean oxygen permeability values, of 3–5 cm/s, were 2 to 4 times lower than bilayers combining the common monounsaturated phospholipid 1-**p**almitoyl-2-**o**leoyl-*sn*-glycero-3-**p**hospho**c**holine (POPC, or PC16:0/18:1) in a 1:1 ratio with cholesterol. As such, this study finds that doubly saturated 14- and 16-carbon phospholipids and cholesterol together can generally reduce membrane oxygen permeability. The extent to which such permeability reductions could promote hypoxia in a biological context requires further investigation.

Keywords

Biophysics · Permeability · Diffusion · Partition coefficient · De novo lipogenesis · Cholesterol · Molecular dynamics simulation

Q. Wang
Department of Chemistry, New Mexico Institute of Mining and Technology (New Mexico Tech), Socorro, NM, USA

Present Address: Department of Chemistry and Biochemistry, Presbyterian College, Clinton, SC, USA

S. C. Pias (✉)
Department of Chemistry, New Mexico Institute of Mining and Technology (New Mexico Tech), Socorro, NM, USA
e-mail: sally.pias@nmt.edu

S. C. Pias et al. (eds.), *Oxygen Transport to Tissue XLVI*, Advances in Experimental Medicine and Biology 1498, https://doi.org/10.1007/978-3-032-10389-5_11

11.1 Introduction

The current study delves into the potential impact of saturated lipids produced through de novo fatty acid synthesis on intracellular hypoxia in breast tumor tissue. Solid tumors frequently experience hypoxia, a major factor promoting cell proliferation, metastasis, and resistance to radiotherapy (Semenza 2016). While vascular abnormalities and inadequate perfusion are well-established contributors to hypoxia (Multhoff et al. 2014), the extent to which lipid abnormalities could also contribute has yet to be clarified.

Cancers generally exhibit dysregulation in processes controlling the levels of cholesterol and lipids in the tissue, including aberrant activation of de novo fatty acid synthesis (Santos and Schulze 2012; Liu et al. 2023). A lipidomics study of human breast tumors by Hilvo et al. found abnormal levels of putative de novo lipid products. Specifically, elevation of the saturated phospholipids 1-**m**yristoyl-2-**p**almitoyl-*sn*-glycero-3-**p**hospho**c**holine ($MPPC_{14,16}$, or PC14:0/16:0) and 1,2-**dip**almitoyl-*sn*-glycero-3-**p**hospho**c**holine ($DPPC_{16,16}$, or PC16:0/16:0) correlated with tumor grade and estrogen receptor status. Above-median $DPPC_{16,16}$ levels were also associated with poor patient survival (Hilvo et al. 2011).

Biophysical studies indicate that variations in lipid composition can modify cell membrane permeability to molecular oxygen (O_2) and its route of diffusion (Krämer et al. 2020; Ghysels et al. 2019; Pias 2021; Wang et al. 2021; Subczynski et al. 2009; Borden and Longo 2004; Dotson et al. 2017; Dotson et al. 2021; Widomska et al. 2007; Dotson and Pias 2018; Angles et al. 2021; Subczynski et al. 1991). Studies to date generally support the assumption that oxygen diffuses passively through tissue after its release into microvessels and that it crosses cell membranes readily through the lipid phase (Pias 2021). At biologically relevant temperatures, saturated phospholipids and cholesterol both tend to reduce lipid bilayer permeability to oxygen (or to the similar gas nitric oxide, NO), relative to unsaturated phospholipids alone (Ghysels et al. 2019; Borden and Longo 2004; Dotson et al. 2017; Dotson et al. 2021; Widomska et al. 2007; Wang et al. n.d.; Wennberg et al. 2012). A notable yet unexplained exception has been reported in liposomes mimicking red blood cell membranes, where the oxygen permeability was found to be anomalously low and to *increase* with cholesterol incorporation (Al-Samir et al. 2021). However, cholesterol has typically been found to lower membrane oxygen permeability, exerting a stronger effect on saturated lipids than on unsaturated ones (Subczynski et al. 2009; Dotson et al. 2017; Dotson et al. 2021; Widomska et al. 2007; Subczynski et al. 1991; Subczynski et al. 1989).

The structure of cholesterol consists of a rigid ring system and a flexible hydrocarbon chain (Fig. 11.1). In membranes, it promotes both fluidity (lateral movement) and order (straightness) of adjacent phospholipids. The cholesterol content of normal plasma membranes varies according to cell type and typically falls in the range 20–50% (by number of molecules), while organelle membranes contain much less cholesterol, typically 5–10% (Mouritsen 2005).

Our group previously found that 50% cholesterol reduces the oxygen permeability of bilayers of the common monounsaturated lipid 1-**p**almitoyl-2-**o**leoyl-*sn*-glycero-3-**p**hospho**c**holine (POPC, or PC16:0/18:1) (Dotson et al. 2017, 2021). Part A of the current work uses atomistic molecular dynamics simulations to examine how breast tumor-associated variations in lipid chain length, headgroup size, and cholesterol incorporation influence model membrane oxygen permeability and related physical properties. It focuses on monounsaturated phospholipids but also identifies lipid saturation as a key parameter warranting further investigation. Notably, $MPPC_{14,16}$ was found in Part A to reduce oxygen permeability in combination with cholesterol, but whether the effect is due to lipid saturation or to the specific combination of lipid chains in $MPPC_{14,16}$ remains to be clarified (Wang et al. n.d.).

Here, we have tested the hypothesis that the permeability-reducing effect of $MPPC_{14,16}$ when combined with cholesterol derives from its 14- and 16-carbon saturated lipid chains but not from

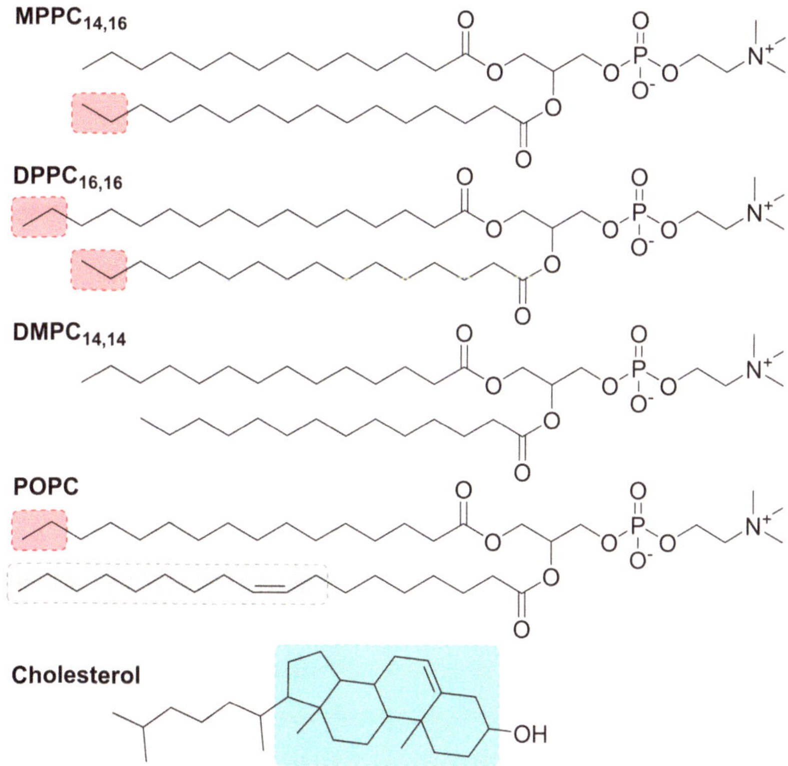

Fig. 11.1 Lipid molecular structures. Differences in saturated chain length are highlighted in red, and the monounsaturated chain of POPC is shaded gray. The rigid cholesterol ring system is shaded blue

their particular structural arrangement. We have examined two structurally similar lipids that each duplicate one of the chains of $MPPC_{14,16}$, namely 1,2-**dim**yristoyl-*sn*-glycero-3-**p**hospho**c**holine ($DMPC_{14,14}$, or PC14:0/14:0) and $DPPC_{16,16}$. All three lipids are pictured in Fig. 11.1, alongside POPC and cholesterol. In order to support the hypothesis, we would expect that $MPPC_{14,16}$/cholesterol (chol), $DPPC_{16,16}$/chol, and $DMPC_{14,14}$/chol would all have similar oxygen permeability values because they incorporate only saturated chains of similar length. Atomistic molecular dynamics simulations have been used to calculate oxygen permeability, using an analytical approach established in our earlier work (Dotson et al. 2017, 2021; Angles et al. 2021). Several other physical parameters affecting oxygen permeation are also presented, including oxygen positional free energy and oxygen lipid-water partition coefficient.

11.2 Methods

Atomistic molecular dynamics simulations model molecular structures in atomic detail and project molecular motion over time using a simplified "force field" representation of atomic interactions based on classical physics. Thermodynamic properties such as oxygen permeability can be calculated from molecular simulation trajectories using statistical analysis. The current study focuses on two hydrated lipid bilayer systems, $DPPC_{16,16}$ and $DMPC_{14,14}$, and compares them with $MPPC_{14,16}$ and POPC systems also studied in Part A (Wang et al. n.d.). For all phospholipid systems compared here, cholesterol has been incorporated at a level of 50 mol%, or a 1:1 molar ratio. This relatively high amount of cholesterol is consistent with our earlier work and occurs in the membranes of red blood cells, eye lens fiber cells, and possibly also in some tumor

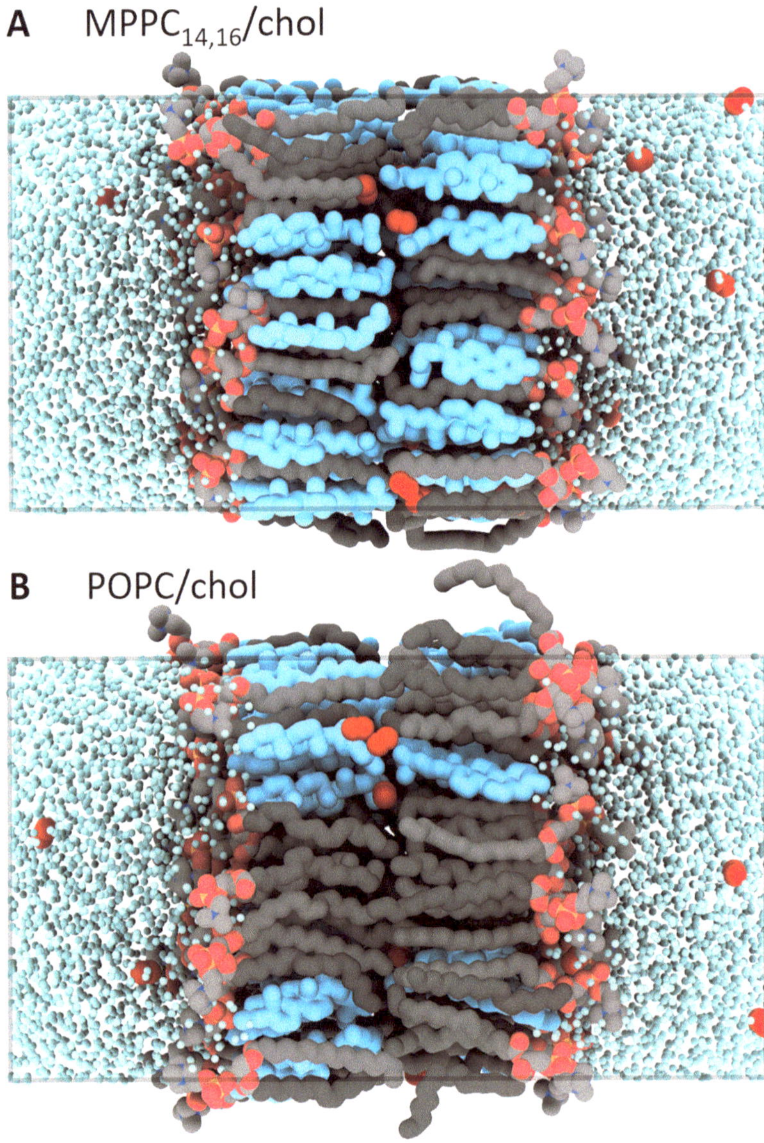

Fig. 11.2 Molecular dynamics simulation snapshot images of three-dimensional hydrated lipid bilayers of (**a**) $MPPC_{14,16}$/chol and (**b**) POPC/chol. Water molecules shown as reduced-size cyan dots; lipid hydrocarbon chains shown in gray, with lipid headgroup atoms colored by element; O_2 molecules shown as paired red spheres; cholesterol molecules shown in light blue. All hydrogen atoms omitted for image clarity

cells (see Dotson et al. 2017, 2021; Angles et al. 2021). 1:1 mixtures of POPC/cholesterol (chol) and $DMPC_{14,14}$/chol have also been used in experimental oximetry studies, providing a basis for validation and comparison (Subczynski et al. 1991, 2009). Representative simulation system images are provided in Fig. 11.2, and the compositional details of the $DPPC_{16,16}$/chol and $DMPC_{14,14}$/chol systems are provided in Table 11.1.

The systems were constructed using the Charmm-GUI server (Jo et al. 2009). GPU-accelerated molecular dynamics simulations were carried out using Amber biomolecular simulation software version 20, along with the Lipid14 force field, TIP3P water model, and an O_2 model developed in our earlier work (Dotson et al. 2021; Götz et al. 2012; Case et al. 2020; Dickson et al. 2014; Jorgensen et al. 1983). Simulations were carried out under constant pressure and constant temperature conditions. The pressure was fixed at 1 bar (~1 atm) using anisotropic coupling with a weak-coupling barostat for 150 nanoseconds (ns) of equilibration, followed

Table 11.1 Simulation system compositions (number of molecules) and average simulation box areas

System	Lipid molecules[a]	Cholesterol[a]	H_2O	O_2	Area (Å^2)[b]
$DPPC_{16,16}$/chol	64 $DPPC_{16,16}$	64	4460	20	2500 ± 20
$DMPC_{14,14}$/chol	64 $DMPC_{14,14}$	64	4460	20	2540 ± 30
$MPPC_{14,16}$/chol	64 $MPPC_{14,16}$	64	4460	20	2610 ± 20
POPC/chol	64 POPC	64	4460	20	2690 ± 40

[a]Molecules equally distributed between the two lipid layers
[b]Calculated over the production phase

by use of a Monte Carlo barostat for 200 ns of additional equilibration and for a subsequent production phase of 300 ns. The production phase used a simulation timestep of 2 femtoseconds, and coordinates were recorded every picosecond (ps). The temperature was held constant at 310 K (37 °C) using Langevin temperature control with a collision frequency of 1 ps^{-1}. The SHAKE algorithm was applied, and bonds to hydrogen were held rigid.

To aid in statistical sampling, the O_2 concentration was enhanced in the range 110–140 mM, roughly 200 times the concentration expected in atmosphere-equilibrated air. We have previously demonstrated that such enhancement avoids physical artifacts and retains O_2 in the dissolved phase, as intended (Dotson et al. 2021; Angles et al. 2021). Two independent simulation replicates and both lipid leaflets of each bilayer were used to assess reproducibility and to estimate error. The membrane oxygen permeability, P_M, was calculated for each lipid bilayer system by tracking O_2 molecule half-permeation events during unrestrained simulations, using an equation validated in our previous work (Dotson et al. 2021; Angles et al. 2021):

$$P_M = \Phi_{esc}/(4N_w) \quad (11.1)$$

Here, Φ_{esc} is the number of O_2 molecule half-permeation events per unit time, and N_w is the average O_2 population in bulk water.

A more general expression for membrane permeability is also relevant, although it was not used to calculate the permeability coefficients reported in this study. The expression, commonly known as Overton's rule, predicts the membrane permeability for a particular substance to be directly proportional to its lipid-water partition coefficient, K_P, times its diffusion coefficient within the membrane, D_M, and also inversely proportional to the membrane permeation thickness, h (Missner and Pohl 2009):

$$P_M = (K_P \cdot D_M)/h. \quad (11.2)$$

Below, we report values for P_M, h, and D_M, as well as the permeability-thickness product $P_M \cdot h$, which corresponds with a solubility-diffusion product (partition coefficient times diffusion coefficient).

Relative positional free energy curves for O_2 were calculated as a function of the depth along the axis of permeation, z, using the relation

$$\Delta G(z) = -RT\, ln(K_P(z)). \quad (11.3)$$

Here, R is the ideal gas constant, and T is the absolute temperature. $K_P(z)$ is the positional lipid-water partition coefficient for O_2, calculated as the time-averaged positional O_2 population, $N(z)$, divided by the time-averaged O_2 population in bulk water, N_w:

$$K_P(z) = N(z)/N_w. \quad (11.4)$$

An overall average O_2 lipid-water partition coefficient, $\langle K_P \rangle$, was calculated for each simulation system by averaging the $K_P(z)$ curve in the range $\pm h/2$ from the bilayer center (midplane).

To characterize the bilayer structure quantitatively, positional electron density curves were calculated from atomic coordinates throughout the production phase, using the Amber analysis program CPPTRAJ (Roe and Cheatham 2013). The same analysis program was used to calculate lipid order parameters for the hydrocarbon chains over the production phase.

11.3 Results and Discussion

Figure 11.2 shows representative simulation snapshot images of $MPPC_{14,16}$/chol and POPC/chol lipid bilayers. Note that these are static representations of highly dynamic systems, in which all atoms move relative to one another and the O_2 molecules (paired red spheres) diffuse throughout the lipid bilayer and surrounding water.

Figure 11.3 provides positional electron density as well as O_2 positional free energy and lipid-water partition coefficient curves. The electron density curves (Fig. 11.3b) provide a quantitative representation of the lipid bilayer structure, reflecting variation in the atomic packing density as a function of distance from the bilayer center (midplane). These curves show highly similar bilayer structures for all systems, with only subtle variations between the saturated and monounsaturated lipid systems that are within the standard deviation ranges (see Fig. 11.3b caption). The general features of all systems include high density in the headgroup regions, low density close to the bilayer center, and plateaus with moderate density in the blue-shaded regions occupied by the rigid cholesterol ring system. These structural features align spatially with features in the O_2 positional free energy curves of Fig. 11.3c, including permeation barriers in the headgroup regions, a sink at the bilayer midplane, and additional permeation barriers along the rigid cholesterol ring system (blue shaded regions). The cholesterol-associated barriers are particularly high in all of the saturated lipid systems, which also show substantial reduction in oxygen permeability, P_M (Table 11.2). These barriers correspond with regions of low local partition coefficient, as shown in the positional O_2 lipid-water partition coefficient curves of Fig. 11.3d, from which the free energy curves were calculated (Eq. 11.3).

Saturated chains tend to pack closely against cholesterol molecules, resulting in reduced O_2 population and increased positional free energy in those regions (Dotson et al. 2017). A physical explanation of this effect has been offered by Wennberg and colleagues, who found that diffusing permeants must disrupt van der Waals interactions between lipid hydrocarbon chains in cholesterol-containing membranes, an effect that is exaggerated for saturated lipids compared to unsaturated ones (Wennberg et al. 2012). Consistent with their finding, the hydrocarbon chains of the $MPPC_{14,16}$/chol bilayer snapshot image in Fig. 11.2a generally appear straighter than those of POPC/chol (Fig. 11.2b). Lipid order parameter values support the visualization, indicating that both hydrocarbon chains of all the saturated-lipid/cholesterol systems are generally more orientationally ordered—and, by implication, straighter—than the corresponding POPC/chol chains (Fig. 11.4). Therefore, the elevated free energy barriers flanking cholesterol in the saturated lipid systems appear to stem from increased hydrocarbon chain ordering in the absence of any double bonds, which introduce kinks and reduce chain order.

We note that the "force field" model used in the current study includes approximation of van der Waals forces (both dispersion and coulombic), such that the model is capable of accounting for the effects of such forces on oxygen partitioning and diffusion. Since dispersion forces depend strongly on the distances between interacting atoms, straighter chains should enable greater numbers of favorable contacts as well as a more extensive contact area among interacting lipid molecules. Our findings are consistent with those of Wennberg and colleagues, who applied a similar molecular simulation model and quantitatively related van der Waals interaction energy among hydrocarbon chains to lipid-water partitioning for multiple permeants—including nitric oxide (NO), which structurally resembles O_2 (Wennberg et al. 2012).

We have previously validated our simulation models for studies of membrane-level oxygen transport and have demonstrated semi-quantitative agreement with experimental measurements (Dotson et al. 2017, 2021; Angles et al. 2021). While there are known discrepancies, the simulation models are able to reproduce expected patterns and trends related to membrane structure and physical properties as well as

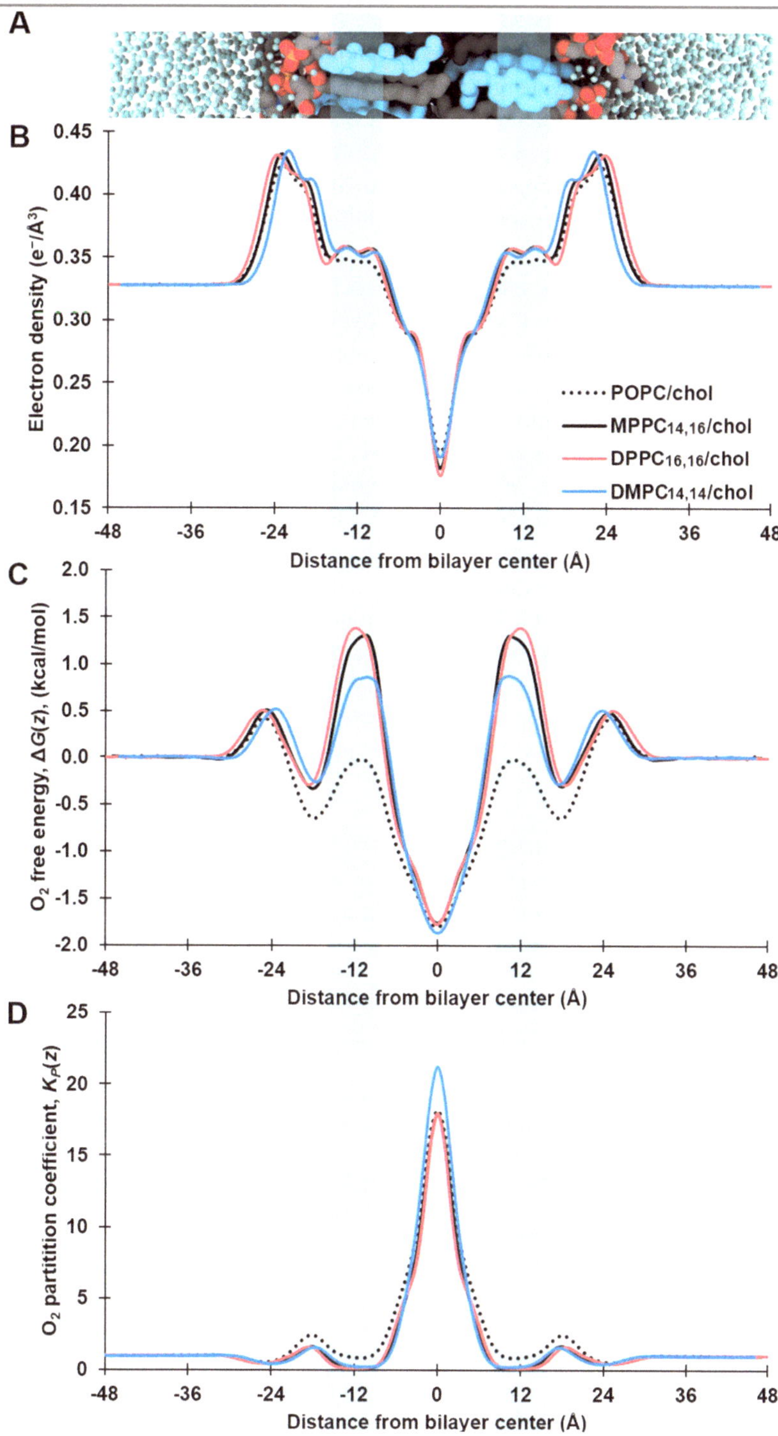

Fig. 11.3 Saturated lipids and cholesterol generate large energy barriers to oxygen permeation. (**a**) Segment of an $MPPC_{14,16}$/chol molecular simulation snapshot image, as in Fig. 11.2a, placed above data plots to show corresponding physical features. Blue shading throughout the figure highlights the regions most affected by the rigid cholesterol rings. (**b**) Electron density curves, plotted as a function of lipid bilayer depth. Standard deviations are fairly consistent throughout the curves, with values of ≤ 0.3 $e^-/Å^3$ for POPC/chol and ≤ 0.2 $e^-/Å^3$ for all other systems. (**c**) Relative positional free energy curves for O_2, $\Delta G(z)$, calculated using Eq. 11.3. Each curve shows an

response to phospholipid and cholesterol composition. In the current work, the free energy curves for $DMPC_{14,14}$/chol and POPC/chol show semi-quantitative agreement with oxygen transport parameter curves and activation energies reported in previous experimental electron paramagnetic resonance (EPR) oximetry studies. Specifically, the experimental curves show a cholesterol-induced oxygen transport barrier for $DMPC_{14,14}$/chol greater than that of POPC/chol (based on oxygen transport parameter data in ref. (Subczynski et al. 2009)). Activation energies calculated from oximetry data collected at various temperatures likewise show a factor of 1.5 increase in the maximum cholesterol-induced barrier in $DMPC_{14,14}$/chol relative to egg yolk phosphatidylcholine (EYPC)/chol (Subczynski et al. 1991). EYPC consists of ~70% POPC, in addition to other unsaturated phospholipids. These patterns compare well with the simulation data reported here, which give a factor of 1.5 increase for $DMPC_{14,14}$/chol relative to POPC/chol in the cholesterol-induced O_2 transport barrier (based on the free energy peaks near ±10 Å from the bilayer center in Fig. 11.3c), although differences in the experimental and simulation methods lead to substantial quantitative differences.

Table 11.2 gives membrane O_2 permeability coefficient, P_M, values calculated from molecular dynamics simulations, along with related membrane physical properties. The common monounsaturated phospholipid POPC is taken as a baseline for comparison because it has been well-studied in our previous work, especially in combination with cholesterol in a 1:1 molar ratio with phospholipid (Dotson et al. 2017, 2021). Part A of the current work found $MPPC_{14,16}$/chol to have an oxygen permeability coefficient four times lower than that of POPC/chol (Wang et al. n.d.). Here, we find that $DPPC_{16,16}$/chol and $DMPC_{14,14}$/chol show similar permeability to $MPPC_{14,16}$/chol, with average P_M values in the range 3–5 cm/s. For these lipids, slight variations

Table 11.2 Membrane permeability and related properties calculated from simulations[a]

Bilayer	P_M (cm/s)[b]	h (Å)[c]	$P_M \cdot h$ ($\times 10^{-6}$ cm^2/s)[d]	$\langle K_P \rangle$[e]	D_M ($\times 10^{-6}$ cm^2/s)[f]
$DPPC_{16,16}$/chol	3.3 ± 0.7	64	2.1 ± 0.4	2.6 ± 0.3	0.8 ± 0.2
$DMPC_{14,14}$/chol	5 ± 1	58	3 ± 1	3.2 ± 0.5	0.9 ± 0.2
$MPPC_{14,16}$/chol[g]	3 ± 1	60	2 ± 1	2.7 ± 0.2	0.7 ± 0.2
POPC/chol[g]	12.1 ± 0.3	60	7.3 ± 0.2	3.7 ± 0.1	2.0 ± 0.1

[a]Where relevant, values are given as mean ± standard deviation
[b]P_M, membrane permeability coefficient for O_2
[c]h, membrane permeation thickness in ångströms (Å), where 1 Å = 0.1 nm
[d]$P_M \cdot h$, permeability-thickness product, with errors propagated from P_M data
[e]$\langle K_P \rangle$, average lipid-water partition coefficient for O_2
[f]D_M, apparent O_2 diffusion coefficient along the permeation axis (z), calculated using Eq. 11.2 with P_M, h, $\langle K_P \rangle$, and their associated errors from the table
[g]Values also reported in ref. (Wang et al. n.d.)

Fig. 11.3 (continued) average over 2 trajectories and both lipid layers (4 values). Standard deviation varies by position and is largest in the cholesterol-spanning region (±18 to ±6 Å from the bilayer center), with peak standard deviations of 0.2 kcal/mol for $MPPC_{14,16}$/chol and 0.1 kcal/mol for all other curves. (**d**) Positional O_2 lipid-water partition coefficient curves, $K_P(z)$, used for calculating the free energy curves in *panel C*. Standard deviation varies by position, with values of ≤0.9 for all curves in the cholesterol-spanning region and peak values of ≤2 near the bilayer center (−6 to +6 Å). *Legend for all plots:* POPC/chol (dotted line), $MPPC_{14,16}$/chol (black solid line), $DPPC_{16,16}$/chol (red line), $DMPC_{14,14}$/chol (blue line). Note that all curves have been truncated on the right and left ends, to remove periodic boundary artifacts generated by simulation box length fluctuations related to the pressure control algorithm

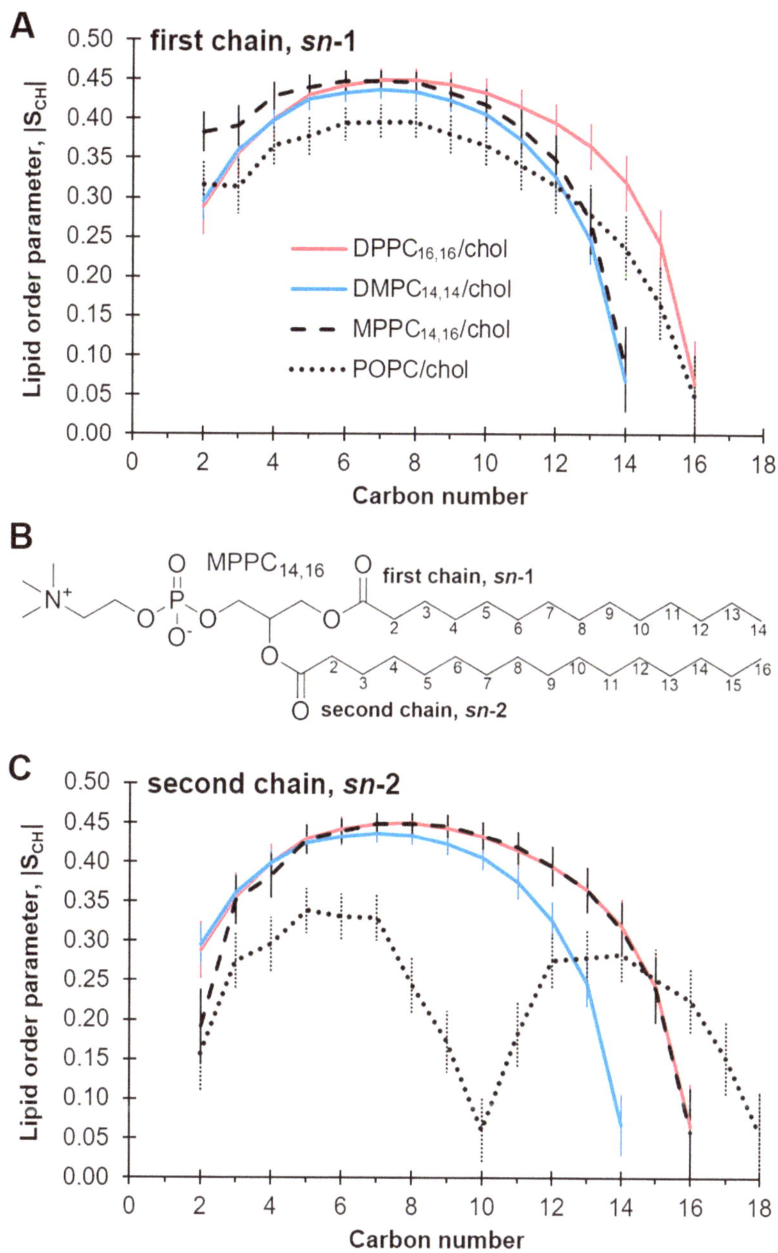

Fig. 11.4 Lipid order parameter curves indicate that the hydrocarbon chains of POPC are generally less ordered than those of the fully saturated lipids. (**a**) Carbon-wise lipid order parameter for the first hydrocarbon chain for each phospholipid. The saturated chain of POPC has lower order than those of the other lipids in the range between carbons #4 and #10; these carbons fall mainly in the region spanned by the cholesterol rings. (**b**) Representative lipid structure, showing chain and carbon numbering. (**c**) Carbon-wise lipid order parameter for the second hydrocarbon chain for each phospholipid. The region around the *cis*-double bond of the POPC monounsaturated chain is apparent as a dip in the curve at carbon #10, indicating that the part of the chain surrounding the double bond is highly disordered. *Legend for both plots:* $DPPC_{16,16}$/chol (red line), $DMPC_{14,14}$/chol (blue line), $MPPC_{14,16}$/chol (black dashed line), POPC/chol (black dotted line). Error bars show standard deviation

in the chain length (14- and/or 16-carbon chains) as well as variations in membrane permeation thickness (58–64 Å) appear not to influence the O_2 permeability significantly—given overlap in the calculated standard deviation ranges. Since membrane permeability depends inversely on the permeation thickness (see Eq. 11.2), we have calculated the permeability-thickness product, $P_M \cdot h$, to account for thickness differences. All the saturated lipid systems share similar $P_M \cdot h$ values, which are at least a factor of 2 lower than the POPC system value. Additionally, the average O_2 lipid-water partition coefficient, $\langle K_P \rangle$, is similar among the three saturated-lipid systems, but the $DPPC_{16,16}$/chol and $MPPC_{14,16}$/chol values are at least ~20% lower than the POPC/chol value (considering standard deviation ranges).

Recall that membrane permeability depends on both the lipid-water partition coefficient and the diffusion coefficient along the axis of permeation, D_M (see Eq. 11.2). Values for D_M given in Table 11.2 are smaller by a factor of 2 for the saturated-lipid/cholesterol systems, compared to POPC/chol. This result is consistent with a model proposed by Subczynski and colleagues, which associates oxygen transport effects of lipid chain unsaturation and cholesterol with alteration of the *dynamics* of chain movement (Subczynski et al. 1991). The model explains the lower diffusion coefficient for the saturated-lipid/cholesterol bilayers as resulting from dampening of the chain rotational dynamics thought to propel oxygen through the bilayer. Namely, the straighter (more ordered) chains of the saturated lipids pack neatly against the cholesterol rings and are less able to rotate to form transient pockets for oxygen to occupy along the diffusion pathway.

All of the findings reported here support the hypothesis that the reduced permeability of $MPPC_{14,16}$/chol derives from the saturation and length of its chains, rather than from their specific arrangement. The current study is limited to simple lipid-bilayer models and does not address many complexities of biological membranes, especially with respect to lipid diversity and the presence of proteins as well as carbohydrates of the glycocalyx. Examination of multicomponent lipid bilayers in Part A of this work indicates that membrane compositional complexity can reduce or eliminate the influence of individual lipid types. However, Part A also found that $MPPC_{14,16}$ and cholesterol together reduced the oxygen permeability of a mixed lipid bilayer including two other monounsaturated lipid types (Wang et al. n.d.), indicating that specific lipids may sometimes influence permeability when part of more complex mixtures.

11.4 Conclusion

This study has used molecular dynamics simulations to test whether the unique chain combination of the saturated phospholipid $MPPC_{14,16}$ drives its oxygen-permeability-reducing effect when combined with membrane cholesterol. The results indicate that the effect is general to saturated phospholipids of similar chain length (14 and/or 16 carbons). We found $DPPC_{16,16}$/chol and $DMPC_{14,14}$/chol to have similar permeability to $MPPC_{14,16}$/chol, falling 2 to 4 times below the permeability of the monounsaturated lipid POPC with cholesterol at 37 °C. Orientational analysis indicates that cholesterol straightens (orders) the hydrocarbon chains of fully saturated phospholipids more than those of monounsaturated POPC, with corresponding elevation of cholesterol-induced oxygen permeation barriers. The data support a model attributing such barriers to favorable contacts among lipid tails that must be disrupted in order for permeation to occur (Wennberg et al. 2012) and a second model attributing slower oxygen diffusion to dampened rotational dynamics of saturated chains flanking rigid cholesterol (Subczynski et al. 1991).

Our prior work has predicted that relatively small changes in membrane oxygen permeability, comparable to those estimated here, may reduce tissue oxygenation if the pathway of oxygen diffusion is lipid-dominated and not primarily through aqueous fluids such as the cytosol and the interstitial fluid (Pias 2021). The primary pathway of oxygen diffusion is yet unknown, although many indicators favor lipid-dominated diffusion (Pias 2020, 2021). As such, the present work is unable to predict with certainty whether saturated membrane lipids can impact tumor hypoxia. Even so, it demonstrates the biophysical impact of 14- and 16-carbon saturated lipid chains associated with de novo fatty acid synthesis on oxygen transport in simplified model systems.

Acknowledgments Research reported in this publication was supported by an Institutional Development Award (IDeA) from the National Institute of General Medical Sciences of the National Institutes of Health (NIH), under grant number P20GM103451. Its contents are solely the responsibility of the authors and do not necessarily represent the official views of the NIH. The authors additionally acknowledge support from the Glendorn

Foundation and the New Mexico Institute of Mining and Technology (New Mexico Tech). Lipid bilayer images were generated using ChimeraX software (Pettersen et al. 2021). The authors thank the reviewers for valuable insights and suggestions.

References

Al-Samir S, Itel F, Hegermann J, Gros G, Tsiavaliaris G, Endeward V (2021) O_2 permeability of lipid bilayers is low, but increases with membrane cholesterol. Cell Mol Life Sci 78(23):7649–7662. https://doi.org/10.1007/s00018-021-03974-9

Angles G, Hail A, Dotson RJ, Pias SC (2021) Atomistic simulations modify interpretation of spin-label oximetry data. part 1: intensified water–lipid interfacial resistances. Appl Magn Reson 52(10):1261–1289. https://doi.org/10.1007/s00723-021-01398-z

Borden MA, Longo ML (2004) Oxygen permeability of fully condensed lipid monolayers. J Phys Chem B 108(19):6009–6016. https://doi.org/10.1021/jp037815p

Case DA, Belfon K, Ben-Shalom IY, Brozell SR, Cerutti DS, Cheatham TE III et al (2020) AMBER 20. University of California, San Francisco

Dickson CJ, Madej BD, Skjevik ÅA, Betz RM, Teigen K, Gould IR, Walker RC (2014) Lipid14: the Amber lipid force field. J Chem Theory Comput 10(2):865–879. https://doi.org/10.1021/ct4010307

Dotson RJ, Pias SC (2018) Reduced oxygen permeability upon protein incorporation within phospholipid bilayers. Adv Exp Med Biol 1072:405–411. https://doi.org/10.1007/978-3-319-91287-5_65. PMCID: PMC6202029

Dotson RJ, Smith CR, Bueche K, Angles G, Pias SC (2017) Influence of cholesterol on the oxygen permeability of membranes: insight from atomistic simulations. Biophys J 112(11):2336–2347. https://doi.org/10.1016/j.bpj.2017.04.046

Dotson RJ, McClenahan E, Pias SC (2021) Updated evaluation of cholesterol's influence on membrane oxygen permeability. In: Advances in experimental medicine and biology, vol 1269, pp 23–30. https://doi.org/10.1007/978-3-030-48238-1_4

Ghysels A, Krämer A, Venable RM, Teague WE, Lyman E, Gawrisch K, Pastor RW (2019) Permeability of membranes in the liquid ordered and liquid disordered phases. Nat Commun 10(1):5616. https://doi.org/10.1038/s41467-019-13432-7

Götz AW, Williamson MJ, Xu D, Poole D, Le Grand S, Walker RC (2012) Routine microsecond molecular dynamics simulations with AMBER on GPUs. 1. Generalized born. J Chem Theory Comput 8(5):1542–1555. https://doi.org/10.1021/ct200909j

Hilvo M, Denkert C, Lehtinen L, Müller B, Brockmöller S, Seppänen-Laakso T et al (2011) Novel theranostic opportunities offered by characterization of altered membrane lipid metabolism in breast cancer progression. Cancer Res 71(9):3236–3245. https://doi.org/10.1158/0008-5472.CAN-10-3894

Jo S, Lim JB, Klauda JB, Im W (2009) CHARMM-GUI membrane builder for mixed bilayers and its application to yeast membranes. Biophys J 97(1):50–58. https://doi.org/10.1016/j.bpj.2009.04.013

Jorgensen WL, Chandrasekhar J, Madura JD, Impey RW, Klein ML (1983) Comparison of simple potential functions for simulating liquid water. J Chem Phys 79(2):926–935. https://doi.org/10.1063/1.445869

Krämer A, Ghysels A, Wang E, Venable RM, Klauda JB, Brooks BR, Pastor RW (2020) Membrane permeability of small molecules from unbiased molecular dynamics simulations. J Chem Phys 153(12):124107. https://doi.org/10.1063/5.0013429

Liu X, Lv M, Zhang W, Zhan Q (2023) Dysregulation of cholesterol metabolism in cancer progression. Oncogene 42(45):3289–3302. https://doi.org/10.1038/s41388-023-02836-x

Missner A, Pohl P (2009) 110 Years of the Meyer–Overton Rule: predicting membrane permeability of gases and other small compounds. ChemPhysChem 10(9–10):1405–1414. https://doi.org/10.1002/cphc.200900270

Mouritsen OG (2005) Life–as a matter of fat: the emerging science of lipidomics. Springer. http://link.springer.com/book/10.1007/b138577/page/1

Multhoff G, Radons J, Vaupel P (2014) Critical role of aberrant angiogenesis in the development of tumor hypoxia and associated radioresistance. Cancer 6(2):813–828. https://doi.org/10.3390/cancers6020813

Pettersen EF, Goddard TD, Huang CC, Meng EC, Couch GS, Croll TI et al (2021) UCSF ChimeraX: structure visualization for researchers, educators, and developers. Protein Sci 30(1):70–82. https://doi.org/10.1002/pro.3943

Pias SC (2020) Pathways of oxygen diffusion in cells and tissues. In: Ryu P-D, LaManna JC, Harrison DK, Lee S-S (eds) Advances in experimental medicine and biology. Springer International Publishing, Cham, pp 183–190. https://doi.org/10.1007/978-3-030-34461-0_23

Pias SC (2021) How does oxygen diffuse from capillaries to tissue mitochondria? Barriers and pathways. J Physiol 599(6):1769–1782. https://doi.org/10.1113/JP278815

Roe DR, Cheatham TE (2013) PTRAJ and CPPTRAJ: software for processing and analysis of molecular dynamics trajectory data. J Chem Theory Comput 9(7):3084–3095. https://doi.org/10.1021/ct400341p

Santos CR, Schulze A (2012, August) Lipid metabolism in cancer. FEBS J. https://doi.org/10.1111/j.1742-4658.2012.08644.x

Semenza GL (2016) The hypoxic tumor microenvironment: a driving force for breast cancer progression. Biochim Biophys Acta 1863(3):382–391. https://doi.org/10.1016/j.bbamcr.2015.05.036

Subczynski WK, Hyde JS, Kusumi A (1989) Oxygen permeability of phosphatidylcholine-cholesterol membranes. Proc Natl Acad Sci 86(12):4474–4478. https://doi.org/10.1073/pnas.86.12.4474

Subczynski WK, Hyde JS, Kusumi A (1991) Effect of alkyl chain unsaturation and cholesterol intercalation on oxygen transport in membranes: a pulse ESR spin labeling study. Biochemistry 30(35):8578–8590. https://doi.org/10.1021/bi00099a013

Subczynski WK, Widomska J, Feix JB (2009) Physical properties of lipid bilayers from EPR spin labeling and their influence on chemical reactions in a membrane environment. Free Radic Biol Med 46(6):707–718. https://doi.org/10.1016/j.freeradbiomed.2008.11.024

Wang Q, Dotson RJ, Angles G, Pias SC (2021) Simulation study of breast cancer lipid changes affecting membrane oxygen permeability: effects of chain length and cholesterol. In: Advances in experimental medicine and biology, vol 1269, pp 15–21. https://doi.org/10.1007/978-3-030-48238-1_3

Wang Q, Dotson RJ, Pias SC (n.d.) Influence of breast tumor-associated lipids on oxygen transport across model cell membranes (Part A: interplay of chain length, headgroup size, and cholesterol)

Wennberg CL, Van Der Spoel D, Hub JS (2012) Large influence of cholesterol on solute partitioning into lipid membranes. J Am Chem Soc 134:5351–5361

Widomska J, Raguz M, Subczynski WK (2007) Oxygen permeability of the lipid bilayer membrane made of calf lens lipids. Biochim Biophys Acta (BBA) – Biomembranes 1768(10):2635–2645. https://doi.org/10.1016/j.bbamem.2007.06.018

12 Simulating Gas Permeation Through the Central Pore of AQP5

Eric Joon Shinn, Ardeshir Vahedi, Walter F. Boron, and Emad Tajkhorshid

Abstract

Aquaporins (AQPs), a family of integral membrane proteins that serve as water channels, have been hypothesized to facilitate the permeation of metabolic gases (O_2 and CO_2) via their central pore (CP) formed in the middle of AQP tetramers. Our examination of aquaporin-5 (AQP5), an isoform expressed in lung epithelial cells, using atomistic molecular dynamics (MD) simulations suggests that gas can indeed permeate through the protein. Free energy calculations by implicit ligand sampling (ILS) predict that gas molecules favorably partition into the hydrophobic interior of the CP. Following up, we performed concentration gradient simulations of membrane-embedded AQP5 with CO_2. Concentration profiles predict that CO_2 partitions favorably within the CP cavity, to greater levels than even the lipid bilayer core. Additional analysis showed that the CP facilitates a greater number of permeations than the four monomeric pores combined and that permeations through the CP occurred with a mean residence time 88.1 ns. Altogether, the results indicate that AQPs may facilitate the permeation of gas across membranes, with the CP playing a key role. Understanding the impact of proteins on gas permeation across membranes contributes to a more complete picture of cellular gas transport and may lead to the discovery of therapeutic targets in disease states involving altered gas transport.

Keywords

Molecular dynamics simulations · Concentration gradient · Permeability · AQP · Central pore

12.1 Introduction

In aerobic metabolism, O_2 and CO_2 must traverse a diverse range of environments, including the membranes of numerous cells in various tissues. Upon entering the lungs, O_2 must cross the membranes of the alveolar epithelial cells, capillary endothelial cells, and red blood cells before finally binding to hemoglobin. These cellular layers present as barriers to the permeation of gas, each with unique properties arising from their macromolecular constituents which consist of lipids, proteins, and carbohydrates.

E. J. Shinn · E. Tajkhorshid (✉)
Department of Biochemistry, and Center for Biophysics and Quantitative Biology, University of Illinois Urbana-Champaign, Urbana, IL, USA
e-mail: shinn3@illinois.edu; emad@illinois.edu

A. Vahedi · W. F. Boron
Department of Physiology and Biophysics, Case Western Reserve University, Cleveland, OH, USA

S. C. Pias et al. (eds.), *Oxygen Transport to Tissue XLVI*, Advances in Experimental Medicine and Biology 1498, https://doi.org/10.1007/978-3-032-10389-5_12

Aquaporins (AQPs) have been hypothesized to facilitate gas transport across the membrane, as they are one of the most abundantly expressed integral membrane proteins in red blood and lung epithelial cell membranes (Boron 2010). Each AQP monomer consists of six transmembrane α-helices forming a water-conducting pore that maintains osmotic homeostasis, and X-ray crystallographic data show that AQPs oligomerize into homo-tetramers (Verkman 2008). In the middle of an AQP tetramer, another pore is formed called the central pore (CP). The contribution of transmembrane helices 2 and 5 by each AQP monomer creates the pore, which is lined mostly by hydrophobic amino acids (Fig. 12.1).

Earlier simulation studies have indicated that gas molecules may permeate through AQPs and prefer to do so by the hydrophobic CP over the monomeric aquapores (MPs) (Wang et al. 2007). Here we report on our recent simulations examining the permeation of CO_2 through AQP5, applying a recently developed method that enables the creation of sustained concentration gradients in simulation (Shinn and Tajkhorshid 2024). Though simulations were conducted for CO_2, we considered the observed dynamical behavior to serve as a general proxy for small, non-polar gas molecules within the context of classical MD simulations, which do not consider quantum mechanics/chemistry. Our results suggest that gases like O_2 and CO_2 indeed prefer to permeate via the CP and that the permeation rate time scale converges upon estimates calculated from biomedical measurements.

12.2 Methods

Atomistic MD simulations of an AQP5 tetramer (PDB: 3d9s) were performed using NAMD2.14, with the CHARMM36m forcefield (Horsefield et al. 2008; Phillips et al. 2005; Phillips et al. 2020; Huang and MacKerell 2013). The AQP5 protein was embedded in a 100 Å × 100 Å POPC (1-palmitoyl-2-oleoyl-sn-glycero-3-phosphocholine) planar lipid bilayer (Fig. 12.1) using CHARMM-GUI (Jo et al. 2008). POPC bilayers are commonly used in MD simulations as a model membrane, as POPC is a major lipid in eukaryotic cell membranes. The AQP5 tetramer was positioned in the membrane with its extracellular domain facing the +z-domain. As such, the sides of the membrane in the +z-domain and –z-domain are referred to as the extracellular and intracellular sides, respectively. For the gradient simulation including CO_2, the molecules were generated by replacing randomly selected water molecules with the software tools VMD and PSFGEN (Humphrey et al. 1996). Analysis and visualization were performed using VMD with user-defined Tcl scripts.

The AQP5-membrane system consisted of 110,000 atoms, which was simulated using an integration time step of 2 fs, a force cutoff distance of 12 Å, a pair list distance of 13.5 Å, and a switching distance of 10 Å. Lennard-Jones forces were calculated every time step, with short-range electrostatics calculated every 2 time steps. PME was used for long-range electrostatics (Pollock and Glosli 1996). For the periodic boundary conditions, the box size measured 100 Å × 100 Å × 110 Å in the x-, y-, and z-dimensions, respectively. The system was simulated at a temperature of $T = 310$ K, maintained using Langevin dynamics (Feller et al. 1995) with a damping coefficient of 0.5/ps, and constant pressure was maintained at $P = 1$ atm using the Langevin piston method (Schneider and Stoll 1978), with a period of 100 fs and damping coefficient of 50 fs.

Implicit ligand sampling (ILS) was used to estimate the potential of mean force (PMF) for CO_2 within the CP and MPs of AQP5 (Cohen et al. 2006). The PMF calculated by ILS reports the Gibbs free energy required to introduce a small ligand, like CO_2, at a given position within a simulation and describes the accessibility of the analyzed region of the simulation to the ligand. ILS was used to analyze a 100 ns simulation of the AQP5 tetramer in which CO_2 was absent.

A concentration gradient simulation of the AQP5 system containing 100 CO_2 molecules was performed for 2 μs with a 3 kcal/mol energy step (Shinn and Tajkhorshid 2024). In this simulation, the motion of CO_2 molecules near the z-periodic boundary was unidirectionally biased

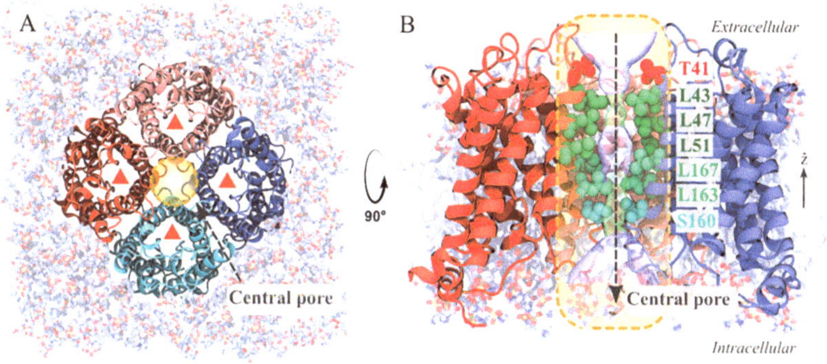

Fig. 12.1 The AQP5-membrane system and central pore. (**a**) The simulated system consisting of an AQP5 tetramer was embedded into a POPC lipid bilayer. Four monomers (orange, pink, blue, and cyan) each contain a water-conducting pore (orange triangles). The CP is formed in the middle at the tetrameric interface (yellow circle). (**b**) A side view of the CP (yellow highlight) shows the amino acids contributed by transmembrane helices 2 and 5. An AQP5 monomer in the foreground is excluded to provide an inside view of the CP. The AQP5 tetramer is positioned in the membrane so that its extracellular domain is oriented toward the +z-domain

using Grid-steered MD (Wells et al. 2007). The bias imposed on the CO_2 molecules effectively was a force applied over a biasing region that produces a ratio of CO_2 concentrations on opposite ends of the simulation that reflects Maxwell-Boltzmann statistics. This method created an enrichment of CO_2 on the extracellular side of the AQP5-membrane and depletion on the intracellular side, generating a sustained concentration gradient in simulation without biasing the CO_2 molecules as they interacted with the protein-membrane system. For a 3 kcal/mol energy step, the theoretical ratio of concentrations is 130-to-1, while for this study, the achieved ratio was 43-to-1 as steady-state flow of CO2 across the membrane dissipates the gradient. The concentration of CO_2 along the transmembrane axis was profiled through the total protein-membrane system, as well as through focused regions, to characterize the transmission of the concentration gradient. The focuses were on the pure lipid phase of the membrane, the AQP5 CP, and MP. These focuses showed the concentration profile through each specific region (i.e., CP focus showed concentration of CO_2 only through the CP, the MP focus only through the MPs of AQP5). An analysis of the permeations of CO_2 that occurred through the AQP5 tetramer was performed, in which permeations were counted and classified as CP events if they entered the CP cavity. Residence times for each permeation event were obtained by recording the initiation and completion of each event. These analyses were performed by user-defined Tcl-scripts in VMD.

12.3 Results

12.3.1 Comparison of Pore Energetics

The Gibbs free energy of partitioning for CO_2 within the CP and MPs was estimated by ILS, which was used to compute the PMF through the pore volumes (Fig. 12.2). Regions of the pore with energy values exceeding that of the bulk solvent, as indicated by the PMF, predict that those regions will be unlikely to be occupied by the ligand, in other words have low concentration. Though the PMF does not provide information about diffusion coefficients, regions of low concentration correspond to low permeability in the view of the inhomogeneous solubility diffusion model (Dotson et al. 2017; Venable et al. 2019; Marrink and Berendsen 1994).

We found the highest energy barrier in the CP, as indicated by the PMF, rose 2.1 kcal/mol (ΔE_{CP}) above that of the bulk solvent (Fig. 12.2). This barrier coincided with the

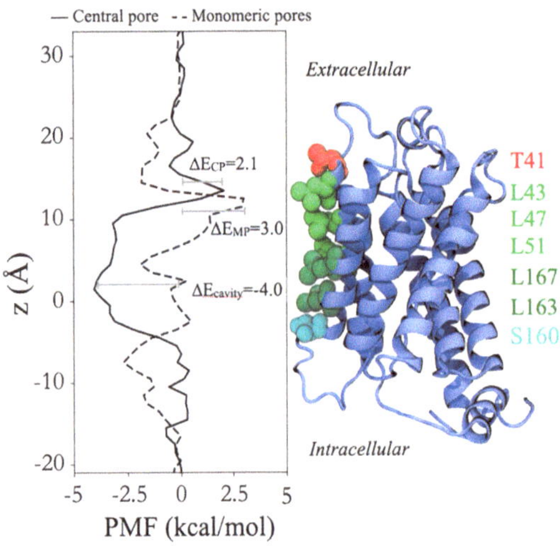

Fig. 12.2 Implicit ligand sampling was used to calculate the PMF for CO_2 within the CP and MPs. The highest energy barrier in the CP (shaded in red) measured 2.1 kcal/mol above that of the bulk solvent at leucine-43, while in the MP the highest barrier was 3.0 kcal/mol at the selectivity filter. The analysis also revealed a low-energy region called the CP cavity (shaded in blue) with a free energy minimum of −4.0 kcal/mol. An AQP5 monomer with CP lining residues is shown on the right as a spatial reference

position of leucine-43 (L43) (z = 13 Å), where the CP is narrow. A smaller barrier was also predicted near threonine-41 (T41) (z = 19 Å). In comparison, the highest barrier calculated in the MP was 3 kcal/mol (ΔE_{MP}), nearly 1 kcal/mol higher than that of the CP. The barrier in the MP was found at the well-studied AQP selectivity filter (Hub and de Groot 2008). In addition, the ILS analysis revealed an extended low-energy region, with a minimum of −4 kcal/mol (ΔE_{cavity}) and length of 13 Å, within the core of the CP. Hydrophobic leucines line this region of the CP which we call the "CP cavity." Overall, the energetics indicate the CP as being more favorable than the MPs for the partitioning of gas, and thus more permeable.

12.3.2 Concentration Profiles

In our concentration gradient simulations, profiles of the CO_2 concentration through the AQP5 CP and MPs reflected estimates from ILS. The concentration profile through the lipids, serving as a useful reference, showed a typical maximum at the center of the bilayer core and distinguishable local minimum at the position of the polar phospholipid head groups on the side of the membrane representing the extracellular side (Fig. 12.3a). Because 100 CO_2 molecules were included in a relatively small simulation volume, the concentration values exceed what would occur in a physiological context. A high concentration, however, was used to achieve adequate sampling and is common practice in MD simulations.

Beginning with the lipid phase from the extracellular side, the dip in CO_2 concentration denoted the beginning of the membrane, with the local minimum at z = 20 Å indicating the position of the phospholipid head groups. Moving down the transmembrane axis, maximal concentration was measured within the hydrophobic lipid tail bilayer core ($z = 0$ Å), followed by a subsequent decrease in CO_2 concentration. Under the gradient condition, the impact of the intracellular headgroups from the bottom leaflet of the lipid bilayer is distinguishable only as a slight deviation in the profile curve from $z = -23$ Å to −27 Å. Beyond z = −30 Å, the CO_2 concentration steadily decreased through the aqueous phase.

For the CP, most notable was the marked elevation in concentration within the CP cavity ($z = 12$ Å to 16 Å), above even that of the lipid tail region of the bilayer, which is known to favorably accommodate small non-polar solutes, like O_2 and CO_2 (Dotson et al. 2017). Meanwhile, the free energy barriers in the CP observed at the extracellular entrance to the CP, near T41 and L43, were represented as consecutive local minima at $z = 20$ Å and $z = 13$ Å, respectively. Another local minimum appears at the position of serine-160 (S160) ($z = -12$ Å), at which point the CP terminates. At either end of the CP cavity, the local CO_2 concentration is significantly lower than other regions and may provide a mechanism of gating for the passage of CO_2.

The MPs show a distinct concentration profile compared to the CP. At the extracellular entrance, the MPs show a local elevation in CO_2 concentration, peaking at $z = 16$ Å, which decreases upon approaching the selectivity filter ($z = 8$ Å).

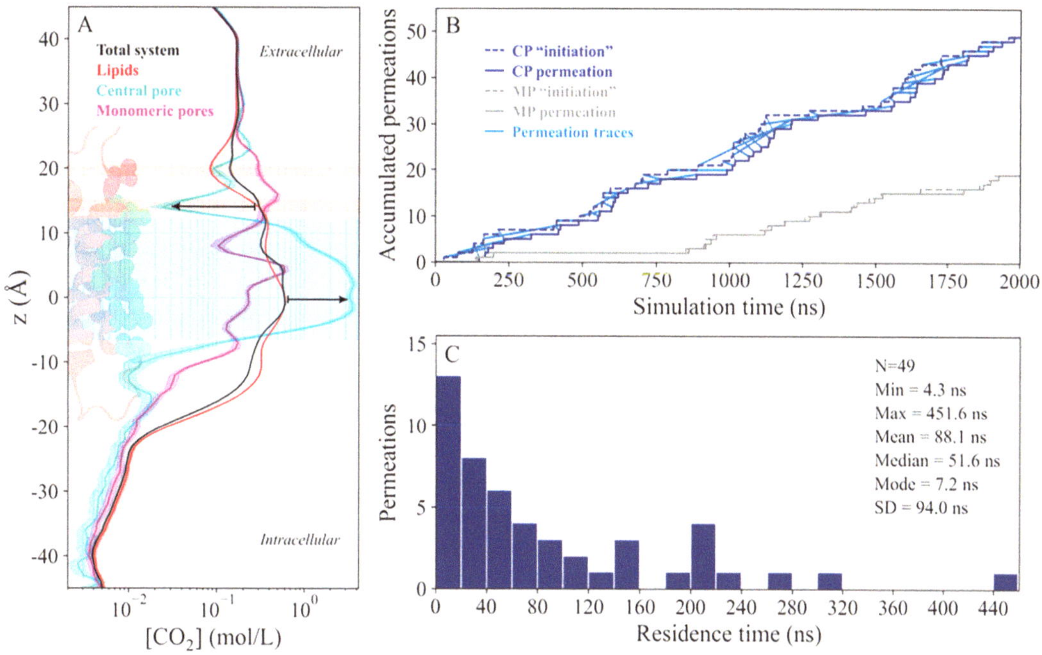

Fig. 12.3 Permeation analysis of AQP5 concentration gradient simulations. (**a**) The concentration of CO_2 was profiled along the transmembrane axis for the total system (black), through the lipid phase (red), AQP5 CP (cyan), and AQP5 MPs (magenta), visualizing the transmission of the gradient across the membrane. The mean concentration is plotted as a line, with the shaded region representing the precision within 1 standard error. The CP exhibited the highest concentration in the CP cavity, exceeding even the lipid bilayer core. (**b**) The cumulative number of permeations that passed through AQP5 were analyzed for the 2 μs simulation. The CP facilitated 49 permeations (blue solid line) while the 4 MPs facilitated 19 (gray solid line), indicating the CP as more permeable. The time at which each permeating CO_2 entered the AQP5 is shown as dotted lines with corresponding color and lines tracing each permeation from initiation to completion are shown in light blue. (**c**) A histogram shows the distribution of residence times for permeations through the CP

The profile exhibits another local maximum at $z = 4$ Å, in between the selectivity filter and the two conserved Asn-Pro-Ala (NPA) motifs that exclude additional protons from being carried through the MPs by water. The elevated CO_2 concentration measured from our simulation not only agrees with our ILS results but also with free energy calculations from a previous MD simulation study predicting favorable partitioning for O_2, CO_2, and NH_3 in the same region (Hub and de Groot 2008). At the depth of the lipid tails and the CP cavity, the overall low CO_2 concentration within the MPs signifies they would be the least permeable to gas, unsurprising given their role as aquapores.

12.3.3 Flow Through the CP

We tracked the accumulation of CO_2 permeations that occurred across AQP5 (Fig. 12.3b). Over the course of the 2 μs gradient simulation, 49 permeations occurred via the CP, while a combined total of 19 permeations occurred through the 4 MPs. While the overall rate of CP permeations appeared steady over the course of the simulation, permeations through the MPs seemed to increase after 1 μs. Of the 49 CP events, 47 were of CO_2 that permeated from the extracellular side to the intracellular and 2 were for the reverse, netting a flow of 45 permeations down the concentration gradient. The greater number of permeations measured through the CP versus the four MPs substantiates the CP as the preferred pathway for gas permeation through AQP5.

12.3.4 Residence Time

The residence time of CO_2 for each permeation event was calculated from the time of entry into and exit from the protein space of each permeating CO_2 molecule, termed the initiation and completion of each event. Traces that are shown in Fig. 12.3b connect the time of initiation to completion for each permeation. A collation of the CP permeation residence times is represented as a histogram in Fig. 12.3c. The mean residence time of CO_2 permeating through the CP was 88.1 ns. The population exhibited a positive skew to the right, having a median of 51.6 ns and mode of 7.2 ns. A visualization of a single CO_2 permeation event from initiation to completion is depicted in Fig. 12.4. Permeations through the MPs occurred more quickly, with a mean residence time of 18.2 ns.

12.4 Conclusion and Discussion

Contention surrounds the idea that integral membrane proteins, like AQPs, contribute significantly to the overall permeability of the membrane to gas molecules, especially since the lipids are presumed to mediate most gas permeation (Hannesschlaeger et al. 2019; Michenkova et al. 2021). The contribution by the lipids, however, may be less than initially expected as their permeability depends on their phase state (Ghysels et al. 2019). Our simulations showed that the AQP5 CP allows for greater partitioning of CO_2 than the MPs and that while both the CP and MP facilitate permeation, the CP facilitates more. Whether the quantity of gas traversing the membrane by AQPs is physiologically significant remains to be answered and will require experimental evidence.

To make a case for gas channels, a principal counterargument that must be overcome is the notion that gases like CO_2 and O_2 partition favorably into the lipid bilayer and are thus expected to easily permeate membranes by the lipid phase even when considering the membrane-tightening effect of cholesterol (Hannesschlaeger et al. 2019; Missner et al. 2008). Moreover, even if a protein possesses a channel that favorably partitions gas molecules, as we showed for the CP of AQP5, physical limitations on the local transport to and away from the pore, known as the access resistance, must also be reconciled (Hannesschlaeger et al. 2019). Because channels formed by a membrane protein represent a small fraction of the membrane surface area, the channel itself must not only support permeant flux greater than the lipid phase, but so too must permeating molecules be able to access the channel rapidly enough to supply it with enough mass to do so. Indeed, the argument for protein gas channels contributing significantly to physiological gas flux across cell membranes is challenging. From our simulations, we observed a total of 1851 permeations across the membrane, only 49 of which occurred exclusively by the CP and 19 by the 4 MPs, 2.6% and 1%, respectively. Furthermore, a prior study using MD simulation concluded that membrane proteins non-specifically reduce gas permeability (Dotson and Pias 2018).

However, the presence of membrane proteins, like AQPs, has been shown to enhance the permeability of cell membranes to gases, like CO_2 and NH_3, by electrophysiological experiments measuring pH (Geyer et al. 2013; Cooper and Boron 1998). Though there has been some disagreement on the ability of the experimental techniques to precisely distinguish the permeability of the membrane from other phenomena, like unstirred layers or conversion between CO_2 and carbonic acid (Hannesschlaeger et al. 2019), the evidence stands in support of the idea that proteins can act as gas channels. More recently, stopped-flow absorbance spectroscopic monitoring of hemoglobin oxygenation in red blood cells indicates that AQP1 and Rh complex account for over half of red blood cell membrane permeability to O_2 (Zhao et al. 2020). Because reasonable theoretical arguments contest experimental reports that the presence proteins like AQPs and the Rh-complex can positively impact membrane gas permeability, the question remains to be

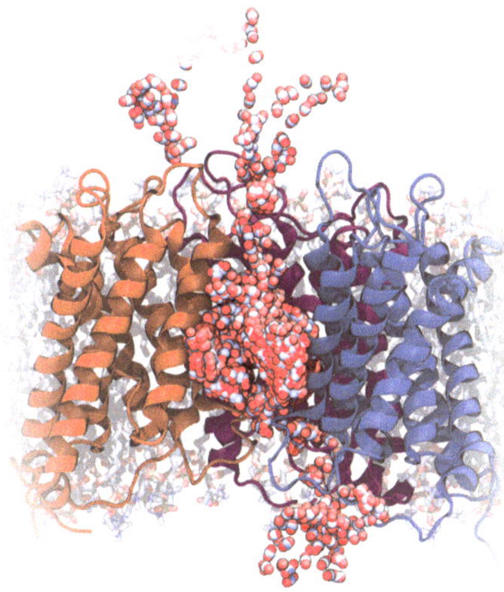

Fig. 12.4 Trajectory of a single CO_2 permeation through the AQP5 CP from initiation to completion. Snapshots of the CO_2 position over time show how a CO_2 molecule diffuses stochastically within the CP for an extended period, 88.1 ns on average

answered by modeling and simulation approaches that can investigate the issue with nanoscopic detail.

Though this work features CO_2, as it is a part of an ongoing collaboration whose outcomes are in preparation for publication, the results obtained from our MD simulations in the present study are to some degree extensible to the membrane permeation of O_2. Within the scope of classical MD where molecules are not subject to the formation or breakage of chemical bonds, O_2 and CO_2 are relatively similar molecules, being small, non-electrolytes, comprised of just two or three atoms. Other common biomolecules, in contrast, may contain many more atoms, be polar, and carry a net electric charge. Even the slightly larger ethanol molecule, a common permeant studied with MD simulations, contains nine atoms and can be expected to enter the CP of AQP5 with considerably greater difficulty if at all. In the aqueous phase of our simulations, we calculated a diffusion coefficient of 3.56×10^{-5} cm^2/s for CO_2 using the Eq. 12.2 $D\Delta t = \langle(r(t + \Delta t) - r(t))^2\rangle$ for a lag time of 40 ps. The resulting value is within a factor-of-2 difference from values calculated for O_2 in simulation (Dotson et al. 2017; Ghysels et al. 2019) and within the range of values obtained experimentally (Polat et al. 2024). For the studied temporo-spatial scale (microseconds and nanometers), we expect O_2 and CO_2 to behave similarly, with both being able to access and permeate through the CP of AQP5. In fact, we might expect O_2 to permeate more readily through the CP, being a smaller molecule. Though we compare O_2 to CO_2 in considering their ability to diffuse through gas channels, we acknowledge that the transport mechanisms for O_2 and CO_2 may differ from each other in a broader physiological context.

The mean residence time measured from our simulations indicates that permeations via AQP5 occur on the scale of tens of nanoseconds. This value provides a useful initial point of comparison with biomedical values. Red blood cells transit the lung alveolar capillary beds in about 0.5 seconds (Petersson and Glenny 2014), contain about 270 million hemoglobin tetramers, enter the lungs at around 75% oxygenation, and leave at 98% (Pierigè et al. 2008). This equates to a delivery of around 230 million O_2 molecules in half a second, 460 molecules per microsecond, to each passing red blood cell. After normalizing flow with the surface area of a red blood cell (134 μm^2) (Houchin et al. 1958), it seems that our prediction of a nanosecond time regime for permeation events involving individual molecules of gas is reasonably accurate, considering the vast difference in scale separating the biomedical values and our molecular simulations.

The outlook for studies of O_2 and CO_2 transport with classical MD simulations appears positive, as the relevant processes occur in accessible temporal and spatial domains. The distance between the epithelium of the alveoli and the membrane of erythrocytes in alveolar capillaries has been measured to be on the order of a single micron (Weibel et al. 1993). With the diffusion rates of O_2 and CO_2 being relatively fast ($\sim 5 \times 10^{-5}$ cm^2/s), it is possible to envision future simulations modeling the entire diffusion process in meticulous detail with advances in computational power. Such detailed modeling would enable scientists and physicians to understand how nanoscopic changes to the diffusion barrier affect the transport of gas in tissue.

References

Boron WF (2010) Sharpey-Schafer lecture: gas channels. Exp Physiol 95:110–1130

Cohen J, Arkhipov A, Braun R et al (2006) Imaging the migration pathways for O_2, CO, NO, and Xe inside myoglobin. Biophys J 91:1844–1857

Cooper GJ, Boron WF (1998) Effect of PCMBS on CO_2 permeability of Xenopus oocytes expressing aquaporin 1 or its C189S mutant. Am J Physiol Cell Physiol 275: C1407–C1682

Dotson RJ, Pias SC (2018) Reduced oxygen permeability upon protein incorporation within phospholipid bilayers. Adv Exp Med Biol 1072:405–411

Dotson RJ, Smith CR, Bueche K et al (2017) Influences of cholesterol on the oxygen permeability of membranes: insight from atomistic simulations. Biophys J 112: 2336–2347

Feller SE, Zhang Y, Pastor RW et al (1995) Constant pressure molecular dynamics simulations: the Langevin piston method. J Chem Phys 103(11): 4613–4621

Geyer RR, Musa-Aziz R, Qin X et al (2013) Relative CO_2/NH_3 selectivities of mammalian aquaporins 0-9. Am J Physiol Cell Physiol 304:C985–C994

Ghysels A, Krämer A, Venable RM et al (2019) Permeability of membranes in the liquid ordered and disordered phases. Nat Commun 10:1–12

Hannesschlaeger C, Horner A, Pohl P (2019) Intrinsic membrane permeability to small molecules. Chem Rev 119:5922–5953

Horsefield R, Nordén K, Fellert M et al (2008) High-resolution x-ray structure of human aquaporin 5. PNAS 105(36):13327–13332

Houchin DN, Munn JI, Parnell BL (1958) A method for the measurement of red cell dimensions and calculation of mean corpuscular volume and surface area. Blood 13(12):1185–1191

Huang J, MacKerell AD (2013) CHARMM36 all-atom additive protein force field: validation based on comparison to NMR data. J Comput Chem 34:2135–2214

Hub JS, de Groot BL (2008) Mechanism of selectivity in aquaporins and aquaglyceroporins. PNAS 105(4): 1198–1203

Humphrey W, Dalke A, Schulten K (1996) VMD: visual molecular dynamics. J Mol Graph 14:33–38

Jo S, Kim T, Iyer VG et al (2008) CHARMM-GUI: a web-based graphical user Interface for CHARMM. J Comput Chem 29:1859–1865

Marrink SJ, Berendsen HJC (1994) Simulation of water transport through a lipid membrane. J Phys Chem 98: 4155–4168

Michenkova M, Taki S, Blosser MC et al (2021) Carbon dioxide transport across membranes. Interface Focus 11:20200090

Missner A, Kugler P, Saparov SM et al (2008) Carbon dioxide transport through membranes. J Biol Chem 283:25340–25347

Petersson J, Glenny RW (2014) Gas exchange and ventilation-perfusion relationships in the lung. Eur Respir J 44:1023–1041

Phillips JC, Braun R, Wang W et al (2005) Scalable molecular dynamics with NAMD. J Comput Chem 26:1781–1802

Phillips JC, Hardy DJ, Maia JDC et al (2020) Scalable molecular dynamics on CPU and GPU architectures with NAMD. J Chem Phys 153:044130

Pierigè F, Serafini S, Rossi L et al (2008) Cell-based drug delivery. Adv Drug Deliv Rev 60(2):286–295

Polat HA, Coelho FM, Vlugt TJH et al (2024) Diffusivity of CO_2 in H_2O: a review of experimental studies and molecular simulations in the bulk and in confinement. J Chem Eng Data 69:3269–3329

Pollock EL, Glosli J (1996) Comments on P^3M, FMM, and the Ewald method for large periodic Coulombic systems. Comput Phys Commun 95:93–110

Schneider T, Stoll E (1978) Molecular-dynamics study of a three-dimensional one-component model for distortive phase transitions. Phys Rev B 17:1302–1322

Shinn EJ, Tajkhorshid E (2024) Generating concentration gradients across membranes for molecular dynamics simulations of periodic systems. Int J Mol Sci 25(7): 3616

Venable RM, Krämer A, Pastor RW (2019) Molecular dynamics simulations of membrane permeability. Chem Rev 119:5954–5997

Verkman AS (2008) Mammalian aquaporins: diverse physiological roles and potential clinical significance. Expert Rev Mol Med 10:1–18

Wang Y, Cohen J, Boron WF et al (2007) Exploring gas permeability of cellular membranes and membrane channels with molecular dynamics. J Struct Biol 157: 534–544

Weibel ER, Federspiel WJ, Fryder-Doffey F et al (1993) Morphometric model for pulmonary diffusion capacity: I. Membrane diffusion capacity. Respir Physiol 93:125–149

Wells DB, Abramkina V, Aksimentiev A (2007) Exploring transmembrane transport through α-Hemolysin with grid-steered molecular dynamics. J Chem Phys 127:125101

Zhao P, Geyer RR, Salameh AI et al (2020) Role of channels in the oxygen permeability of red blood cells. bioRxiv

Part IV

Hypoxia and Metabolic Regulation

13 Role of the Dual-Specificity Phosphatase 6 (DUSP6) in the Activation of ERK1/2 Signaling by Hypoxia or Acidosis in Tumor Cells

Anne Riemann, Sarah Reime, and Oliver Thews

Abstract

Important pathophysiological characteristics of human tumors are the presence of hypoxia and of an extracellular acidosis. Both metabolic parameters can affect the malignant behavior of tumors but also the response to treatments. Here the modulation of the ERK1/2 signaling plays a relevant role. ERK1/2 activation is controlled by the dual-specificity phosphatase 6 (Dusp6) and by this modulates cellular processes. For this reason, the study analyzed the impact of hypoxia and/or extracellular acidosis on the expression of Dusp6. Several tumor cell lines were exposed to control conditions (room air, pH = 7.4), hypoxia (pO_2 = 1.5 mmHg), and/or an extracellular pH of 6.6 for 24 h. Gene expression was measured by qPCR, RNA sequencing, and Western blot. In addition, the expression and activation of signaling molecules (ERK1/2, mTOR) were analyzed. Most of the tumor lines showed a significant reduction in the Dusp6 expression during acidosis and hypoxia. The combination of hypoxia+acidosis simultaneously had the strongest impact on Dusp6 mRNA expression. The ERK1/2 phosphorylation showed an inverse behavior and was increased during acidosis and hypoxia. Therefore, the ERK1/2 activation might result (at least partially) from reduced phosphatase activity. ROS generation during acidosis was not responsible for Dusp6 changes. However, mTor activation correlated well with Dusp6 expression. Additionally, other regulating proteins, like the transcription factor Ets2 and the p53, which were regulated by hypoxia and acidosis as well, correlated strongly with Dusp6 expression. It can be concluded that ERK1/2 activation during acidosis and/or hypoxia results from Dusp6 repression, which could be induced by mTor signaling or the expression of other regulating proteins.

Keywords

Tumor acidosis · Cell signaling · mTOR · p53 expression · Ets2 expression

13.1 Introduction

The metabolic microenvironment of tumors is characterized by the fact that there is often a pronounced O_2 deficiency (hypoxia) and the extracellular pH is significantly reduced (acidosis). Both changes can influence the biological behavior of tumor cells (e.g., cell migration, metastasis, cell survival) and thus promote

A. Riemann · S. Reime · O. Thews (✉)
Julius-Bernstein-Institute of Physiology, University of Halle, Halle, Germany
e-mail: oliver.thews@medizin.uni-halle.de

S. C. Pias et al. (eds.), *Oxygen Transport to Tissue XLVI*, Advances in Experimental Medicine and Biology 1498, https://doi.org/10.1007/978-3-032-10389-5_13

malignancy. It has long been known that hypoxia can influence the expression of relevant genes via HIF-1 and hypoxia-responsive elements (HRE). A functional influence is also possible via the activation of various intracellular signaling cascades. For example, acidosis leads to increased ERK1/2 phosphorylation, which in turn causes functional changes in tumor cell biology (Riemann et al. 2011, 2014). However, it has also been shown that hypoxia can lead to ERK1/2 activation (Minet et al. 2000).

ERK1/2 phosphorylation is typically induced via the Ras-Raf-MEK cascade. However, there are also mechanisms of dephosphorylation of ERK1/2 via specific phosphatases, whereby the ERK1/2 phosphorylation status results from the balance of the kinases and phosphatases involved. An important ERK1/2 phosphatase is dual-specificity phosphatase 6 (DUSP6), the activation or inhibition of which can lead to a functional influence on the cell (Ahmad et al. 2018; Yoo et al. 2023). For ERK1/2 phosphorylation induced by acidosis and/or hypoxia, the question therefore arises, whether it is caused by activation of the kinases (e.g., MEK) or inhibition of the phosphatase involved (e.g., Dusp6).

For this reason, the aim of the present study was to investigate the influence of acidosis and/or hypoxia on the expression of Dusp6 and to unravel which signaling pathways could be responsible for a modulation of Dusp6 by the metabolic microenvironment.

13.2 Methods

13.2.1 Cell Lines

The experiments were performed in three rat tumor cell lines (subline AT1 of the R-3327 Dunning prostate carcinoma, Walker-256 mammary carcinoma, Yoshida sarcoma). All cell lines were grown in RPMI medium supplemented with 10% fetal calf serum (FCS) at 37 °C under humidified 5% CO_2 atmosphere. For acidotic conditions, cells were incubated in FCS-lacking medium buffered with 10 mM HEPES/10 mM MES adjusted to 6.6 for 24 h. To induce hypoxia, cells were incubated at a pO_2 of 1.5 mmHg in a hypoxia chamber (HypoxyLab, Oxford Optronix, Oxford, UK) for 24 h.

13.2.2 Experimental Setting

After incubation under different conditions, cells were harvested, lysed, and total RNA was extracted with BlueZol reagent according to the manufacturer's instructions. After removing genomic DNA contamination, RNA concentration was determined by NanoDrop (Biochrom, Germany). For qPCR validation, 1 μg RNA was subjected to reverse transcription with SuperScript II reverse transcriptase (Thermo Fisher Scientific, Waltham, MA, USA) and analyzed by qPCR using the Platinum SYBR Green qPCR Supermix (Thermo Fisher Scientific, Waltham, MA, USA). The primers for Dusp6-qPCR were forward TGAGGATCAGCTCCGACTCT and reverse TTTGCCTCGGGCTTCATCTAT. The obtained data were normalized to 18S or Hprt1 and were related to the respective control. Whole genome RNA-sequencing in AT1 cells was performed by Novogene Co., Ltd. (Cambridge, UK) preparing the sequencing libraries and running the paired-end sequencing (2 x 150 bp) on a Nova-Seq6000 Illumina system. Adaptor clipping and data quality control were provided by the service company as well. HISAT2 (v. 2.1.0) served for read mapping to the rat genome rn7.

Western blotting was performed according to standard protocols. In brief, cells were lysed (0.5 M Tris-HCl, pH 6.8; 10% SDS; 10% 2-mercaptoethanol; 20% glycerol; 0.01% bromophenol blue), separated by sodium dodecyl sulfate polyacrylamide gel electrophoresis, and transferred to a nitrocellulose membrane. Subsequently, membranes were incubated with antibodies specific for ERK1/2 (#4696), pERK1/2 (#9101), mTOR (#4517), or p-mTOR (#2971, all Cell Signaling). The bound primary antibody was visualized by IRDye secondary antibodies (Licor Biosciences, Lincoln, NE, USA) using the imaging system Odyssey (Licor Biosciences, Lincoln, NE, USA). Quantitative analysis was performed with Image Studio Lite software (Licor Biosciences).

13.3 Results

In all three tumor cell lines, 24-h incubation with a lowered pH led to an increase in ERK1/2 phosphorylation (Fig. 13.1). Hypoxia also led to a slight (statistically insignificant) ERK1/2 activation. The changes in the combination of acidosis +hypoxia corresponded approximately to the hypoxic incubation. Dusp6 mRNA expression showed an almost mirror-image behavior to ERK1/2 phosphorylation. Both acidosis and especially hypoxia showed a decrease in Dusp6 mRNA expression in most cell lines (Fig. 13.1). Analysis of the correlation between Dusp6 expression and ERK1/2 phosphorylation across all cell lines and experimental conditions showed a negative correlation ($r = -0.308$).

Since it was known from other studies that ROS and mTOR signaling can influence Dusp6 expression (Ahmad et al. 2018; Cheung et al. 2020), the influence of acidosis and hypoxia was analyzed in AT1 cells. Acidosis reduced the Dusp6 mRNA expression to 68.0 ± 1.7% of control ($p < 0.001$). Adding tiron as a ROS scavenger for 24 h did not change the pH effect (74.9 ± 8.1%; $p = 0.324$ vs. acidosis alone). Also, incubation of the cells at pH 7.4 and adding H_2O_2 (25 μM, 24 h) had only a minor impact on Dusp6 (86.0 ± 8.9%; $p = 0.193$), indicating that ROS seems not to be responsible for the acidosis-induced change in Dusp6 mRNA expression. In contrast, mTOR phosphorylation was reduced to 74.7 ± 2.1% by acidosis and to 81.6 ± 4.5% by hypoxia in this cell line. The combination of

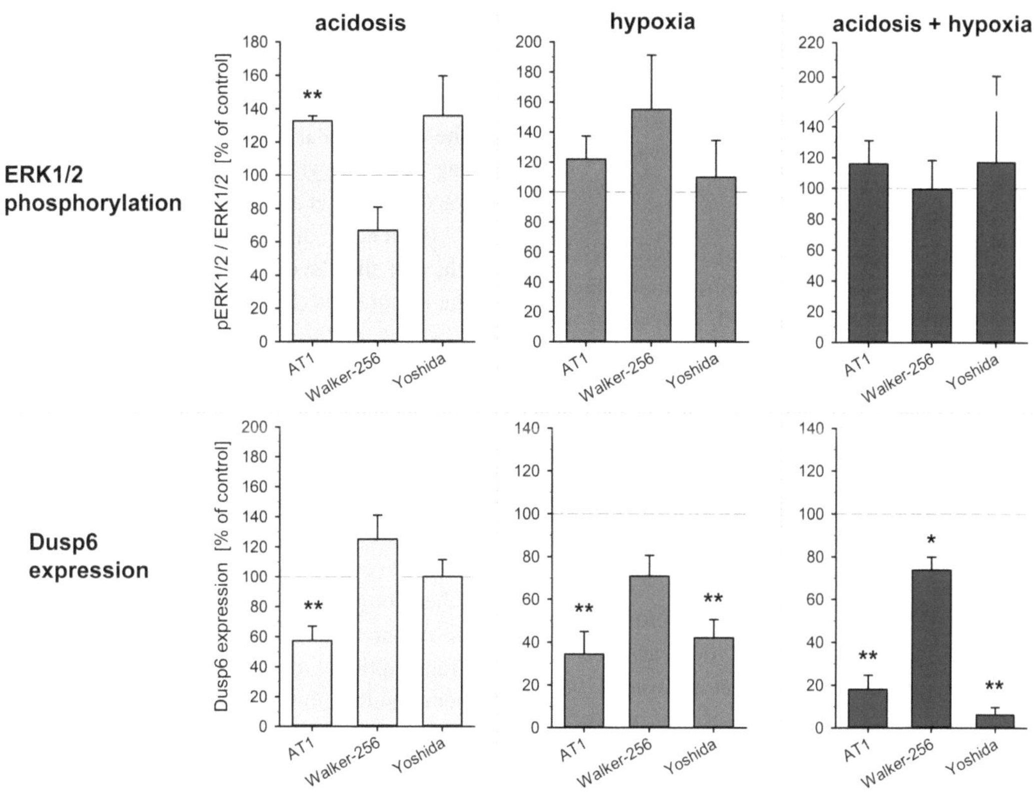

Fig. 13.1 ERK1/2 phosphorylation and Dusp6 mRNA expression under acidic (pH 6.6) and hypoxic (pO_2 = 1.5 mmHg) conditions or the combination of acidosis+hypoxia for 24 h in different tumor cell lines. $n = 4$–7; (*) $p < 0.05$, (**) $p < 0.01$ versus control

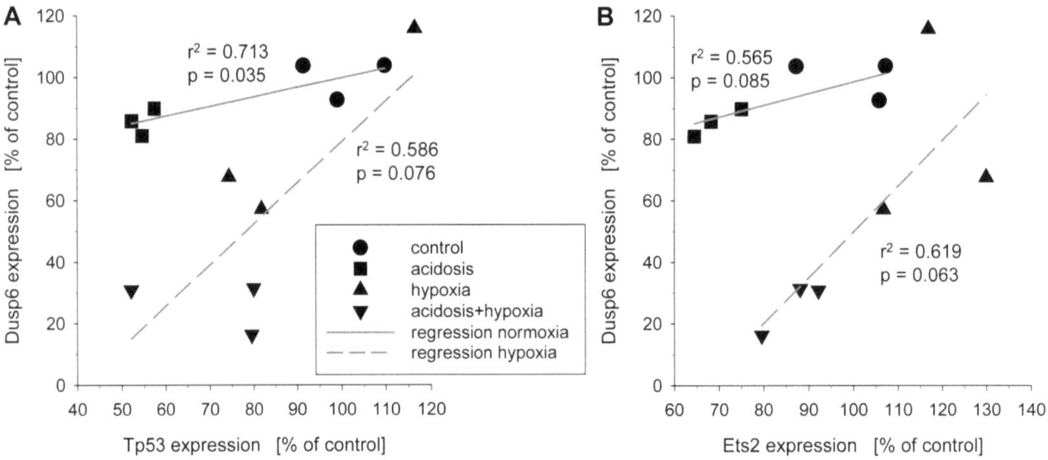

Fig. 13.2 Correlation of mRNA expression of Dusp6 and (**a**) Tp53 or (**b**) Ets2 in AT1 cells under acidic, hypoxic conditions or the combination of acidosis+hypoxia for 24 h. Regression lines were calculated for experiments under normoxia (room air) or hypoxia ($pO_2 = 1.5$ mmHg), respectively

acidosis+hypoxia even led to a decrease to 46.1 ± 7.8%. In parallel, in these cells, the mRNA expression of Dusp6 was reduced to 85.5 ± 2.6% by acidosis or to 80.1 ± 18.0% by hypoxia and to 26.1 ± 4.9% by acidosis+hypoxia, indicating a strong correlation between mTOR phosphorylation and Dusp6 expression ($r^2 = 0.930$; $p = 0.170$).

Dusp6 expression can also be influenced by various other proteins (transcription factors, RNA-binding proteins, growth factors) (Ahmad et al. 2018; Piya et al. 2012; Zhang et al. 2010). In the RNA-seq expression analysis, it was seen that fibroblast growth factor 2 (Fgf2) and tristetraprolin (Zfp36) were upregulated by acidosis, but hypoxia had no effect. ETS proto-oncogene 1 (Ets1) was only upregulated by hypoxia, and pumilio homolog 2 (Pum2) was practically unaffected by metabolic factors. Only tumor protein p53 (Tp53) and ETS proto-oncogene 2 (Ets2) were modulated by both acidosis and hypoxia, and their expression showed a linear correlation with Dusp6 expression under both normoxia and hypoxia (Fig. 13.2), with the slope of the correlation being much more pronounced under hypoxic conditions.

13.4 Conclusions

Previous studies have shown that in various tumor cell lines, a short-term (3 h) acidification of the extracellular space very quickly led to a strong phosphorylation of ERK1/2 by about 200% (Riemann et al. 2011). This ERK1/2 activation influenced migration, proliferation, cell death, and the development of chemoresistance of the tumor cells (Riemann et al. 2014; Sauvant et al. 2008). At the same time, it was shown that these short-term changes were caused by increased ROS formation due to acidosis, but extracellularly added ROS had no effect (Riemann et al. 2011). These results from short-term studies may explain rapid functional changes in biological behavior caused by acidosis. However, the question also arises as to whether longer-term acidification also induces ERK1/2 phosphorylation, which in turn can lead to transcriptional changes in tumor cells. The present results show that both acidosis and hypoxia activate ERK1/2 after incubation for 24 h, but to a much lesser extent than short-term incubation (Fig. 13.1) (Riemann et al. 2011). ERK1/2 phosphorylation results from the balance of activating kinases and deactivating phosphatases (e.g., Dusp6), whereby the phosphatases are ascribed an important function in the feedback regulation of ERK1/2. The current study shows that metabolic factors such as hypoxia and/or acidosis over 24 h can also induce ERK1/2 activation via repression of phosphatases and thus may induce both functional and transcriptional changes in tumor cells. The results on hypoxia contradict data from Bermudez et al. (2011). They described that incubation of cells at a pO_2 of approx. 8 mmHg for 24 h led to a stabilization of Dusp6 mRNA. The reason for these different findings could be due to the level of hypoxia. The situation is further complicated by the fact that the authors were able to show that the hypoxia-induced change in Dusp6 expression was dependent on both HIF-1 stabilization and the activation of MEK-ERK1/2 (Bermudez et al. 2011).

With regard to potential signaling mechanisms that lead to acidosis- or hypoxia-induced changes in Dusp6 expression, ROS, which themselves lead to ERK1/2 phosphorylation (Riemann et al. 2011), do not appear to play a role. On the other hand, the present data show that mTOR phosphorylation under acidosis and/or hypoxia correlates well with Dusp6 mRNA expression. Since other studies have also described an influence of mTor signaling on Dusp6 (Ahmad et al. 2018; Bermudez et al. 2008), further studies should investigate the exact mechanism. The present results also indicate that other regulatory proteins that have a direct influence on Dusp6 expression are affected by acidosis and hypoxia. In particular, the expression of Tp53 and Ets2 correlated very well with Dusp6 (Fig. 13.2). A fundamental correlation between these genes has been described by others (Piya et al. 2012; Ekerot et al. 2008). With regard to metabolic parameters, however, the present study shows for the first time that both acidosis and hypoxia influence Tp53 and Ets2 in a comparable way to Dusp6, but that acidosis and hypoxia apparently influence the cooperation of Tp53 and Ets2 with Dusp6 expression independently of each other. Further investigations must unravel the signaling pathways through which the interaction of these proteins occurs.

References

Ahmad MK, Abdollah NA, Shafie NH et al (2018) Dual-specificity phosphatase 6 (DUSP6): a review of its molecular characteristics and clinical relevance in cancer. Cancer Biol Med 15:14–28

Bermudez O, Marchetti S, Pages G et al (2008) Post-translational regulation of the ERK phosphatase DUSP6/MKP3 by the mTOR pathway. Oncogene 27: 3685–3691

Bermudez O, Jouandin P, Rottier J et al (2011) Post-transcriptional regulation of the DUSP6/MKP-3 phosphatase by MEK/ERK signaling and hypoxia. J Cell Physiol 226:276–284

Cheung EC, DeNicola GM, Nixon C et al (2020) Dynamic ROS control by TIGAR regulates the initiation and progression of pancreatic cancer. Cancer Cell 37: 168–182 e164

Ekerot M, Stavridis MP, Delavaine L et al (2008) Negative-feedback regulation of FGF signalling by DUSP6/MKP-3 is driven by ERK1/2 and mediated by Ets factor binding to a conserved site within the DUSP6/MKP-3 gene promoter. Biochem J 412:287–298

Minet E, Arnould T, Michel G et al (2000) ERK activation upon hypoxia: involvement in HIF-1 activation. FEBS Lett 468:53–58

Piya S, Kim JY, Bae J et al (2012) DUSP6 is a novel transcriptional target of p53 and regulates p53-mediated apoptosis by modulating expression levels of Bcl-2 family proteins. FEBS Lett 586:4233–4240

Riemann A, Schneider B, Ihling A et al (2011) Acidic environment leads to ROS-induced MAPK signaling in cancer cells. PLoS One 6:e22445

Riemann A, Schneider B, Gündel D et al (2014) Acidic priming enhances metastatic potential of cancer cells. Pflügers Arch 466:2127–2138

Sauvant C, Nowak M, Wirth C et al (2008) Acidosis induces multi-drug resistance in rat prostate cancer cells (AT1) in vitro and in vivo by increasing the activity of the p-glycoprotein via activation of p38. Int J Cancer 123:2532–2542

Yoo DH, Im YS, Oh JY et al (2023) DUSP6 is a memory retention feedback regulator of ERK signaling for cellular resilience of human pluripotent stem cells in response to dissociation. Sci Rep 13:5683

Zhang Z, Kobayashi S, Borczuk AC et al (2010) Dual specificity phosphatase 6 (DUSP6) is an ETS-regulated negative feedback mediator of oncogenic ERK signaling in lung cancer cells. Carcinogenesis 31:577–586

14 Blood Flow Regulates Metabolism in Hematopoietic Development

Pamela L. Wenzel

Abstract

Blood flow modifies oxygen availability and biomechanical forces within the vasculature of the embryo as the hematopoietic system develops. The aorta-gonad-mesonephros (AGM) envelops the largest artery in the body and is a critical site for the emergence of hematopoietic stem cells (HSCs). Herein, I discuss the role of hypoxia-inducible factors (HIFs) and force as determinants of metabolism and fate determination. To address the effects of blood flow on hematopoietic development, I employ mouse embryo models and biomimetic culture. Real-time cell metabolic analyses show that oxygen consumption rates (OCR) and extracellular acidification rates (ECAR) are altered by flow in cultures of AGM cells. Additionally, mitochondrial permeability transition pore (mPTP) activity assays indicate that mPTP opening is regulated as precursors commit to a hematopoietic fate. Moreover, transcriptome data suggest that expression of *Ppif,* the gene that encodes the mPTP regulator Cyclophilin D, may be downregulated by fluid force. Together, these data implicate blood flow in regulation of metabolic activity in the AGM region, influencing bioenergetics and serving as a cue for HSC emergence. The broader significance of this research lies in its potential to accurately recapitulate the hematopoietic niche in vitro for study of blood disorders and improvement of cellular therapies.

Keywords

Blood development · Blood flow · Hematopoietic stem cells · Hemogenic endothelium · Metabolism · Mitochondrial permeability transition pore

14.1 Introduction

Oxygen deprivation is a natural cue in developing mammalian embryos (Simon and Keith 2008). Indeed, oxygen gradients provide critical spatial information necessary to coordinate development of the embryonic blood system, cardiovascular system, nervous system, and a multitude of other organs. Stem cells, particularly of the blood system, are known to occupy hypoxic niches, even into adulthood (Spencer et al. 2014). Yet, oxygen is the primary electron acceptor in countless biochemical reactions required for

P. L. Wenzel (✉)
Department of Integrative Biology & Pharmacology, McGovern Medical School, The University of Texas Health Science Center at Houston, Houston, TX, USA

Immunology Program, The University of Texas MD Anderson UTHealth Graduate School of Biomedical Sciences, Houston, TX, USA
e-mail: Pamela.L.Wenzel@uth.tmc.edu

S. C. Pias et al. (eds.), *Oxygen Transport to Tissue XLVI*, Advances in Experimental Medicine and Biology 1498, https://doi.org/10.1007/978-3-032-10389-5_14

life, and abundant evidence indicates that insufficient oxygen delivery to embryonic tissues results in developmental defects and lethality. Thus, it is not surprising that stem cell niches are frequently found in close proximity to the vasculature (Putnam 2014).

In embryogenesis, hematopoietic stem cells (HSCs) and progenitors are specified from unique endothelial cells that line the arterial vasculature, termed hemogenic endothelium (HE) (Yoshimoto and Yoder 2009). Within developing arterial blood vessels, cells change from an endothelial to hematopoietic identity in a process known as the endothelial-to-hematopoietic transition (EHT). HSCs capable of durable renewal of erythrocytes, platelets, and the majority of cells of the immune system arise from within an evolutionarily conserved "hotspot" of HSC fate specification—the aorta-gonad-mesonephros (AGM)—at approximately 10.5–11.5 days of murine (E10.5-E11.5) and 5 weeks of human embryogenesis (Lensch and Daley 2004). We and others have shown that blood flow is required for differentiation of HE capable of producing HSCs (Diaz et al. 2015; Adamo et al. 2009; Kim et al. 2015; North et al. 2009; Wang et al. 2011).

14.2 Flow-Induced Cellular Signaling in Hematopoiesis

How blood flow promotes blood development is of great interest since this mechanistic information could aid in recapitulating the embryonic hematopoietic niche in culture. In fact, accurate biomimetic modeling of the hematopoietic niche remains elusive but, if successful, would redefine strategies available for hematologic disease modeling and therapeutic use of HSCs. In 2009, two reports were published demonstrating that blood flow in zebrafish (North et al. 2009) and biomechanical force caused by blood flow in mice (Adamo et al. 2009) drive the emergence of hematopoietic cells. Both studies showed that flow-induced nitric oxide generation contributes to blood development. In continuation of these studies, we examined the transcriptome for clues to the intracellular signaling that regulates hematopoiesis in response to blood flow (Diaz et al. 2015). We found that flow triggers calcium oscillations in the cytosol to elevate production of prostaglandin E_2, which thereby activates a signaling cascade through cAMP-PKA-CREB. Inhibition of this signaling pathway by blocking antibodies or chemical inhibitors blunted hematopoietic activity. We also reported that flow upregulates Notch, Wnt, and Hedgehog (Diaz et al. 2015; Lee et al. 2013; Horton et al. 2020; Horton et al. 2021), developmental pathways critical for hematopoiesis (Clements and Traver 2013). Interestingly, flow can be mimicked pharmacologically via nitric oxide donors (Adamo et al. 2009; North et al. 2009; Wang et al. 2011) and a synthetic prostaglandin E_2 derivative (Diaz et al. 2015; Kim et al. 2015) to enhance hematopoietic activity from pluripotent stem cells and cells from the AGM. This information provides potential druggable targets to enhance hematopoiesis in vitro, yet we believe additional flow-responsive signaling is important for blood development.

Initiation of circulation within the embryo alters oxygen availability at the site of HSC emergence on the arterial wall. In fact, examination of gene expression from Ncx1-deficient embryos without a heartbeat in an independent study (Azzoni et al. 2021) and AGM cultures from our published work (Diaz et al. 2015) suggests that flow alters signaling indicative of change in oxygen within the aorta. Using a platform that permits culture of embryonic cells with precise intensities of wall shear stress (WSS) modeled after blood flow in the mouse embryo, we find that hypoxia-inducible factor 1 (HIF-1) is a topmost enriched pathway in WSS compared to static (no flow) cultures of AGM. Likewise, HIF-1 is enriched in comparisons of several EHT cell types in wild-type embryos and cardiac mutant littermates. Contributing to this enrichment are transcripts for HIF-1α and factors implicated in HIF regulation, including various PI3K subunits. How and whether this signaling is important for hematopoiesis remains largely unaddressed.

Mouse studies manipulating genes encoding HIF protein subunits suggest that HIF-1 and

HIF-2 are both critical for proper formation of the vasculature and for development of some blood cell lineages. The alpha subunits of HIF-1 and HIF-2 transcription factors are regulated by oxygen tension; whereas, the related HIF-1β protein, also known as aryl hydrocarbon receptor nuclear translocator (ARNT), forms heterodimers with HIF-1α and HIF-2α and is believed to be constitutively expressed (Mandl and Depping 2014). Evidence from *Arnt* genetic knockout studies suggested that HIF signaling is required for proliferation of multilineage hematopoietic progenitors derived from embryos and embryonic stem cells (Adelman et al. 1999). Impaired hematopoiesis resulting from loss of HIF-α-Arnt signaling could be rescued by exogenous vascular endothelial growth factor (VEGF). Similarly, deletion of *Hif1a*, encoding the hypoxia-responsive alpha subunit of HIF-1, reduced myeloid multilineage and committed erythroid progenitors in HIF-1α-deficient embryos (Yoon et al. 2006). Unlike the prior study, defects in erythroid colonies and decreased levels of hemoglobin in erythroid cells could not be rescued by the addition of cytokines like VEGF and erythropoietin. Instead, erythropoiesis could be improved by the addition of a synthetic iron chelate that alleviates defective iron utilization, suggesting that HIF-1 is an important regulator of iron homeostasis which is required for erythropoiesis. These studies raise the possibility that HIF-1 regulates a panoply of intracellular processes in hematopoiesis, including metabolic pathways.

HIF-1 is widely recognized for its strong association with metabolic adaptations affecting glucose metabolism via glycolysis, the pentose phosphate pathway (PPP), and oxidative phosphorylation (OXPHOS). Interestingly, evidence shows that HIF-1α protein is stabilized by hypoxia, but only when glucose is available (Osada-Oka et al. 2010). Accumulation of HIF-1α leads to increase in glucose-6-phosphate (G6P), a key substrate utilized in the PPP for the synthesis of NADPH, ribonucleotides, deoxyribonucleotides, and amino acid precursors (Nagao et al. 2019). These molecules are the building blocks for cycling cells, and evidence in cancer cells suggests that HIF-1α plays a predominantly pro-proliferative role. Yet, the effect of HIF-1 on cell cycling in healthy cells is more complex, especially in non-hypoxic environments (Dreute et al. 2024).

14.3 Metabolic Adaptation in Blood Precursors

Blood flow increases the availability of oxygen to the arterial vasculature and can also facilitate increased utilization of oxygen by cells lining the vessel walls. In our studies, we find that consumption of oxygen in the AGM region is elevated by flow and that the force generated by blood flow triggers this adaptation (Fig. 14.1a). Conversely, extracellular acidification is reduced in the presence of WSS (Fig. 14.1b). We speculate that initiation of blood circulation in the embryo not only increases supply of oxygen to tissues where hematopoiesis begins but also that signaling driving intracellular oxygen consumption is stimulated by the frictional force generated by blood flow. A study analyzing single-cell transcriptomics showed that cells undergoing EHT in Ncx1 heartbeat mutant embryos fail to undergo a metabolic switch from glycolysis to OXPHOS that is found in wild-type cells. The researchers then mimicked the effects of hypoxia via stabilization of HIF-1α by inhibiting prolyl hydroxylases with dimethyloxalylglycine (DMOG) in embryo explant cultures. Upon chemical stabilization of HIF-1α, which presumably elevated glycolysis, intraembryonic hematopoiesis was decreased (Azzoni et al. 2021). The authors speculate that severe hypoxia associated with a lack of circulation in the Ncx1 mutants prevents this metabolic switch toward elevated OXPHOS and suggests that the HE to pro-HSC transition is the stage of EHT that is sensitive to effects of circulation on the path to HSC commitment.

Switch in metabolism appears to be important for blood development based upon other studies in mice and humans. Indeed, one report shows that HE and other EHT populations have high basal respiration levels and utilize glutamine for

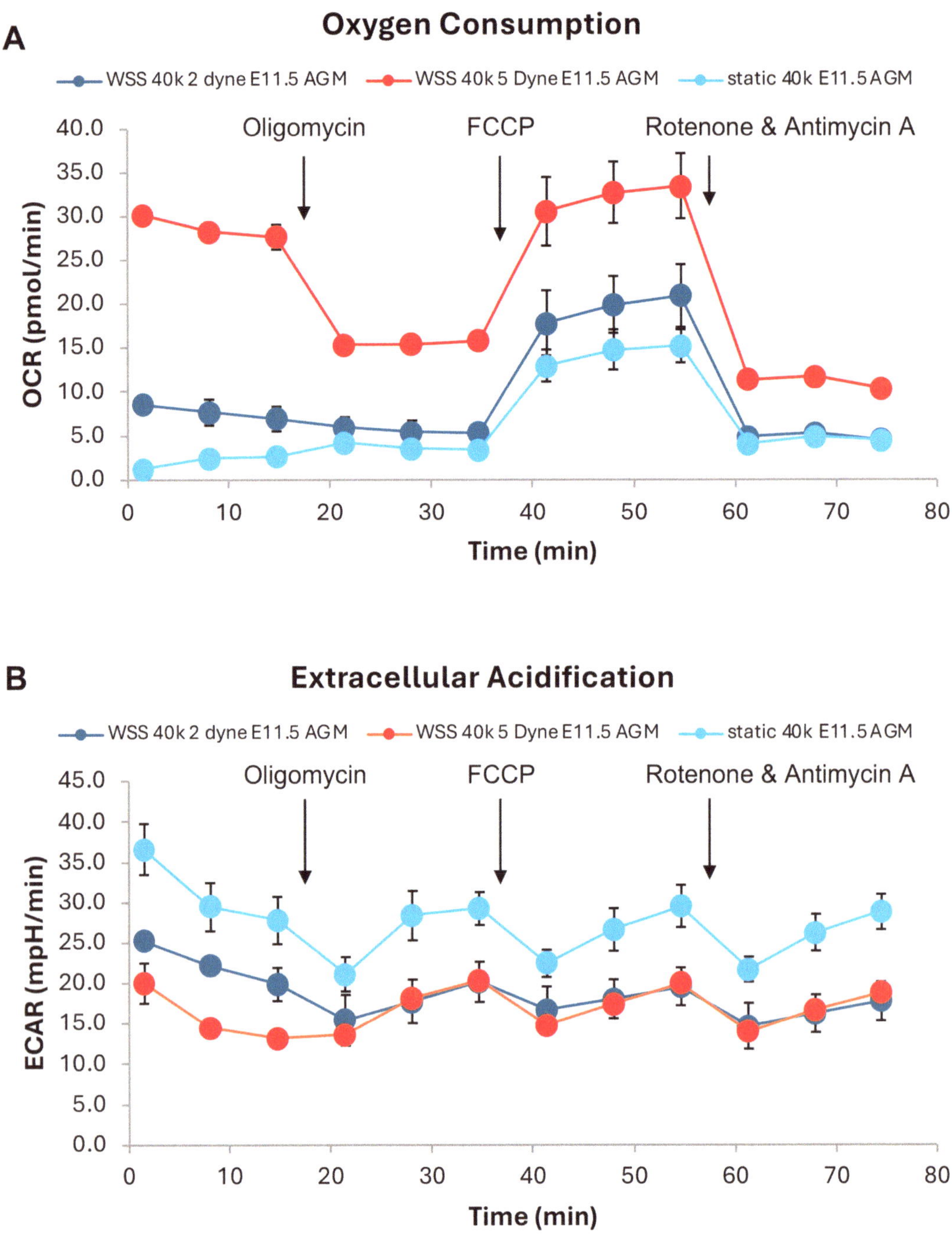

Fig. 14.1 Oxygen consumption and extracellular acidification under wall shear stress (WSS) or static (no flow) conditions in embryonic day 11.5 (E11.5) aorta-gonad-mesonephros (AGM) cultures. Three replicates are shown with bars indicating the standard error of the mean at each time point. (**a**) Oxygen consumption was determined in real-time cell metabolic assays using the Seahorse Mito Stress test assay following 36-h treatment of cells with one of two intensities of WSS (2 or 5 dyne cm^{-2}) or static conditions. OCR: Oxygen consumption rate. (**b**) Concurrent measurement of acidification of media serves as an indicator of glycolytic activity in the Seahorse MitoStress test assay. ECAR: Extracellular acidification rate

the TCA cycle and mitochondrial respiration (Oburoglu et al. 2021). Respiration rate in EHT cells is elevated relative to HE cells, regardless of the presence or absence of glucose, supporting that OXPHOS increases during EHT and that glutamine is sufficient to fuel metabolism in EHT cell types. In fact, glutamine deprivation produced > 80% reduction in surface expression of the CD43 hematopoietic cell marker in differentiating HE cells, further supporting a potential role for mitochondrial metabolism in commitment to the hematopoietic fate. In another report by the same researchers, they observe that differentiation of HE toward primitive erythroid hematopoiesis is dependent on glycolysis in human induced pluripotent stem cell cultures (Oburoglu et al. 2022). In contrast, differentiation of HE toward definitive hematopoiesis is dependent upon OXPHOS-mediated cholesterol metabolism.

We recently showed that mitochondrial ultrastructure is important for blood development as well (Dumbali et al. 2024). The mitochondrial permeability transition pore (mPTP) is a supramolecular channel that regulates the exchange of solutes across the inner mitochondrial membrane. The inner mitochondrial membrane is highly impermeable but has been shown under specific conditions to lose barrier function, such as in hypoxia-induced ischemia, which can induce excess reactive oxygen species (ROS), or other conditions that drive a drop in pH. We have found that mPTP conductance shapes metabolic machinery as cells transit toward hematopoietic identity in the embryo. The pore demonstrates modest activity in early arterial endothelial (eAE) cells, followed by restriction of pore opening just prior to commitment to a hematopoietic fate (Fig. 14.2a). The transient closing of the mPTP appears to aid in stimulating biogenesis of cristae, or folds in the inner mitochondrial membrane (Dumbali et al. 2024). Enforced closure of the mPTP by pharmacological or genetic means enhances hematopoiesis in vitro. HSCs also acquire an increased capacity to engraft, repopulate, and expand in the bone marrow of adult transplant recipient mice (Dumbali et al. 2024). We also found that WSS applied to E10.5 AGM suppresses expression of one of the regulatory molecules of the mPTP, Cyclophilin D, encoded by the *Ppif* gene (Fig. 14.2b). By extension, one might predict that force could contribute to regulation of mPTP opening and could modulate some of the same biology impacted by the pore, including mitochondrial phenotypes.

Multiple aspects of blood flow, including increased oxygen availability and generation of WSS, contribute to a reprogramming of energy metabolism as cells commit to the hematopoietic fate (Fig. 14.3). An active area of research is centered on understanding how mechanotransduction regulates pathways that dictate how energetic and anabolic demands are met during EHT and whether signaling beyond the mPTP plays a role in that process. Future research should be aimed at understanding the effects of flow and the force it generates on metabolic adaptation and mitochondrial functions in blood development and disease.

14.4 Experimental Methods

14.4.1 Mouse Embryos and Culture of AGM

All animal experiments were approved under the Animal Welfare Committee protocol AWC-23-0111 by the UTHealth Houston Animal Welfare Committee. Copulation plugs were identified following pairing of C57BL/6 male and female pairs, defining day 0.5 of embryonic gestation. Embryos were collected on embryonic day E10.5 or E11.5 for microdissection of the AGM region, dissociation of cells, and culture under static or WSS conditions. Embryos of both sexes were analyzed in parallel and/or combined into single samples.

Embryonic tissue was dissociated into single-cell suspension with Accutase (STEMCELL Technologies) at 37 °C with periodic agitation. Cells were resuspended in M5300 Myelocult medium (STEMCELL Technologies) enriched with 10 ng/ml murine SCF (Peprotech), 24 μM HEPES (Invitrogen), non-Essential Amino Acids

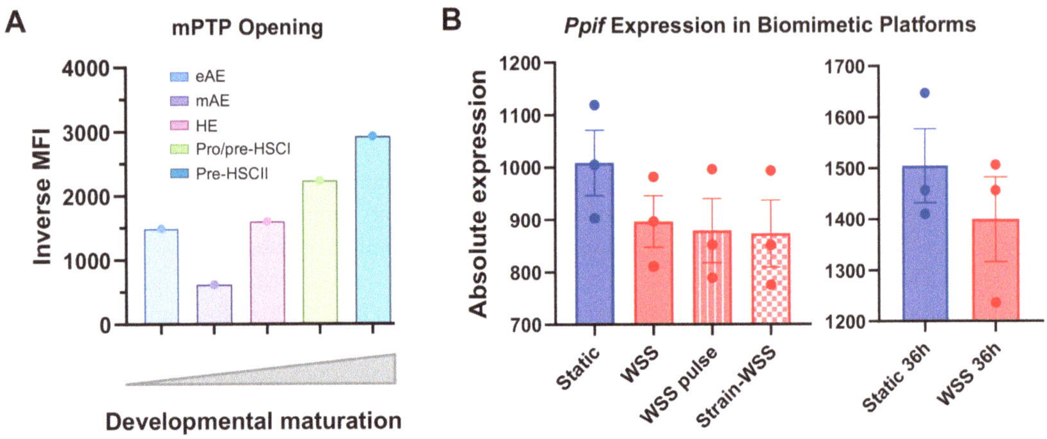

Fig. 14.2 Mitochondrial permeability transition pore (mPTP) is regulated as cells commit to a hematopoietic fate and in response to force generated by blood flow. (**a**) Dye efflux from mitochondria via pore opening reveals high mPTP activity in E11.5 embryo pre-HSCs and low activity in mAE (One-way ANOVA, $p < 0.05$). Low inverse mean fluorescence intensity (MFI) indicates more dye retention and less mPTP opening in mPTP MitoProbe assay. This suggests that pore conductance is suppressed as arterial endothelium matures and subsequently increased with acquisition of hematopoietic fate in hemogenic endothelium and pre-HSCs. eAE: early arterial endothelium; mAE: mature arterial endothelium; HE: hemogenic endothelium; Pro/pre-HSCI: pro-HSCs and pre-HSCs pooled; Pre-HSCII: pre-HSC II cells purported to precede definitive adult-type HSCs. (**b**) WSS of 5 dyne cm^{-2} in E10.5 AGM cells produces a non-significant trend in suppression of *Ppif* expression with or without cyclic strain (stretching) and regardless of pulsatility of flow, suggesting that blood flow could participate in downregulation of mPTP activity vital for differentiation of eAE to mAE. Absolute values on the y-axis are from Illumina beadchip intensities after background correction and quantile normalization to housekeeping genes. One-way ANOVA, n.s. for 16-h force type comparisons on left; T-test, n.s. for 36-h static versus WSS comparison on right. Each condition represents three independent biological replicates. Bars indicate standard error of the mean

(Gibco), Sodium Pyruvate (Gibco), and Penicillin-Streptomycin (Gibco). Cells were strained through a 70-micron filter and seeded into μ-Slide VI 0.4 channel slides (ibidi GmbH) at 400,000 cells cm^{-2}. Following cell attachment for 6 h, cells were cultured under static conditions or in the presence of WSS (5 dyne cm^{-2}). WSS was applied using a previously described microfluidic flow system for either 16 or 36 h (Li et al. 2015). Cells were retrieved by dissociation with Accutase, or RNA was prepared by direct lysis on ibidi slides with Buffer RLT Plus (RNeasy Plus Micro, Qiagen) and processed for analysis of *Ppif* expression in Illumina BeadChip assays according to our prior report (Diaz et al. 2015). Briefly, transcriptome data were generated on Illumina MouseWG-6 v2.0 R2 expression beadchips, followed by background correction and quantile normalization to housekeeping genes with GenomeStudio (Illumina). Normalized values from the beadchips are reported as absolute values for *Ppif* expression.

14.4.2 Measurement of Metabolism by Real-Time Cell Metabolic Assay

The Seahorse XFe96 Analyzer was used to measure oxygen consumption rate (OCR) and extracellular acidification rate (ECAR) as surrogates of oxidative phosphorylation and glycolysis, respectively, using 40,000 cells per well dissociated from the AGM exposed to static or WSS conditions. Chemical compounds injected were components of the Seahorse XFp Cell Mito Stress Test (Agilent), and included oligomycin, carbonyl cyanide-4 (trifluoromethoxy) phenylhydrazone (FCCP), rotenone, and antimycin A.

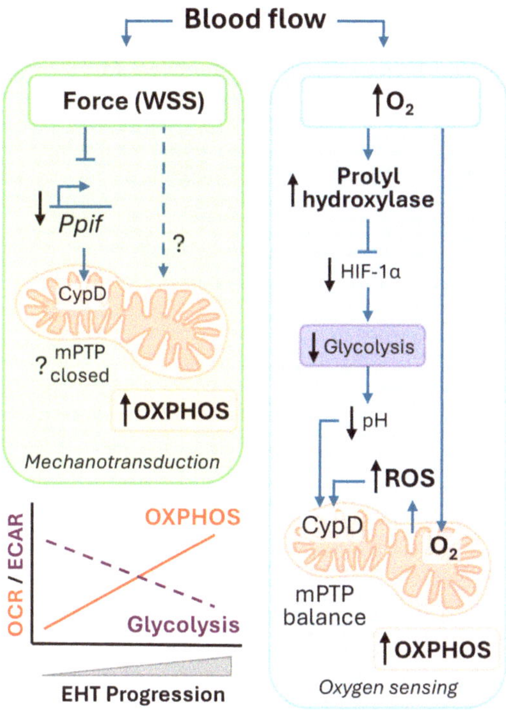

Fig. 14.3 Blood flow regulates energy metabolism via cooperative signaling downstream of mechanotransduction and oxygen sensing in cells undergoing transition to the hematopoietic fate. As cells differentiate under the effects of blood flow, WSS and increased oxygen availability support a switch from glycolysis to OXPHOS that is essential for the acquisition of hematopoietic identity. Transient closing of the mPTP is important for the mitochondrial maturation necessary for hematopoietic stem cells, but a role for force in regulating that closure remains an active area of research

14.4.3 Quantification of mPTP Opening

Following dissociation of E11.5 AGM cells to single-cell suspension, dead cells were labeled by incubation for 15 min in PBS with Zombie NIR viability dye (1:1000, Biolegend). Cells were subsequently incubated on ice for 30 min with antibodies against EHT surface markers EPCR/CD201 (EPCR-APC, 1:100, clone eBio1560, eBiosciences), cKit (cKit-PerCP-Cy5.5, 1:100, clone 2B8, Biolegend) VE-Cadherin (VEC-BV421, 1:100, clone 7B4, BD Biosciences), CD44 (CD44-PE, 1:100, clone IM7, Biolegend), and CD45 (CD45-BV510, 1: 100, clone 30-F11, Biolegend). The opening of the mPTP was quantified by the MitoProbe Transition Pore Assay Kit (ThermoFisher). MitoProbe reagents were then applied according to manufacturer instructions and included labeling of cells at 37 °C with 0.01 μM calcein AM, 0.4 mM $CoCl_2$, and 0.5 μM ionomycin reagents in calcium-containing Hank's Buffered Saline Solution. Cells were analyzed immediately following loading of calcein dye by flow cytometry. Fluorescence of all events acquired was averaged using FlowJo version 10.10.0 software, and inverse values were determined by subtraction from maximum fluorescence.

14.4.4 Statistical Analysis

Numerical data were plotted using GraphPad Prism 10 and analyzed by SigmaPlot 12.5 for statistical significance. Differences were determined by one-way ANOVA for multiple comparisons or the unpaired t-test for two groups. Significance was defined as $p < 0.05$ and n.s. indicates non-significant difference. Pooled data are reported as ± standard error of the mean (SEM) and individual biological replicates depicted as individual data points, when feasible.

Acknowledgments Financial support for this project was provided by the National Institutes of Health to P.L.W. (R01DK111599). I would like to thank Tyler Hagan for feedback regarding the drafting of the manuscript and Andre Palmer and Allan Doctor for helpful discussion regarding oxygen delivery.

Competing Interests The author declares no potential or real conflict of interest in relation to the research presented in this manuscript.

Ethics Approval All animal experiments were conducted according to the Animal Welfare Committee protocol AWC-23-0111 and approved by the UTHealth Houston Animal Welfare Committee.

References

Adamo L, Naveiras O, Wenzel PL et al (2009) Biomechanical forces promote embryonic haematopoiesis.

Nature 459:1131–1135. https://doi.org/10.1038/nature08073
Adelman DM, Maltepe E, Simon MC (1999) Multilineage embryonic hematopoiesis requires hypoxic ARNT activity. Genes Dev 13:2478–2483. https://doi.org/10.1101/gad.13.19.2478
Azzoni E, Frontera V, Anselmi G et al (2021) The onset of circulation triggers a metabolic switch required for endothelial to hematopoietic transition. Cell Rep 37: 110103. https://doi.org/10.1016/j.celrep.2021.110103
Clements WK, Traver D (2013) Signalling pathways that control vertebrate haematopoietic stem cell specification. Nat Rev Immunol 13:336–348. https://doi.org/10.1038/nri3443
Diaz MF, Li N, Lee HJ et al (2015) Biomechanical forces promote blood development through prostaglandin E2 and the cAMP-PKA signaling axis. J Exp Med 212: 665–680. https://doi.org/10.1084/jem.20142235
Dreute J, Pfisterer M, Schmitz ML (2024) A reductionist perspective on HIF-1α's role in cell proliferation under non-hypoxic conditions. Biochim Biophys Acta Mol Cell Res 1871:119683. https://doi.org/10.1016/j.bbamcr.2024.119683
Dumbali SP, Horton PD, Moore TI, Wenzel PL (2024) Mitochondrial permeability transition dictates mitochondrial maturation upon switch in cellular identity of hematopoietic precursors. Commun Biol 7:967. https://doi.org/10.1038/s42003-024-06671-y
Horton PD, Dumbali S, Wenzel PL (2020) Mechanoregulation in hematopoiesis and hematologic disorders. Curr Stem Cell Rep 6:86–95. https://doi.org/10.1007/s40778-020-00172-4
Horton PD, Dumbali SP, Bhanu KR et al (2021) Biomechanical regulation of hematopoietic stem cells in the developing embryo. Curr Tissue Microenviron Rep 2:1–15. https://doi.org/10.1007/s43152-020-00027-4
Kim PG, Nakano H, Das PP et al (2015) Flow-induced protein kinase A–CREB pathway acts via BMP signaling to promote HSC emergence. J Exp Med 212:633. https://doi.org/10.1084/jem.20141514
Lee HJ, Li N, Evans SM et al (2013) Biomechanical force in blood development: Extrinsic physical cues drive pro-hematopoietic signaling. Differentiation 86:92–103. https://doi.org/10.1084/jem.20142235
Lensch MW, Daley GQ (2004) Origins of mammalian hematopoiesis: in vivo paradigms and in vitro models. Curr Top Dev Biol 60:127–196. https://doi.org/10.1016/S0070-2153(04)60005-6
Li N, Diaz MF, Wenzel PL (2015) Application of fluid mechanical force to embryonic sources of hemogenic endothelium and hematopoietic stem cells. Methods Mol Biol 1212:183–193. https://doi.org/10.1007/7651_2014_95
Mandl M, Depping R (2014) Hypoxia-inducible aryl hydrocarbon receptor nuclear translocator (ARNT) (HIF-1β): is it a rare exception? Mol Med 20:215–220. https://doi.org/10.2119/molmed.2014.00032
Nagao A, Kobayashi M, Koyasu S et al (2019) HIF-1-dependent reprogramming of glucose metabolic pathway of cancer cells and its therapeutic significance. Int J Mol Sci 20:238. https://doi.org/10.3390/ijms20020238
North TE, Goessling W, Peeters M et al (2009) Hematopoietic stem cell development is dependent on blood flow. Cell 137:736–748. https://doi.org/10.1016/j.cell.2009.04.023
Oburoglu L, Mansell E, Woods NB (2021) Glutamine metabolism regulates endothelial to hematopoietic transition and hematopoietic lineage specification. Sci Rep 11:17589. https://doi.org/10.1038/s41598-021-97194-7
Oburoglu L, Mansell E, Canals I et al (2022) Pyruvate metabolism guides definitive lineage specification during hematopoietic emergence. EMBO Rep 23:e54384. https://doi.org/10.15252/embr.202154384
Osada-Oka M, Hashiba Y, Akiba S et al (2010) Glucose is necessary for stabilization of hypoxia-inducible factor-1α under hypoxia: Contribution of the pentose phosphate pathway to this stabilization. FEBS Lett 584: 3073–3079. https://doi.org/10.1016/j.febslet.2010.05.046
Putnam AJ (2014) The instructive role of the vasculature in stem cell niches. Biomaterials 2:1562–1573. https://doi.org/10.1038/jid.2014.371
Simon MC, Keith B (2008) The role of oxygen availability in embryonic development and stem cell function. Nat Rev Mol Cell Biol 9:285–296. https://doi.org/10.1038/nrm2354
Spencer JA, Ferraro F, Roussakis E et al (2014) Direct measurement of local oxygen concentration in the bone marrow of live animals. Nature 508:269–273. https://doi.org/10.1038/nature13034
Wang L, Zhang P, Wei Y et al (2011) A blood flow-dependent klf2a-NO signaling cascade is required for stabilization of hematopoietic stem cell programming in zebrafish embryos. Blood 118:4102–4110. https://doi.org/10.1182/blood-2011-05-353235
Yoon D, Pastore YD, Divoky V et al (2006) Hypoxia-inducible factor-1 deficiency results in dysregulated erythropoiesis signaling and iron homeostasis in mouse development. J Biol Chem 281:25703–25711. https://doi.org/10.1074/jbc.M602329200
Yoshimoto M, Yoder MC (2009) Developmental biology: birth of the blood cell. Nature 457:801–803. https://doi.org/10.1038/457801a

Part V

Hemodynamic Physiology

Erythrocytes as Active Regulators of Blood Flow: A Brief Overview

15

Felix Scholkmann, Martin Wolf, and Ursula Wolf

Abstract

While the primary function of red blood cells (RBCs) in the human body is the transport of gases (oxygen and carbon dioxide) in the vascular system, research conducted over the past few decades has demonstrated that RBCs also play a role in the active regulation of blood flow by releasing vasoactive substances (adenosine triphosphate, ATP, and nitric oxide, NO). This chapter provides an overview of the process involved in this ATP- and NO-mediated vasodilation of RBCs, and its relevance to human physiology and pathophysiology.

Keywords

Erythrocytes · Blood flow · ATP · Nitric oxide · Vasodilation · Physiology · Pathophysiology

15.1 Introduction

Among the smallest cell types in the human body (with a typical diameter of 6–8 μm), erythrocytes (red blood cells, RBCs) represent the most abundant cell type, comprising ~84% of the total human cell count (Sender et al. 2016) and ~40% of the human blood volume. While the primary function of RBCs is the transport of gases (oxygen, O_2, and carbon dioxide, CO_2), research conducted over the past decades has demonstrated that RBCs have another important function: the regulation of blood flow through the release of vasoactive substances (adenosine triphosphate, ATP, and nitric oxide, NO) (Richardson et al. 2020a).

This chapter provides a summary of this particular function of RBCs and examines its relevance for different aspects of human physiology and pathophysiology.

F. Scholkmann (✉)
Biomedical Optics Research Laboratory, Department of Neonatology, University Hospital Zurich, University of Zurich, Zurich, Switzerland

Institute of Complementary and Integrative Medicine, University of Bern, Bern, Switzerland

Neuroscience Center Zurich, University of Zurich and ETH Zurich, Zurich, Switzerland
e-mail: Felix.Scholkmann@usz.ch

M. Wolf
Biomedical Optics Research Laboratory, Department of Neonatology, University Hospital Zurich, University of Zurich, Zurich, Switzerland

Neuroscience Center Zurich, University of Zurich and ETH Zurich, Zurich, Switzerland

U. Wolf
Institute of Complementary and Integrative Medicine, University of Bern, Bern, Switzerland

S. C. Pias et al. (eds.), *Oxygen Transport to Tissue XLVI*, Advances in Experimental Medicine and Biology 1498, https://doi.org/10.1007/978-3-032-10389-5_15

15.2 Erythrocyte ATP- and NO-Mediated Vasodilation

In a seminal work, published in 1992, Bergfeld and Forrester (1992) showed that RBCs release ATP in response to a brief exposure to a hypoxic/hypercapnic environment. A subsequent study conducted by this research group demonstrated that (i) ATP is released by RBCs in response to both low partial pressure of oxygen (pO_2) and low pH, and (ii) ATP induces a vasodilatory response when applied intraluminally (Ellsworth et al. 1995). The erythrocyte was identified as an O_2 sensor and regulator of vascular tone.

A number of factors have been identified as potential triggers for the release of ATP by RBCs. To date, the following have been experimentally proven: mechanical deformation of RBCs (Sprague et al. 1996; Faris and Spence 2008; Williams and Miller 1980; Sprague et al. 1998), decreased pO_2 (hypoxia) (Bergfeld and Forrester 1992; Ellsworth et al. 1995; Faris and Spence 2008), increased partial pressure of CO_2 (pCO_2) (hypercapnia) (Bergfeld and Forrester 1992), increased temperature (Kalsi and Gonzalez-Alonso 2012), elevated lactate concentration (hyperlactatemia) (Rozier et al. 2007), bacterial toxins (Skals et al. 2014), adenosine (Misiti et al. 2022), centrifugation (Connes et al. 2018), haemolysis (Sikora et al. 2014), and a reduction in the extracellular pH (acidosis) (Ellsworth et al. 1995) (Fig. 15.1a).

Due to the absence of mitochondria (and thus the ability for oxidative phosphorylation), RBCs produce ATP by the anaerobic glycolytic Embden-Meyerhof pathway (EMP), which breaks down glucose to generate lactate and 2 moles of ATP (Melo et al. 2019) (Fig. 15.1b). As a side-arm of the EMP, the Luebering-Rapoport shunt is also active, producing 2,3-diphosphoglycerate (2,3-DPG), which plays a crucial role in regulating the affinity between haemoglobin and O_2 (Duc and Engel 2009). Additionally, 2,3-DPG acts as an antioxidant, protecting haemoglobin oxygenation, chelating ferrous ions, and scavenging hydroxyl radicals, peroxyl radicals and cation radicals (Tellone et al. 2019). While 80–90% of the glucose is metabolised via the EMP, 10% is metabolised via the pentose phosphate pathway.

The release of ATP from RBCs is triggered by at least five mechanisms: (i) activation of the membrane-bound calcium (Ca^{2+}) channel Piezo 1, (ii) activation of membrane-bound adenylyl cyclase (AC) through interaction with guanine nucleotide-binding protein (G protein), (iii) phosphorylation of the voltage-dependent anion channel (VDAC), (iv) formation of pore-forming toxins (PFTs), and (v) haemolysis (Richardson et al. 2020a). Figure 15.2a provides a visualisation of these processes.

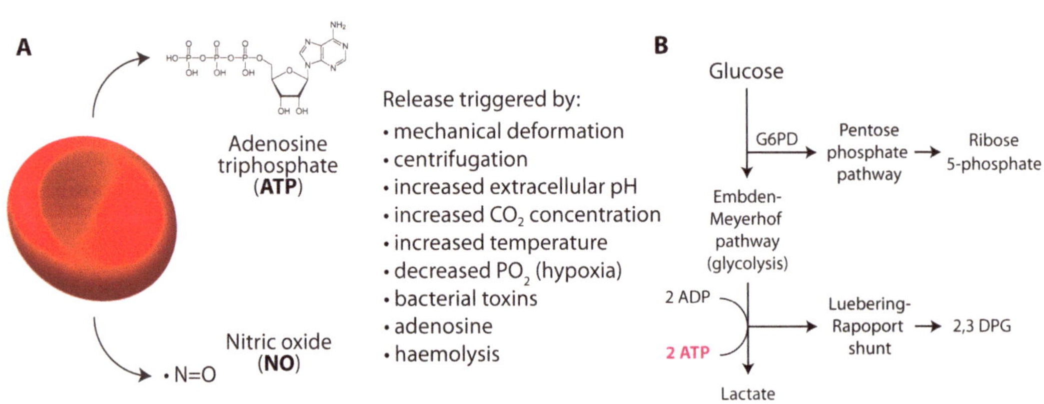

Fig. 15.1 (**a**) Listing of processes triggering the release of ATP and NO from RBCs. (**b**) The main erythrocyte metabolic pathways (Melo et al. 2019). G6PD: phosphogluconate dehydrogenase, 2,3 DPG: 2,3-diphosphoglycerate

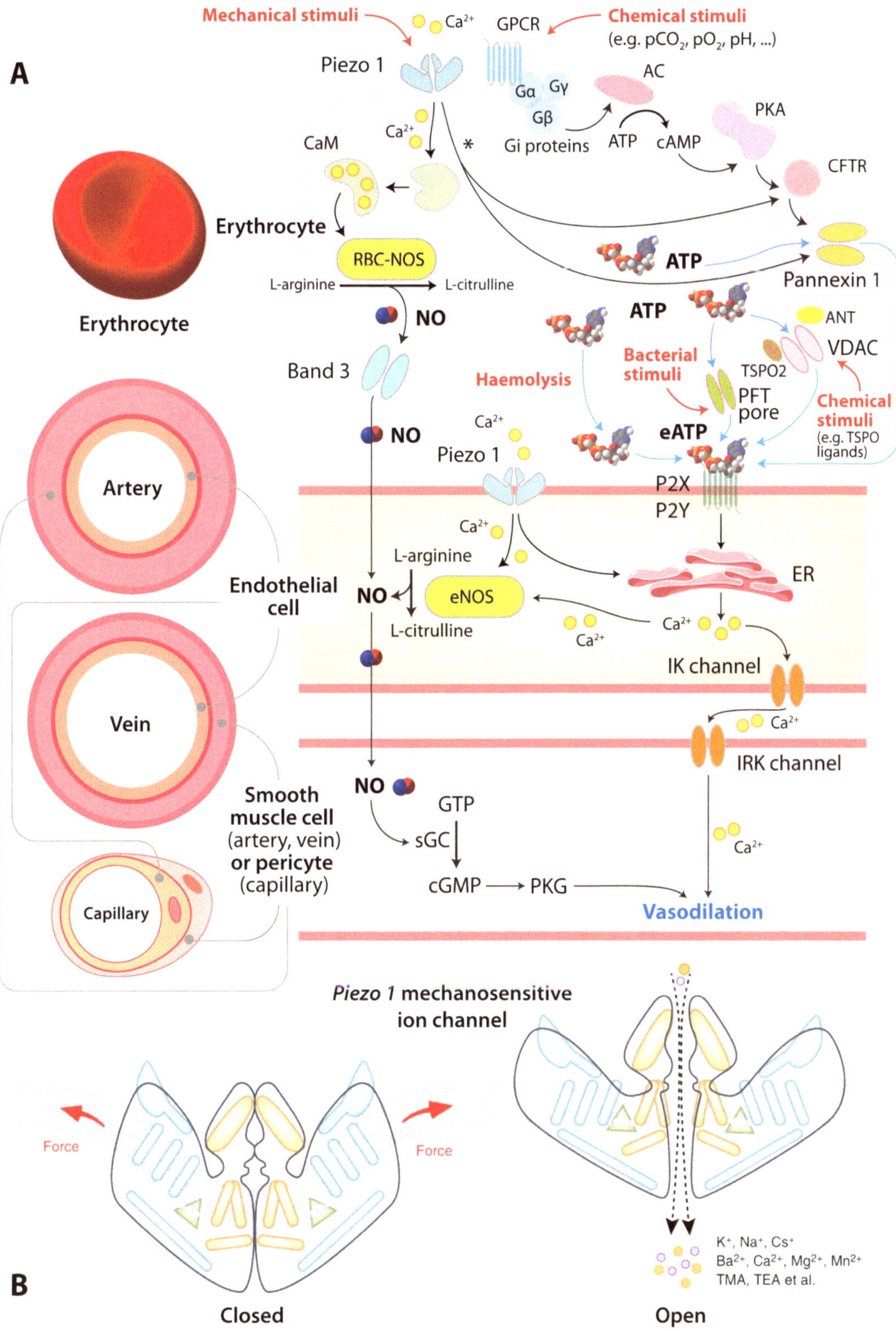

Fig. 15.2 (**a**) Visualisation of the five main mechanisms of ATP- and NO-mediated vasodilation. Figure based on Richardson et al. (2020a), with addition of the pathways involving VDAC, the PFT pore, and haemolysis. The asterisk indicates that the exact steps of this pathway are unknown yet. The mechanism involving the TSPO2-

The first process is triggered by mechanical stimuli, i.e. mechanical deformation of the RBC membrane, which results in the opening of the Piezo1 ion channel (Fig. 15.2b). This results in an influx of Ca^{2+}, which initiates the production of NO by the RBC nitric oxide synthase (RBC-NOS) (LoBue et al. 2023) and the release of ATP through the ATP-permeable ion channel Pannexin1 (PANX1) (Richardson et al. 2020a). This process also requires the involvement of the cystic fibrosis transmembrane conductance regulator (CFTR), which belongs to the ATP-binding cassette transporter superfamily (Sprague et al. 1998). NO is then released via the erythroid anion exchanger-1 (Band 3) (Hsu 2024), diffuses through the endothelial cell layer of the vessels, and triggers vasodilation. Extracellular ATP (eATP) then binds to purinergic receptors of the P2X- and PXY-subfamily on endothelial cells, which stimulates the release of Ca^{2+} into the cytoplasm, triggering the activity of endothelial nitric oxide synthase (eNOS) to produce NO and, in parallel, the release of Ca^{2+} through the intermediate Ca^{2+}-activated potassium (IK) channel into the extracellular space. NO then diffuses into smooth muscle cells (SMCs) (in arteries and veins) or pericytes (in capillaries), resulting in vasodilation. In parallel, the Ca^{2+} influx in SMCs and/or pericytes through the Ca^{2+}-sensitive inward-rectifier potassium (IRK) channel also triggers vasodilation (as a propagation phenomenon along the vessel, i.e. the "conducted vasomotor response" (Mironova et al. 2024) with an electrical conduction speed of ~1500 μm/s and a mechanical one of ~500 μm/s (Dietrich et al. 2002)).

The second process is initiated by chemical stimuli (e.g. pO_2, pCO_2, pH, lactate, adenosine) which activate the membrane-bound AC (also known as ATP diphosphate-lyase) through interaction with an as yet unidentified G protein, i.e. a G protein-coupled receptor (GPCR). AC catalyses the conversion of ATP to 3′,5′-cyclic AMP (cAMP), which then activates protein kinase A (PKA). The cAMP-PKA pathway is therefore triggered. PKA then stimulates the release of ATP by phosphorylating CFTR. The released ATP through PANX1 (Sridharan et al. 2010) then follows the same path as in case of the first process described. The triggering of the β-adrenergic receptor (βAR), a member of the GPCR family, has been shown, for instance, to result in the release of ATP from RBCs (Montalbetti et al. 2011).

The third process entails the release of ATP through the VDAC, which forms a complex with the translocator protein 2 (TSPO2), ANT, and other as yet unidentified proteins: the TSPO2-VDAC-ANT supramolecular complex (Marginedas-Freixa et al. 2018a). This pathway is triggered by the phosphorylation of VDAC induced by translocator protein (TSPO) ligands (Marginedas-Freixa et al. 2018a; b), such as porphyrins or endozepines. The eATP released via VDAC then initiates vasodilation.

The fourth process involves PFTs (Dal Peraro and van der Goot 2016) that are secreted from both Gram-negative and Gram-positive bacteria, which can trigger the release of ATP through transmembrane pores generated (Skals et al. 2014), resulting in vasodilation.

Fig. 15.2 (continued) VDAC-ANT also involves interactions (not shown in the figure) with A, PKC, and PKA. (**b**) Visualisation of the open and closed state of the Mammalian Piezo1 ion channel. Reproduced with permission from Qin et al. (2021). AC: adenylyl cyclase, ATP: adenosine triphosphate, Band 3: erythroid anion exchanger-1, CaM: calmodulin, cAMP: cyclic adenosine monophosphate, cGMP: cyclic guanosine monophosphate, eATP: extracellular ATP, eNOS: endothelial nitric oxide synthase, ER: endoplasmic reticulum, G proteins: guanine nucleotide-binding proteins, GPCR: G-protein-coupled receptors, GTP: guanosine triphosphate, IK channel: intermediate Ca^{2+}-activated potassium, IRK channel: Ca^{2+}-sensitive inward-rectifier potassium channel, NO: nitric oxide, P2Y: purinergic G protein-coupled receptors, Pannexin 1: Pannexin 1 ion channel, PFT pore: pore-forming toxins pore, Piezo 1: Piezo 1 ion channel, PKA: protein kinase A, PKG: protein kinase G, PKS: protein kinase S, CFTR: cystic fibrosis transmembrane conductance regulator, RBC-NOS: red blood cell nitric oxide synthase, sGC: soluble guanylate cyclase, TSPO: translocator protein, TSPO2: translocator protein 2, VDAC: voltage-dependent anion channel

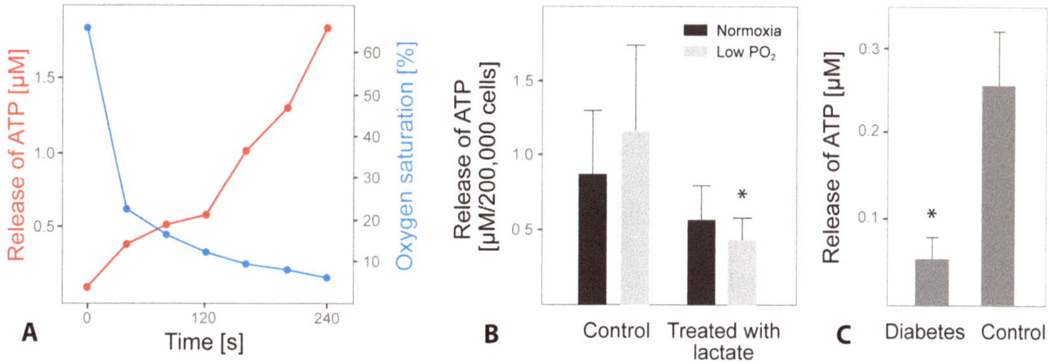

Fig. 15.3 (**a**) RBC ATP release as a function of oxygen saturation according to data from Faris and Spence (2008). (**b**) Suppression of RBC ATP release by lactate under hypoxia according to data from Rozier et al. (2007). * $p < 0.05$, significant difference from normoxia. (**c**) Diminished RBC ATP release during RBC deformation according to data from Subasinghe and Spence (2008). * $p < 0.05$, significant difference from controls

The fifth process of ATP release is initiated by cell lysis, i.e. haemolysis (Sikora et al. 2014; Ferguson et al. 2021). Sikora et al. concluded that haemolysis "is a primary ATP-release mechanism in human erythrocytes" (Sikora et al. 2014). However, this conclusion was subsequently questioned by Kirby et al. (2015) (but see also the reply of Sikora et al. (2015)). In addition to the release of ATP, haemolysis instigates a number of vasoactive processes, including the scavenging of NO and the release of arginase-1 and microvesicles. Moreover, the haemoglobin breakdown product haem is a damage-associated molecular pattern (DAMP) (Dutra and Bozza 2014; Bozza and Jeney 2020) which is vasoactive by influencing vascular endothelial function and interacting with vascular signalling. Extracellular ATP itself is also a DAMP (Roh and Sohn 2018). Furthermore, haemolysis can trigger the formation of neutrophil extracellular traps (NETs) (Kono et al. 2014) which are also vasoactive (by, for example, stimulating the release of endothelin-1) (Aldabbous et al. 2016).

15.3 Significance for Human Physiology and Pathophysiology

The capacity of RBCs to facilitate ATP- and NO-mediated vasodilation through diverse mechanisms can be regarded as a reliable indication of their crucial active involvement in regulating blood flow and O_2 delivery, particularly in situations such as haemolysis, mechanical deformation of RBCs (e.g. through shear stress), hypoxia, or hypercapnia.

It seems reasonable to conclude that, under physiological or pathophysiological conditions in vivo, the five main mechanisms of RBC ATP release are active and also interact with each other. It has been shown that the hypoxia-induced ATP release and the RBC deformation-induced ATP release are non-additive, resulting in a non-linear relationship between pO_2 and ATP release (whereas hypoxia appears to lead to RBC membrane deformation due to conformational changes in membrane-bound haemoglobin) (Faris and Spence 2008) (Fig. 15.3a).

RBC ATP- and NO-mediated vasodilation seems to be important, for example, to regulate microvascular skeletal muscle perfusion (Ellsworth 2004; Ellsworth and Sprague 2012). In addition, as pointed out by Dietrich et al. (2002), the mechanism seems to be also relevant for neurovascular coupling where the ATP release by RBCs due to neuronal activity triggers a local microvascular vessel dilation involving a conducted vasomotor response in pial and cortical microvessels.

There are several factors that can interfere with the RBC ATP release mechanisms. One such factor is lactate, which has been shown to reduce the ability of RBCs to release ATP under hypoxia, thereby impairing the vascular regulatory capacity of the RBCs (Fig. 15.3b). This may be pathophysiologically relevant, as elevated blood lactate levels have been observed in several chronic human diseases, such as in multiple sclerosis (MS) (Amorini et al. 2014; Cerexhe et al. 2022; Esmael et al. 2021), post-acute COVID-19 syndrome (PACS) (van der Togt and Rossman 2023; Lopez-Hernandez et al. 2023), diabetes (Juraschek et al. 2013; Crawford et al. 2010), migraine (Altunkaynak et al. 2013), myalgic encephalomyelitis/chronic fatigue syndrome (ME/CFS) (Ghali et al. 2019; Fluge et al. 2016), mild cognitive impairment (MCI) (Pan et al. 2019), and cancer (Sitter et al. 2022; Cheung et al. 2020). Furthermore, serum lactate is an "independent predictor of hospital mortality in critically ill patients in the emergency department" (Bou Chebl et al. 2017) and "predicts poor prognosis of acute ischemic stroke" (Jin et al. 2022). In these cases, lactate-induced impairment of ATP- and NO-mediated vasodilation may play a role, e.g. by exacerbating existing problems with blood perfusion and tissue oxygenation, as observed in MS (Martinez Sosa and Smith 2017; Lassmann 2003; D'Haeseleer et al. 2015), PACS (Adingupu et al. 2023a, b), diabetes (Arden and Sivaprasad 2011; Wang et al. 2021), migraine (Frank et al. 2020; Dalkara et al. 2010), ME/CFS (van Campen et al. 2021; Campen et al. 2020), MCI (Snyder et al. 2021; Kim et al. 2020), and cancer (Ruan et al. 2009; Vaupel and Mayer 2016). Thus far, direct evidence of RBC ATP release impairment has been obtained for the case of diabetes (Subasinghe and Spence 2008) (Fig. 15.3c). Furthermore, impaired RBC ATP release has been found in cystic fibrosis patients (Sprague et al. 1998) and in RBC treated with amyloid β peptide (indicating a significance for Alzheimer's disease) (Misiti et al. 2008).

How much ATP is released by RBCs is also determined by the flexibility of the cell membrane: stiffer (less deformable) RBCs release less ATP (Fischer et al. 2003).

RBC deformability is altered in numerous conditions such as increased RBC oxidative stress (Becatti et al. 2016) (Fig. 15.4a), increased inflammation (Tsuda 2012) (Fig. 15.4b), increased blood glucose concentration (Keymel et al. 2011) (Fig. 15.4c), and increased glycated haemoglobin concentration (Keymel et al. 2011) (Fig. 15.4d). Furthermore, the RBC deformability has been found to be reduced in abnormal states of iron availability and metabolism, i.e. in anaemia (Anderson et al. 1999; Pavliukovych and Pavliukovych 2021) or iron overload (e.g. in case of hemochromatosis (Richardson et al. 2020b)). There seems to be a U-shaped correlation between the iron status (probably primarily the ferritin level) and the erythrocyte deformability). In all these conditions, an effect on RBC ATP-and NO-mediated vasodilation is likely.

15.4 Discussion, Conclusion, and Outlook

In this chapter, we provided an overview of the main pathways of ATP- and NO-mediated vasodilation, as well as their significance for human physiology and pathophysiology.

Further research is required to gain a deeper understanding of the biochemical pathways involved and to investigate the relationship between the active role of blood flow regulation by RBCs and various physiological processes, such as neurovascular coupling, as well as its role in disease.

In light of the potential pivotal role of ATP- and NO-mediated vasodilation by RBCs in human physiology, the development of a blood test to quantify this directly (or indirectly, for example, by accessing the deformability of the RBC) based on individual blood samples could represent a novel approach for individualised medical diagnostics or disease monitoring,

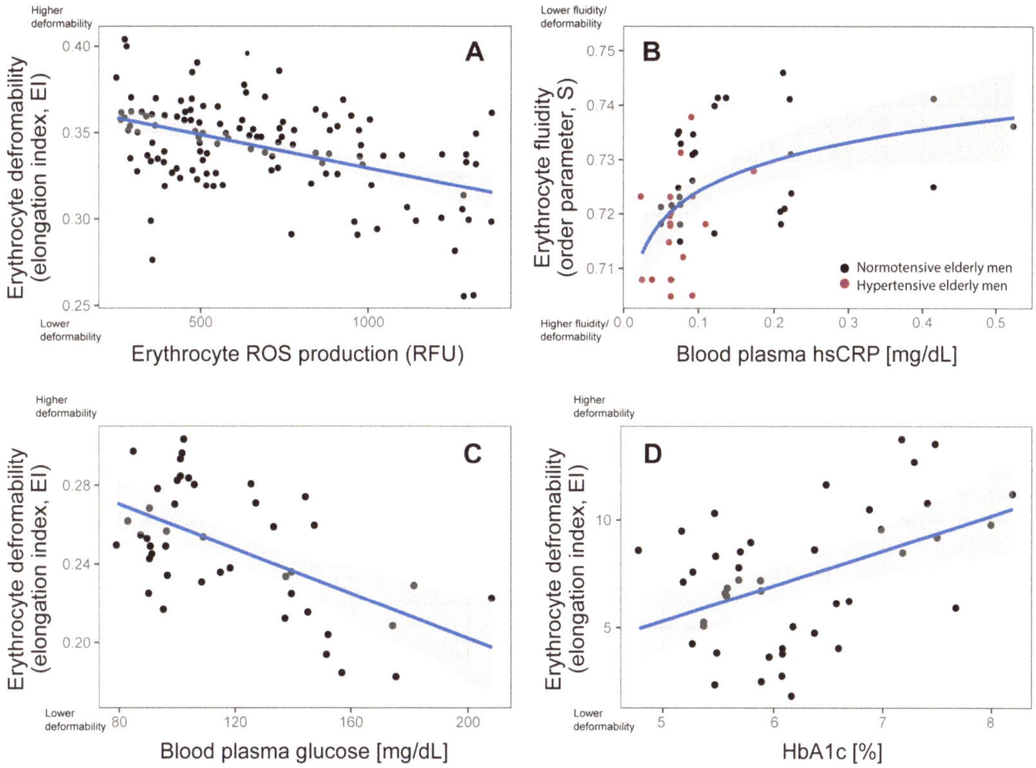

Fig. 15.4 Correlations of RBC deformability with (**a**) reactive oxygen species (ROS) production (indicating oxidative stress) (according to data from Becatti et al. (2016)) (fit with a linear function: $r = -0.461$, $p < 0.001$), (**b**) blood plasma C-reactive protein (CRP) concentration (according to data from Tsuda (2012)) (fit with a power-law function: $r = 0.488$, $p < 0.01$), (**c**) blood plasma glucose concentration (according to data from Keymel et al. (2011)) (fit with a linear function: $r = 0.570$, $p < 0.01$), and (**d**) glycated haemoglobin (HbA1c) concentration (according to data from Keymel et al. (2011)) (fit with a linear function: $r = 0.452$, $p < 0.01$). hsCRP: high-sensitive CRP, RFU: relative fluorescence unit. The grey area indicates the 95% confidence interval of the fitted function. For the analysis, the original data from the respective publications were digitised and the visualisations and regression analyses were carried out using R (version 4.3.0) and Matlab (R2024b)

including metabolic, inflammatory or neurodegenerative diseases.

Acknowledgements The authors thank Rachel Scholkmann for proofreading the manuscript. MW declares that he is co-founder and president of the board of OxyPrem AG.

References

Adingupu DD et al (2023a) Brain hypoxia, neurocognitive impairment, and quality of life in people post-COVID-19. J Neurol 270(7):3303–3314

Adingupu DD et al (2023b) Reduced cerebrovascular oxygenation in individuals with post-acute COVID-19 syndrome (PACS) ("long COVID"). In: Adv Exp Med Biol, vol 1438, pp 211–216

Aldabbous L et al (2016) Neutrophil extracellular traps promote angiogenesis: evidence from vascular pathology in pulmonary hypertension. Arterioscler Thromb Vasc Biol 36(10):2078–2087

Altunkaynak Y et al (2013) Serum lactic acid and pyruvic acid levels in patients with migraine and tension type headache. Dusunen Adam J Psychiatry Neurol Sci 26: 276–280

Amorini AM et al (2014) Serum lactate as a novel potential biomarker in multiple sclerosis. Biochim Biophys Acta 1842(7):1137–1143

Anderson C, Aronson I, Jacobs P (1999) Erythrocyte deformability is reduced and fragility increased by iron deficiency. Hematology 4(5):457–460

Arden GB, Sivaprasad S (2011) Hypoxia and oxidative stress in the causation of diabetic retinopathy. Curr Diabetes Rev 7(5):291–304

Becatti M et al (2016) Erythrocyte oxidative stress is associated with cell deformability in patients with retinal vein occlusion. J Thromb Haemost 14(11): 2287–2297

Bergfeld GR, Forrester T (1992) Release of ATP from human erythrocytes in response to a brief period of hypoxia and hypercapnia. Cardiovasc Res 26(1):40–47

Bou Chebl R et al (2017) Serum lactate is an independent predictor of hospital mortality in critically ill patients in the emergency department: a retrospective study. Scand J Trauma Resusc Emerg Med 25(1):69

Bozza MT, Jeney V (2020) Pro-inflammatory actions of Heme and other hemoglobin-derived DAMPs. Front Immunol 11:1323

Campen C, Rowe PC, Visser FC (2020) Reductions in cerebral blood flow can be provoked by sitting in severe Myalgic encephalomyelitis/chronic fatigue syndrome patients. Healthcare (Basel) 8(4):394

Cerexhe L et al (2022) Blood lactate concentrations during rest and exercise in people with Multiple Sclerosis: a systematic review and meta-analysis. Mult Scler Relat Disord 57:103454

Cheung SM et al (2020) Lactate concentration in breast cancer using advanced magnetic resonance spectroscopy. Br J Cancer 123(2):261–267

Connes P et al (2018) Centrifugation-induced release of ATP from red blood cells. PLoS One 13(9):e0203270

Crawford SO et al (2010) Association of blood lactate with type 2 diabetes: the atherosclerosis risk in communities carotid MRI study. Int J Epidemiol 39(6):1647–1655

D'Haeseleer M et al (2015) Cerebral hypoperfusion: a new pathophysiologic concept in multiple sclerosis? J Cereb Blood Flow Metab 35(9):1406–1410

Dal Peraro M, van der Goot FG (2016) Pore-forming toxins: ancient, but never really out of fashion. Nat Rev Microbiol 14(2):77–92

Dalkara T, Nozari A, Moskowitz MA (2010) Migraine aura pathophysiology: the role of blood vessels and microembolisation. Lancet Neurol 9(3):309–317

Dietrich HH, Ellsworth ML, Dacey RG (2002) The red blood cell, ATP and integrated vascular responses to neuronal stimulation. Int Congr Ser 1235:277–287

Duc G, Engel K (2009) Effect of 2,3-DPG concentration on hemoglobin-oxygen affinity of whole blood. Scand J Clin Lab Invest 24(4):405–412

Dutra FF, Bozza MT (2014) Heme on innate immunity and inflammation. Front Pharmacol 5:115

Ellsworth ML (2004) Red blood cell-derived ATP as a regulator of skeletal muscle perfusion. Med Sci Sports Exerc 36(1):35–41

Ellsworth ML, Sprague RS (2012) Regulation of blood flow distribution in skeletal muscle: role of erythrocyte-released ATP. J Physiol 590(20): 4985–4991

Ellsworth ML et al (1995) The erythrocyte as a regulator of vascular tone. Am J Phys 269(6 Pt 2):H2155–H2161

Esmael A et al (2021) Mitochondrial dysfunction and serum lactate as a biomarker for the progression and disability in MS and its correlation with the radiological findings. Neurol Res 43(7):582–590

Faris A, Spence DM (2008) Measuring the simultaneous effects of hypoxia and deformation on ATP release from erythrocytes. Analyst 133(5):678–682

Ferguson BS et al (2021) Red blood cell ATP release correlates with red blood cell hemolysis. Am J Physiol Cell Physiol 321(5):C761–C769

Fischer DJ et al (2003) Determination of erythrocyte deformability and its correlation to cellular ATP release using microbore tubing with diameters that approximate resistance vessels in vivo. Analyst 128(9):1163–1168

Fluge O et al (2016) Metabolic profiling indicates impaired pyruvate dehydrogenase function in myalgic encephalopathy/chronic fatigue syndrome. JCI Insight 1(21): e89376

Frank F et al (2020) Migraine and aura triggered by normobaric hypoxia. Cephalalgia 40(14):1561–1573

Ghali A et al (2019) Elevated blood lactate in resting conditions correlate with post-exertional malaise severity in patients with Myalgic encephalomyelitis/Chronic fatigue syndrome. Sci Rep 9(1):18817

Hsu K (2024) Erythroid anion transport, nitric oxide, and blood pressure. Front Physiol 15:1363987

Jin XX et al (2022) Elevated lactate dehydrogenase predicts poor prognosis of acute ischemic stroke. PLoS One 17(10):e0275651

Juraschek SP et al (2013) Plasma lactate and diabetes risk in 8045 participants of the atherosclerosis risk in communities study. Ann Epidemiol 23(12): 791–796 e4

Kalsi KK, Gonzalez-Alonso J (2012) Temperature-dependent release of ATP from human erythrocytes: mechanism for the control of local tissue perfusion. Exp Physiol 97(3):419–432

Keymel S et al (2011) Impaired red blood cell deformability in patients with coronary artery disease and diabetes mellitus. Horm Metab Res 43(11): 760–765

Kim CM et al (2020) Associations between cerebral blood flow and structural and functional brain imaging measures in individuals with neuropsychologically defined mild cognitive impairment. Neurobiol Aging 86:64–74

Kirby BS et al (2015) Liberation of ATP secondary to hemolysis is not mutually exclusive of regulated export. Blood 125(11):1844–1845

Kono M et al (2014) Heme-related molecules induce rapid production of neutrophil extracellular traps. Transfusion 54(11):2811–2819

Lassmann H (2003) Hypoxia-like tissue injury as a component of multiple sclerosis lesions. J Neurol Sci 206(2):187–191

LoBue A et al (2023) Red blood cell endothelial nitric oxide synthase: a major player in regulating

cardiovascular health. Br J Pharmacol. https://doi.org/10.1111/bph.16230
Lopez-Hernandez Y et al (2023) The plasma metabolome of long COVID patients two years after infection. Sci Rep 13(1):12420
Marginedas-Freixa I et al (2018a) Human erythrocytes release ATP by a novel pathway involving VDAC oligomerization independent of pannexin-1. Sci Rep 8(1):11384
Marginedas-Freixa I et al (2018b) Induction of ATP release, PPIX transport, and cholesterol uptake by human red blood cells using a new family of TSPO ligands. Int J Mol Sci 19(10):3098
Martinez Sosa S, Smith KJ (2017) Understanding a role for hypoxia in lesion formation and location in the deep and periventricular white matter in small vessel disease and multiple sclerosis. Clin Sci (Lond) 131(20): 2503–2524
Melo D et al (2019) Interplay between erythrocyte peroxidases and membrane. IntechOpen
Mironova GY et al (2024) The conducted vasomotor response and the principles of electrical communication in resistance arteries. Physiol Rev 104(1):33–84
Misiti F et al (2008) Amyloid peptide inhibits ATP release from human erythrocytes. Biochem Cell Biol 86(6): 501–508
Misiti F, Carelli-Alinovi C, Rodio A (2022) ATP release from erythrocytes: a role of adenosine1. Clin Hemorheol Microcirc 80(2):61–71
Montalbetti N et al (2011) Homeostasis of extracellular ATP in human erythrocytes. J Biol Chem 286(44): 38397–38407
Pan H et al (2019) Association among plasma lactate, systemic inflammation, and mild cognitive impairment: a community-based study. Neurol Sci 40(8):1667–1673
Pavliukovych N, Pavliukovych O (2021) Erytrocytes deformability in patients with coronary artery disease and anemia. Atherosclerosis 331:e226
Qin L et al (2021) Roles of mechanosensitive channel Piezo1/2 proteins in skeleton and other tissues. Bone Res 9(1):44
Richardson KJ, Kuck L, Simmonds MJ (2020a) Beyond oxygen transport: active role of erythrocytes in the regulation of blood flow. Am J Physiol Heart Circ Physiol 319(4):H866–H872
Richardson KJ, McNamee AP, Simmonds MJ (2020b) Hemochromatosis alters the sensitivity of red blood cells to mechanical stress. Transfusion 60(12): 2982–2990
Roh JS, Sohn DH (2018) Damage-associated molecular patterns in inflammatory diseases. Immune Netw 18(4):e27
Rozier MD, Zata VJ, Ellsworth ML (2007) Lactate interferes with ATP release from red blood cells. Am J Physiol Heart Circ Physiol 292(6):H3038–H3042
Ruan K, Song G, Ouyang G (2009) Role of hypoxia in the hallmarks of human cancer. J Cell Biochem 107(6): 1053–1062
Sender R, Fuchs S, Milo R (2016) Revised estimates for the number of human and bacteria cells in the body. PLoS Biol 14(8):e1002533
Sikora J et al (2014) Hemolysis is a primary ATP-release mechanism in human erythrocytes. Blood 124(13): 2150–2157
Sikora J et al (2015) Response: hemolysis is a primary and physiologically relevant ATP release mechanism in human erythrocytes. Blood 125(11):1845–1846
Sitter B, Forsmark A, Solheim O (2022) Elevated serum lactate in glioma patients: associated factors. Front Oncol 12:831079
Skals M et al (2014) Bacterial RTX toxins allow acute ATP release from human erythrocytes directly through the toxin pore. J Biol Chem 289(27):19098–19109
Snyder B et al (2021) Cerebral hypoxia: its role in age-related chronic and acute cognitive dysfunction. Anesth Analg 132(6):1502–1513
Sprague RS et al (1996) ATP: the red blood cell link to NO and local control of the pulmonary circulation. Am J Phys 271(6 Pt 2):H2717–H2722
Sprague RS et al (1998) Deformation-induced ATP release from red blood cells requires CFTR activity. Am J Phys 275(5):H1726–H1732
Sridharan M et al (2010) Pannexin 1 is the conduit for low oxygen tension-induced ATP release from human erythrocytes. Am J Physiol Heart Circ Physiol 299(4):H1146–H1152
Subasinghe W, Spence DM (2008) Simultaneous determination of cell aging and ATP release from erythrocytes and its implications in type 2 diabetes. Anal Chim Acta 618(2):227–233
Tellone E et al (2019) New role for an old molecule: the 2,3-diphosphoglycerate case. Biochim Biophys Acta Gen Subj 1863(10):1602–1607
Tsuda K (2012) Associations between high-sensitivity C-reactive protein and membrane fluidity of red blood cells in hypertensive elderly men: an electron spin resonance study. Int J Hypertens 2012:292803
van Campen C, Rowe PC, Visser FC (2021) Cerebral blood flow remains reduced after tilt testing in myalgic encephalomyelitis/chronic fatigue syndrome patients. Clin Neurophysiol Pract 6:245–255
van der Togt V, Rossman JS (2023) Hypothesis: inflammatory acid-base disruption underpins Long Covid. Front Immunol 14:1150105
Vaupel P, Mayer A (2016) Tumor hypoxia: causative mechanisms, microregional heterogeneities, and the role of tissue-based hypoxia markers. Adv Exp Med Biol 923:77–86
Wang Y et al (2021) Cerebral perfusion alterations in type 2 diabetes mellitus – a systematic review. Front Neuroendocrinol 62:100916
Williams AR, Miller DL (1980) Photometric detection of ATP release from human erythrocytes exposed to ultrasonically activated gas-filled pores. Ultrasound Med Biol 6(3):251–256

Capillary Red Blood Cell Velocity Is Decreased in Skeletal Muscle of Septic Rats: In Vivo Imaging Using Videocapillaroscopy

16

Rin Kataoka, Naoki Hitosugi, Kazuki Hotta, Narumi Fukuzaki, Kota Izawa, Shuri Arai, Tomotaka Koike, Kentaro Kamiya, and Atsuhiko Matsunaga

Abstract

This study aimed to investigate the impact of sepsis on red blood cell (RBC) velocity, diameter, and categories of skeletal muscle capillaries 3–4 h after sepsis onset. Male Wistar rats (9–11 weeks old, $n = 14$) were randomly classified into sham and cecal ligation and perforation (CLP) groups, respectively. In vivo imaging of capillaries in the spinotrapezius muscle was performed using stain-free videocapillaroscopy. RBC velocity and capillary diameter were analyzed. Capillaries were classified into three groups: (1) continuously flowing capillaries defined as maintaining continuous flow for >30 s; (2) intermittently flowing, showing brief stagnation lasting <10 s; (3) stopped-flow, showing no flow for >10 s during 30 s of observation periods. The number of capillaries in each category is compared between the groups. The RBC velocity of the CLP group was significantly lower than that of the sham group ($p = 0.01$), but the capillary diameter was not different between the groups. The CLP group showed a trend of a weak negative correlation between capillary diameter and RBC velocity ($r^2 = 0.16, p = 0.08$), but the sham group did not. No significant difference was observed between the groups in each capillary category. These results suggest that sepsis decreases RBC velocity in resting skeletal muscle capillaries but has no effect on capillary category or diameter.

Keywords

Sepsis · Skeletal muscle · Microcirculation · Red blood cell velocity · Capillary imaging

16.1 Introduction

Sepsis is a life-threatening organ dysfunction caused by a dysregulated host response to infection (Singer et al. 2016). Approximately 48.9 million people worldwide are estimated to have contracted sepsis, resulting in 11 million sepsis-related deaths, accounting for almost 20% of global deaths (Rudd et al. 2020).

Critically ill patients have been reported to experience an acute and persistent loss of muscle mass following intra-abdominal sepsis

R. Kataoka · K. Hotta (✉) · N. Fukuzaki · K. Izawa · S. Arai · K. Kamiya · A. Matsunaga
Graduate School of Medical Sciences, Kitasato University, Sagamihara, Japan
e-mail: hotta.kazuki@kitasato-u.ac.jp

N. Hitosugi
Department of Rehabilitation, St. Marianna University Yokohama Seibu Hospital, Yokohama, Japan

T. Koike
Department of Intensive Care Center, Kitasato University Hospital, Sagamihara, Japan

S. C. Pias et al. (eds.), *Oxygen Transport to Tissue XLVI*, Advances in Experimental Medicine and Biology 1498, https://doi.org/10.1007/978-3-032-10389-5_16

(Cox et al. 2021). Additionally, microcirculatory disturbances in skeletal muscle occur 24 h after the onset of sepsis in a rat model (Lam et al. 1994). Early detection and management of sepsis and septic shock in the initial hours after onset improve outcomes. Microcirculatory parameters are more reliable predictors of outcomes than systemic hemodynamic measures in patients with sepsis (De Backer et al. 2013). However, whether sepsis directly affects skeletal muscle microcirculation remains unclear.

Capillary blood flow is strictly regulated according to the oxygen demand of the muscle. High red blood cell (RBC) velocity indicates better oxygen diffusion capacity of skeletal muscle. Capillaries are classified into continuous, intermittent, and stopped-flow categories. In septic animals, researchers have observed decreased capillary flow velocity and an increased number of stopped-flow capillaries in muscles (Ellis et al. 2002). Shock (hypotension) can temporarily slow or completely halt blood flow, resulting from vascular inflammation or leukocyte adhesion. However, the effects of septic shock on skeletal muscle microcirculation remain unknown. We aimed to investigate the impact of sepsis on microcirculatory parameters, including RBC velocity, diameter, and categories of skeletal muscle capillaries 3–4 h after sepsis onset.

16.2 Material and Methods

16.2.1 Animal Preparation

Fourteen male Wistar rats (9–11 weeks old; body weight 315 g [95% confidence interval (CI): 295–355]; $n = 13$, CLEA Japan, Tokyo) were used in this study. The rats were maintained on a 12 h:12 h light-dark cycle and had free access to food and water. The animal experimental protocol was approved by the Animal Care and Use Committee (Approval number Ei-ken 22-17) and was performed in accordance with the guidelines established by the Physiological Society of Japan. The rats were assigned to the sham-operated control group (sham; $n = 6$) and the CLP group (CLP; $n = 7$). No differences in body weight or age were noted between the two groups.

16.2.2 Surgical Preparation

This CLP model simulates peritonitis due to cecal necrosis and extruded feces. The CLP procedure was performed according to a previously described moderate-grade CLP model (Rittirsch et al. 2009). During the entire experimental procedure, 1.5–2.0% isoflurane inhalation anesthesia was administered to spontaneously breathing rats. The anesthetized rats were maintained at a body temperature of 37 °C. In the CLP groups, the cecum was exposed, ligated with 4-0 silk, perforated with an 18-gauge needle, and then returned to the abdominal cavity. In the sham group, the cecum was exposed without ligation or puncture. Subcutaneous administration of saline (3 mL/100 g) was performed as resuscitation after abdominal closure (Hubbard et al. 2005). After the CLP or sham procedure, a catheter was placed in the right common carotid artery. The catheter was connected to blood pressure transducers (model MLT0670; ADInstruments Japan Inc.) and blood pressure amplifiers (model FE117; ADInstruments Japan Inc.) to obtain mean arterial pressure (MAP) and heart rate (HR) (Fujii et al. 2021). Blood samples were collected from a catheter 3–4 h after the CLP or sham procedure to measure lactate and blood glucose.

For capillary imaging, the skin and fascia on the left spinotrapezius muscle were gently removed, avoiding bleeding and muscle damage (Hitosugi et al. 2024). The exposed muscle tissue was moistened by continuously superfusing with warmed Krebs-Henseleit buffer (KHB (in mM): 132 NaCl, 4.7 KCl, 21.8 $NaHCO_3$, 2 $MgSO_4$, 2 $CaCl_2$) adjusted to pH 7.4 (Sonobe et al. 2008).

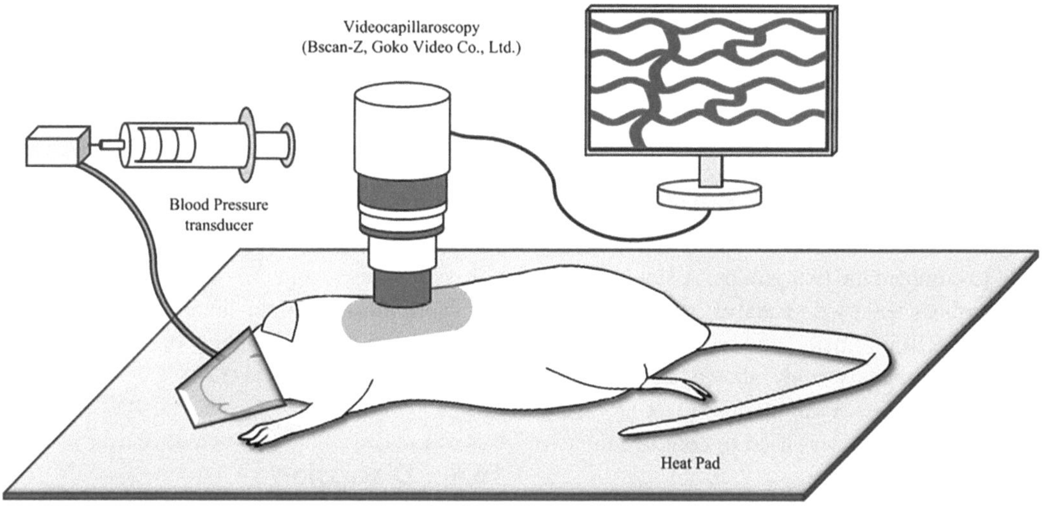

Fig. 16.1 Schematic diagram of in vivo imaging of capillaries in the spinotrapezius muscle using videocapillaroscopy. The anesthetized rats were maintained at a body temperature of 37 °C. The spinotrapezius muscle was gently exposed, and videocapillaroscopy was positioned directly above the spinotrapezius muscle, recording for 180 s at rest. Additionally, mean arterial pressure and heart rate were continuously measured during the measurement

16.2.3 Capillary Imaging

Figure 16.1 shows the schematic diagram of capillary imaging using videocapillaroscopy (Bscan-Z, Goko Imaging Devices Co., Ltd.) (Neubauer-Geryk et al. 2019). Imaging was performed 3–4 h after the CLP or sham procedure. The videocapillaroscopy focused on the superficial capillaries of the skeletal muscle. In vivo imaging of capillary blood flow was conducted for 180 s with a frame rate of 30 frames/s and a resolution of 1280 × 960 pixels (646 × 485 μm). After completing all experimental protocols, the rats were euthanized by heart extraction under 5% isoflurane anesthesia. Muscles were carefully removed and weighed in all groups.

16.2.4 Image Analysis and Statistics

Movie data obtained by videocapillaroscopy were saved as stack images and analyzed using image analysis software ImageJ (NIH, Bethesda, MD, USA) by blinded image analysts who were not informed about the group. The RBC velocity was analyzed in unbranched capillaries parallel to muscle fibers, and RBCs could be tracked for 10–15 frames. The three most in-focus capillaries were selected by analysts and analyzed for each movie. The RBC velocity was calculated by dividing the distance of RBC movement within 10–15 frames by the time. RBC velocity and capillary diameter were calculated at three points for each capillary and averaged for statistical analysis. In this study, capillaries were categorized into three groups as follows: continuous, intermittent, and stopped-flowing capillaries. Capillaries in which RBCs were observed for 900 frames (30 s) were included in the categorization. The capillaries with a continuous flow of >30 s were classified as continuously flowing. The capillaries with blood flow stagnation observed for <10 s and > 10 s out of 30 s of observation were classified as intermittent and stopped-flowing capillaries, respectively (Schulze et al. 2022). For each group, the number of capillaries in each category relative to the total number of capillaries analyzed was expressed as a percentage.

All statistical analyses were performed using GraphPad Prism (Version 10.0.2), with a significance level of <5%. Median values and 95% CIs for animal characteristics and blood data were calculated, and Mann-Whitney U test was used to compare the two groups. Median values and 95% CIs for RBC velocity and capillary diameter were calculated, and Mann-Whitney U test was used to compare the two groups. A single regression analysis was used to analyze the relationship between RBC velocity and capillary diameter because a previous study showed slower RBC velocity in narrower capillaries (Li et al. 2019). The chi-square test was used to compare the two groups of categories.

16.3 Results

Table 16.1 shows animal characteristics. Body weight and muscle weight were not different between the groups. The CLP group showed a significant decrease in MAP but not in HR and a significant increase in lactate levels compared to the sham group.

Figure 16.2 shows the results of the RBC velocity and capillary diameter. One video was excluded from the analysis due to being out of focus. Consequently, RBC velocity and diameter measurements were obtained from 5 rats in the sham group (15 capillaries) and 7 in the CLP group (21 capillaries). The RBC velocity of the CLP group was significantly lower than that of the sham group (CLP 400 [244–521] versus sham 708 [383–1037] μm/s; median [95%CI]; $p = 0.01$). No significant difference was observed in capillary diameter between the sham and CLP groups. Figure 16.3 shows the results of the relationship between RBC velocity and capillary diameter. A trend of a weak negative correlation was observed between capillary diameter and RBC velocity in the CLP group ($r^2 = 0.16$, $p = 0.08$).

Table 16.2 shows the percentage of capillaries in each category. Category measurements were obtained from 6 rats in the sham group (29 capillaries) and 7 in the CLP group (52 capillaries). The chi-square test showed no difference in the percentage of each category between the groups ($p = 0.76$).

16.4 Discussion

To the best of our knowledge, this study is the first to report on the effects of sepsis on capillary RBC velocity, diameter, and categories of skeletal muscle 3–4 h after the onset of sepsis. We

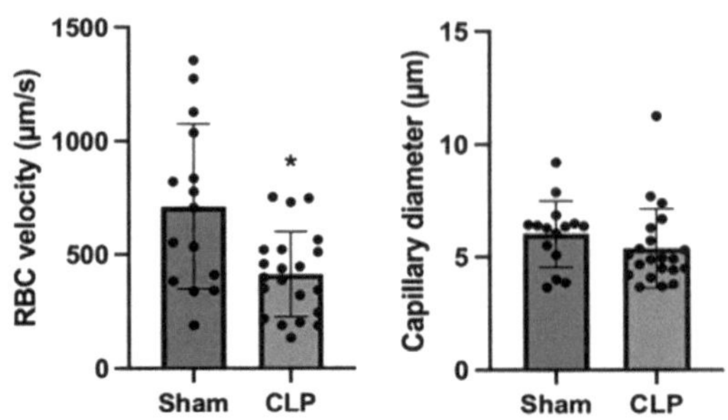

Fig. 16.2 Red blood cell (RBC) velocity (left) and capillary diameter (right) in rat spinotrapezius muscle. Sham sham-operated control group, CLP cecal ligation and puncture. Three capillaries were selected from each rat. The dots represent individual capillaries. Data are presented as median ± 95% CI. Sham group: $n = 5$ rats (15 capillaries), CLP group: $n = 7$ rats (21 capillaries). * represents p-value less than 0.05 vs. sham

Table 16.1 Animal characteristics

	Sham($n = 6$)	CLP($n = 7$)	P value
Body weight (g)	315 [291–362]	315[272–370]	0.91
Spinotrapezius muscle weight (mg)	175 [145–223]	151 [122–170]	0.07
Mean arterial pressure (mmHg)	90 [82–108]	77 [73–112]	0.03
Heart rate (bpm)	402 [355–416]	367 [332–404]	0.13
Lactate (mmol/L)	2.6 [1.6–2.7]	3.1 [2.1–3.6]	0.02
Blood glucose (mg/dL)	160 [151–175]	176 [153–225]	0.06

Data are presented as median [95%CI]

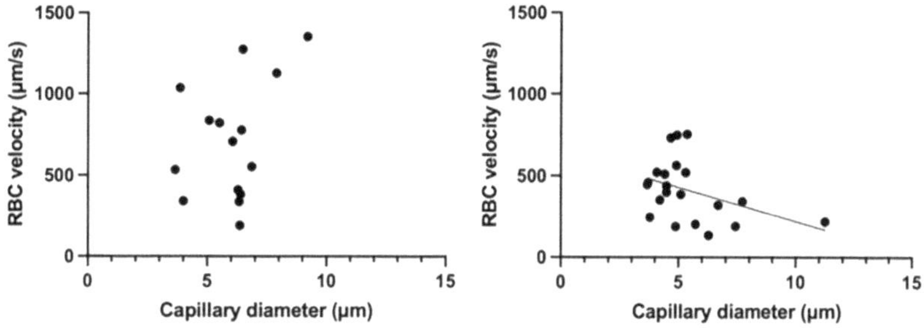

Fig. 16.3 The correlation between RBC velocity and vessel diameter in capillaries of Sham (left) and CLP (right) groups. Sham: $r^2 = 0.14$, $p = 0.18$; CLP: $r^2 = 0.17$, $p = 0.08$

Table 16.2 Capillary categories

	Continuous (%)	Intermittent (%)	Stopped (%)
Sham	80.0	2.9	17.1
CLP	78.9	5.3	15.8

Each category's capillary count relative to the total number of analyzed capillaries is expressed as a percentage. Six rats (29 capillaries) in sham group and 7 rats (52 capillaries) were included for the category analysis. The p-value of the chi-square test was 0.76

performed in vivo blood flow imaging using stain-free videocapillaroscopy for both CLP and sham groups. The main findings of this study are as follows: (1) lower RBC velocity in the CLP group with hypotension, (2) no difference in capillary categories and diameter between the groups, and (3) a weak negative correlation trend between RBC velocity and capillary diameter in the CLP group. Overall, this study suggests that sepsis acutely alters skeletal muscle microcirculation.

The RBC velocity is an indicator of oxygen diffusion capacity in skeletal muscle. The decreased RBC velocity in the CLP group suggests an oxygen diffusion incapacity 3–4 h after the onset of sepsis. Arterial blood pressure is a crucial indicator of systemic circulatory dynamics and is determined by cardiac output and systemic vascular resistance. The lower MAP in the CLP group may indicate septic shock. Consistent with previous studies, severe hypotension induced by septic shock has been reported 3 h after CLP (Eyenga et al. 2014). Blood lactate was significantly higher in the CLP group than in the sham group. Lactate is a marker indicating tissue hypoxia induced by septic shock. Therefore, septic shock could reduce RBC velocity in skeletal muscle capillaries. This study did not directly evaluate the extent to which the reduction in red blood cell (RBC) velocity affects oxygen delivery. Indeed, Poole et al. suggested that a decrease in RBC velocity may limit oxygen diffusing capacity (DO_2), as capillary hematocrit and the longitudinal recruitment of capillary surface area depend on flow dynamics (Poole et al. 2020). In sepsis as well, future studies are warranted to quantitatively assess the relationship between reduced RBC velocity and oxygen diffusing capacity. In addition, although both MAP and RBC velocity were decreased in the CLP group, their direct relationship was not clarified in this study. Future investigations evaluating changes in RBC velocity under corrected MAP conditions may help elucidate the causal relationship between systemic blood pressure and microcirculatory flow dynamics. In our previous study using the same CLP model, no significant deviations were observed in systemic variables such as PaO_2, $PaCO_2$, MAP, and SaO_2 (Hitosugi et al. 2024), suggesting that systemic disturbances were unlikely to have substantially influenced the microvascular flow observed in the present study. Under septic conditions, systemic hematocrit is generally reported to decrease, primarily due to hemodilution (Souza et al. 2012). Although transient local increase in hematocrit may occur due to red blood cell aggregation in low-flow microvascular regions, the predominant systemic trend remains a reduction. Therefore, while changes in hematocrit and other hemorheological factors may influence RBC velocity, their exact contribution to the observed velocity reduction in this study remains unclear and warrants future investigations.

Capillaries were classified into three categories, and we observed no significant differences between groups. The proportion of continuously flowing capillaries in the sham group was 80% in this study, supporting previous research (Ye et al. 2022). Microcirculation is tightly regulated to supply oxygen to muscle mitochondria. About three capillaries around muscle cells supply oxygen to the mitochondria near the vessels. However, stopped-flowing capillaries cannot supply oxygen, resulting in an oxygen diffusion incapacity throughout the muscle. Reduced RBC velocity inhibits oxygen transport over greater distances, limiting the oxygen diffusion capacity. These results suggest that although no impact was observed on the capillary category, oxygen diffusion capacity was partially limited in the skeletal muscle of septic rats. This apparent discrepancy may suggest that, during the early stages of sepsis, the spatial distribution of capillary flow patterns remains preserved, whereas decreased RBC velocity has already occurred within individual capillaries. This implies that oxygen delivery to skeletal muscle may be compromised despite an unaltered categorical distribution of capillary flow. This study focused on resting skeletal muscle. However, evaluating microcirculatory dynamics during muscle activity remains a subject for future investigation.

16.5 Conclusions

This study performed in vivo blood flow imaging using videocapillaroscopy in septic rats. Sepsis decreased the RBC velocity in skeletal muscle capillaries; however, it did not affect the capillary categories or diameter.

Acknowledgments This work was supported by JSPS KAKENHI Grant Number 24K14254.

Competing Interests The authors have no conflicts of interest to declare that are relevant to the content of this chapter.

Ethics Approval The animal experimental protocol was approved by the Animal Care and Use Committee (Approval number Ei-ken 22-17) and was performed in accordance with the guidelines established by the Physiological Society of Japan.

References

Cox MC, Booth M, Ghita G et al (2021) The impact of sarcopenia and acute muscle mass loss on long-term outcomes in critically ill patients with intra-abdominal sepsis. J Cachexia Sarcopenia Muscle 12(5):1203–1213

De Backer D, Donadello K, Sakr Y et al (2013) Microcirculatory alterations in patients with severe sepsis: impact of time of assessment and relationship with outcome. Crit Care Med 41(3):791–799

Ellis CG, Bateman RM, Sharpe MD et al (2002) Effect of a maldistribution of microvascular blood flow on capillary O(2) extraction in sepsis. Am J Physiol Heart Circ Physiol 282(1):H156–H164

Eyenga P, Roussel D, Morel J et al (2014) Early septic shock induces loss of oxidative phosphorylation yield plasticity in liver mitochondria. J Physiol Biochem 70(2):285–296

Fujii Y, Sumikura H, Nagahama D (2021) Establishment of a novel miniature veno-venous extracorporeal membrane oxygenation model in the rat. Artif Organs 45(1):63–67

Hitosugi N, Hotta K, Taketa Y et al (2024) The effect of sepsis and reactive oxygen species on skeletal muscle interstitial oxygen pressure during contractions. Microcirculation 31(1):e12833

Hubbard WJ, Choudhry M, Schwacha MG et al (2005) Cecal ligation and puncture. Shock 1:52–57

Lam C, Tyml K, Martin C et al (1994) Microvascular perfusion is impaired in a rat model of normotensive sepsis. J Clin Invest 94(5):2077–2083

Li B, Lu X, Moeini M et al (2019) Atherosclerosis is associated with a decrease in cerebral microvascular blood flow and tissue oxygenation. PLoS One 14(8):e0221547

Neubauer-Geryk J, Hoffmann M, Wielicka M et al (2019) Current methods for the assessment of skin microcirculation: part 1. Postepy Dermatol Alergol 36(3):247–254

Poole DC, Pittman RN, Musch TI et al (2020) August Krogh's theory of muscle microvascular control and oxygen delivery: a paradigm shift based on new data. J Physiol 598(20):4473–4507

Rittirsch D, Huber-Lang MS, Flierl MA et al (2009) Immunodesign of experimental sepsis by cecal ligation and puncture. Nat Protoc 4(1):31–36

Rudd KE, Johnson SC, Agesa KM et al (2020) Global, regional, and national sepsis incidence and mortality, 1990-2017: analysis for the Global Burden of Disease Study. Lancet (London, England) 395(10219): 200–211

Schulze KM, Weber RE, Horn AG et al (2022) Effects of pulmonary hypertension on microcirculatory hemodynamics in rat skeletal muscle. Microvasc Res 141: 104334

Singer M, Deutschman CS, Seymour CW et al (2016) The third international consensus definitions for sepsis and septic shock (Sepsis-3). JAMA 315:801–810

Sonobe T, Inagaki T, Poole DC et al (2008) Intracellular calcium accumulation following eccentric contractions in rat skeletal muscle *in vivo*: role of stretch-activated channels. Am J Physiol Regul Integr Comp Physiol 294(4):R1329–R1337

Souza AC, Volpini RA, Shimizu MH et al (2012) Erythropoietin prevents sepsis-related acute kidney injury in rats by inhibiting NF-κB and upregulating endothelial nitric oxide synthase. Am J Physiol Renal Physiol 302(8):F1045–F1054

Ye C, Kawasaki M, Nakano K et al (2022) Acquisition and analysis of microcirculation image in septic model rats. Sensors (Basel) 22(21):8471

The Role of Nitric Oxide in Regulating Microvascular Oxygen Pressure During Diaphragm Contraction in Rats

17

Kazuki Hotta, Narumi Fukuzaki, Rin Kataoka, Kota Izawa, Shuri Arai, Kentaro Kamiya, and Atsuhiko Matsunaga

Abstract

Background The diaphragm muscle is the primary muscle involved in inspiration and is unique to mammals. Because the diaphragm is a constantly active muscle, its contraction may depend on oxidative phosphorylation and O_2 supply to create adenosine triphosphate. This study aimed to evaluate the role of intrinsic vasodilator nitric oxide (NO) in diaphragm O_2 dynamics.

Methods Wistar male rats ($n = 6$, 10 wks old, 311 ± 14 g) were mechanically ventilated under isoflurane anesthesia. The diaphragm was exposed to measure microvascular partial O_2 pressure (PO_2mv) by phosphorescence quenching technique during electrical stimulation-induced diaphragm contractions (6 V, 2 ms, 2 Hz, 180 s). The 180 s of muscle contractions and PO_2mv measurement were repeated 10 min after endothelial NO synthase inhibition by the intra-arterial infusion of nitro-l-arginine methyl ester (L-NAME; eNOS inhibitor). The PO_2mv change during the transition from rest to contraction was fitted to a nonlinear regression model. Time delay, time constant (tau), rate constant (K), and nadir in PO_2mv were calculated and compared before and after eNOS inhibition. [Results] Diaphragm PO_2mv decreased during contractions both before and after eNOS inhibition. The time delay and K were not different before and after L-NAME. However, the L-NAME administration decreased the tau and nadir in PO_2mv (tau, 7.61 [5.18–13.41] vs. 3.36 [1.81–5.57] s; nadir, 7.45 [1.41–14.85] vs. 3.93 [0.89–9.47] mmHg; before vs. after eNOS inhibition, median [25–75 percentiles], $P < 0.05$, respectively).

Discussion The time constant in PO_2mv is determined by oxygen supply to the capillaries and oxygen utilization. The faster time constant and lower nadir in PO_2mv possibly indicate the slower and lower oxygen supply to diaphragm capillaries after eNOS inhibition.

Conclusion The NO has a crucial role in the diaphragm oxygen dynamics of intact rats.

K. Hotta (✉) · K. Kamiya · A. Matsunaga
Department of Rehabilitation, School of Allied Health Sciences, Kitasato University, Sagamihara, Japan

Department of Rehabilitation Sciences, Graduate School of Medical Sciences, Kitasato University, Sagamihara, Japan
e-mail: hotta.kazuki@kitasato-u.ac.jp

N. Fukuzaki · R. Kataoka · K. Izawa · S. Arai
Department of Rehabilitation Sciences, Graduate School of Medical Sciences, Kitasato University, Sagamihara, Japan

S. C. Pias et al. (eds.), *Oxygen Transport to Tissue XLVI*, Advances in Experimental Medicine and Biology 1498, https://doi.org/10.1007/978-3-032-10389-5_17

Keywords

Respiratory muscle · eNOS · Microvascular oxygen pressure · L-NAME · Capillary

17.1 Introduction

The diaphragm is unique to mammals. No mammal could survive without it. The diaphragm is a muscular membrane between the abdominal and thoracic cavities and is the primary muscle of inspiration. Metabolic features of the diaphragm are high levels of oxidative metabolism (diaphragm, citrate synthase activity of 39 mol/min/g compared with spinotrapezius muscle, citrate synthase activity of 14 mol/min/g) (Delp et al. 1998), mitochondrial density in type I and IIa fibers (Sieck et al. 1998), and blood flow (50% higher than tibialis anterior muscle) (Delp et al. 1998). However, the mechanism of blood flow control in the diaphragm is still vague. Recently, diaphragmatic blood flow at rest and during contractions was shown to be severely reduced with prolonged mechanical ventilation, which is associated with diaphragm vascular dysfunction, which occurs through both endothelium-dependent and endothelium-independent mechanisms (Horn et al. 2019). Therefore, diaphragmatic blood flow is affected by various pathological conditions.

It has long been known from experiments using animals that diaphragmatic blood flow is nitric oxide (NO)-dependent. Chang, et al. measured blood flow of anesthetized and mechanically ventilated Sprague-Dawley rats (8–10 weeks old) using Laser Doppler at rest (Chang et al. 1997). They showed an increase in endothelium-dependent blood flow stimulated by acetylcholine (ACh) but a lack of blood flow reduction by NO synthase (NOS) inhibition (Chang et al. 1997). They suggested that N omega-nitro-L-arginine methyl ester (L-NAME), known as NOS inhibitor, should not be used as a specific NO synthase inhibitor in the rat diaphragm. On the other hand, the diaphragm is a skeletal muscle that continually contracts and relaxes as long as we live (just like the heart muscle). Therefore, it is necessary to understand the mechanism of diaphragm blood flow control during contraction. This is because the diaphragm's workload and oxygen (O_2) demand are extremely reduced during mechanical ventilation, which is not physiological. Poole, et al. utilized phosphorescence quenching techniques for determining microvascular oxygen pressure (PO_2mv) across the transition to electrically stimulated contractions (1 Hz) within the rat diaphragm (Geer et al. 2002). They demonstrated that diaphragmatic PO_2mv decreased during contraction, and the kinetics (responses to contractions) were significantly faster in the highly oxidative diaphragm compared with the low-oxidative spinotrapezius muscle (Geer et al. 2002). However, the impact of NO on PO_2mv kinetics is still unknown.

The diaphragmatic PO_2mv measurement using phosphorescence quenching technique has already been validated (Poole et al. 1995). The PO_2mv is determined by the balance between O_2 supply and utilization in capillaries. In the diaphragm, where oxygen demand is increased by contraction, PO_2mv decreases during the transition from rest to contraction. We hypothesized that NOS inhibition impedes the blood flow response to the rat diaphragm, thereby affecting the O_2 supply-demand balance and decreasing the PO_2mv more quickly. To test the hypothesis, this study aimed to determine the effect of NOS inhibition on PO_2 dynamics of contracting diaphragm.

17.2 Methods

Animals and Experimental Design Wistar male rats (10 weeks old, $n = 6$) were obtained from Japan CLEA and were housed in two per cage. The rats were maintained on a 12 h:12 h light-dark cycle and had free access to food and water. After an acclimation period of at least 1 week, electrical stimulation-induced diaphragmatic contraction, PO_2mv, mean arterial blood pressure (MABP), and heart rate (HR) measurements were performed simultaneously under light isoflurane anesthesia and mechanical ventilation. The PO_2mv measurement was initiated 60 s before the 180 s of muscle contraction. After confirming that the PO_2mv returned to baseline values, the L-NAME (10 mg/kg) was administered into the carotid artery. The diaphragmatic contraction and PO_2mv measurement were resumed 15 min after the L-NAME administration. After completion of all experimental protocols, the rats were euthanized by heart extraction under 5% isoflurane anesthesia. The heart and diaphragm were carefully removed and weighed. This study was reviewed and approved by the animal care and use committee at Kitasato University (Ei-ken 22–17). and was performed in accordance with the guidelines established by the Physiological Society of Japan.

Surgical Procedure All rats were anesthetized with isoflurane mixed with 40% O_2 during surgery (Univentor 410 Anaesthesia Unit, Univentor, Malta). The isoflurane concentration was set at 3% for induction and then lowered to 1.2–1.5% for maintenance anesthesia. Following tracheotomy intubation using a 14-G catheter (Terumo, Tokyo, Japan), the rats were mechanically ventilated (SAR-830AP, CWE, USA) at a respiratory rate of 70/min, and a minute volume of 500–600 mL/min (experimental setup is shown in Fig. 17.1). All surgical procedures were performed with rectal temperature monitoring, and the rectal temperature was maintained at 37.0 °C using a heating pad (Fig. 17.1). The right carotid artery was cannulated with polyethylene (PE) tubing (SP-31, Natsume, Tokyo, Japan) for arterial blood pressure monitoring using a PowerLab system (PowerLab 8/35, ADInstruments, Nagoya, Japan). The MABP and HR were analyzed every 1 second. Heparinized saline (10 IU/mL) was injected into the cannulas to prevent blood clotting. The diaphragm was exposed by removing the skin and peritoneum to expose the xiphoid cartilage. The xiphoid cartilage was connected to a cotton thread. A thread was passed through the xiphoid cartilage and pulled upward to expose the ventral diaphragm. The diaphragm was moisturized with warmed Krebs–Henseleit buffer (in mM); 132 NaCl, 4.7 KCl, 21.8 $NaHCO_3$, 2$MgSO_4$, and 2$CaCl_2$] adjusted to pH 7.35–7.45 with 100% CO_2. The exposed diaphragm was attached to silver wire electrodes, which were sutured to the sternum (cathode) and left ribs (anode) regions of the muscle. Blood samples were collected from the carotid artery and used for gas analysis (arterial PO_2, PCO_2, pH).

PO_2mv Measurement The PO_2mv of the diaphragm was measured using a phosphorescence quenching technique during muscle contraction. The PO_2mv was measured via the phosphorescent quenching method using a phosphorometer (OxyLED, Oxygen Enterprises, Havertown, PA, USA) and the phosphorescent probe Oxyphor PdG4 (Pd-meso-tetra-(3,5-dicarboxyphenyl)-tetrabenzoporphyrin; 10μM solution; Oxygen Enterprises) (Hitosugi et al. 2024; Takamizawa et al. 2024). The Oxyphor PdG4 has been used to measure interstitial PO_2 in the skeletal muscle in previous studies (Hirai et al. 2018, 2019). Our previous study had already shown the validity of Oxyphor PdG4 for measuring PO_2mv (see supplemental figure 1 published in (Takamizawa et al. 2024)). The phosphorescent probe was administered into the carotid artery (Oxyphor PdG4, 200μM, 0.5 mL/kg) (Takamizawa et al. 2024). The principle of the phosphorescence quenching technique has been described previously (Behnke et al. 2001). This probe is highly soluble in fluid and cannot diffuse across vessel walls, making it a suitable choice for biological tissue applications (Esipova et al. 2011). Briefly, this technique utilizes the Stern-Volmer equation, which provides a quantitative description of the oxygen dependence of the phosphorescent probes, expressed as:

$$PO_2 = [(\tau_0/\tau) - 1]/(kQ \times \tau_0) \qquad (17.1)$$

where kQ is the quenching constant and τ_0 and τ are the phosphorescence lifetimes in the absence of O_2 and at a given PO_2, respectively. Phosphorescence lifetime τ is independent of the local probe concentration and insensitive to the

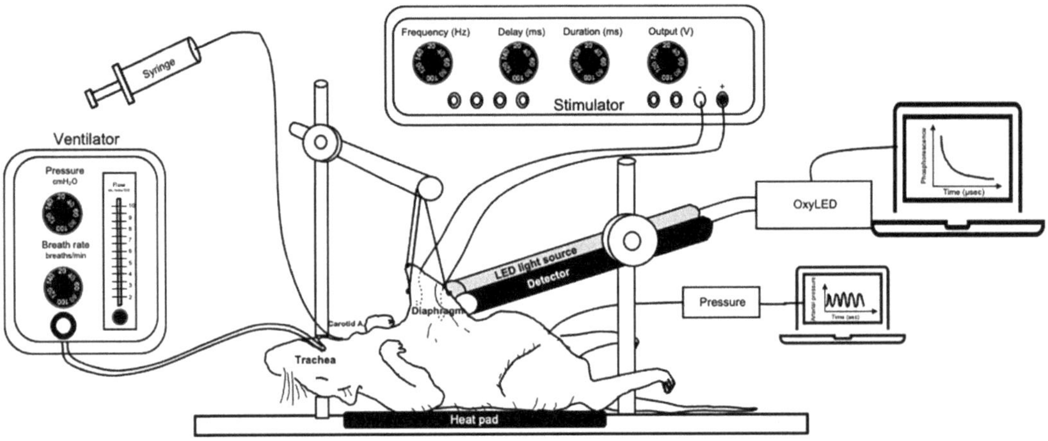

Fig. 17.1 Schematic diagram of experimental setup. The diaphragm was exposed under isoflurane anesthesia and mechanical ventilation. The LED light source and detector were placed 5–10 mm from the left costal diaphragm to measure the PO_2mv

presence of endogenous chromophores and fluorophores (Wilson et al. 2006). The light guide was placed 2–5 mm superficial to the ventral surface of the exposed diaphragm. The phosphorometer allows phosphorescence lifetime measurements. The excitation light (635 nm wavelength) was focused on the diaphragm. The PO_2mv was recorded at intervals of 1 s during the entire protocol. In previous studies, the decrease of PO_2 during contraction has been fitted to nonlinear regression to analyze time delay, time constant, and rate constant (Hirai et al. 2018; Colburn et al. 2020). A total of 60 s was used for the fitting to nonlinear regression, including 20 s of baseline before the start of contraction and 40 s after the start of contraction. This analysis interval was chosen because PO_2 reached its lowest point approximately 30–40 s after the start of contraction. The kinetics of PO_2mv and PO_2 at the onset of contractions were fitted to a nonlinear regression model using GraphPad Prism, version 10.0 (GraphPad Software, La Jolla, CA, USA):

$$PO_{2(t)} = PO_{2(BL)} - APO_2\left[1 - e^{-(t - TD/\tau)}\right] \tag{17.2}$$

where $PO_{2(t)}$ is PO_2mv at a given time point (t), $PO_{2(BL)}$ is baseline PO_2 (i.e., precontraction), APO_2 is the amplitude of PO_2 from baseline to the lowest point, TD is the time delay, e is the exponential function, and τ is the time constant (i.e., time taken to achieve 63% of the response). The goodness of fit and model parsimony were determined using three criteria: the coefficient of determination, the sum of squared residuals, and visual inspection.

Electrical Muscle Stimulation Diaphragmatic contractions were induced using an electrical stimulator (Fig. 17.1; 6 V, 2 ms, 2 Hz, for 180 s; model SEN-2101; Nihon Kohden Corporation, Tokyo, Japan). During mechanical ventilation, spontaneous diaphragm contraction was observed at 42 ± 3 per minute. In this study, the diaphragm was electrically stimulated at 120 per minute, approximately three times higher than the rate of spontaneous diaphragm contraction.

Statistical Analysis All statistical analyses were performed using GraphPad Prism 10.1.1. Two-way analyses of variance (ANOVA) with Tukey's post-hoc analysis was performed to analyze the effect of time and L-NAME on the PO_2mv, MABP, and HR. Data were presented as mean $\pm$ standard deviation. Significance was set at $p < 0.05$.

Table 17.1 Animal characteristics

Body weight (g)	311 ± 14
PaO_2 (mmHg)	93 ± 15
$PaCO_2$ (mmHg)	45 ± 4
pH	7.33 ± 0.05
MAP (mmHg) at rest (pre/post L-NAME)	98 ± 9 / 112 ± 11 *
MAP (mmHg) during contraction (pre/post L-NAME)	77 ± 9 # / 102 ± 11 *#
HR (bpm) at rest (pre/post L-NAME)	344 ± 23 / 353 ± 19
HR (bpm) at pre/post L-NAME	346 ± 15 / 352 ± 23
Diaphragm weight (mg)	549 ± 79

$n = 6$; PaO_2 arterial partial pressure of oxygen, $PaCO_2$ arterial partial pressure of carbon dioxide, *MAP* mean arterial pressure, *HR* heart rate, *L-NAME* endothelial NO synthase inhibition by the intra-arterial infusion of nitro-l-arginine methyl ester; $*P < 0.01$ vs. pre; # $P < 0.01$ vs. at rest in same condition; mean ± SD

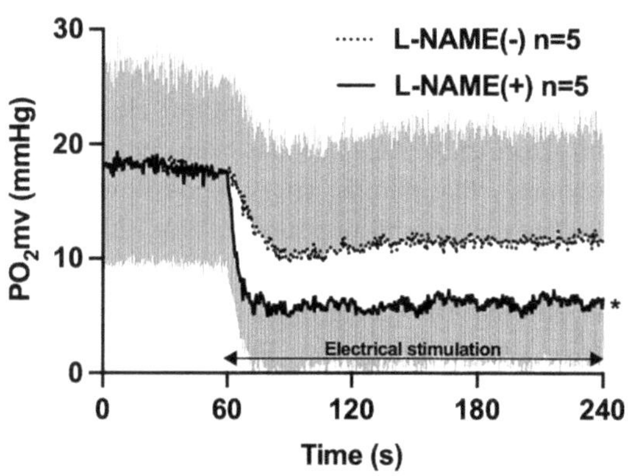

Fig. 17.2 PO_2mv. Muscle contraction was initiated at 60 seconds. The dotted and solid lines indicate before (L-NAME (−)) and after (L-NAME (+)) the L-NAME administration, respectively. Microvascular PO_2 during contraction was lower in the L-NAME (+) than in the L-NAME (−) group. Data were presented as mean ± standard deviation. * represents p-values less than L-NAME (−)

17.3 Results

Table 17.1 shows animal characteristics. Blood gas analysis showed no obvious hypoxemia or hypercapnia (Table 17.1). The MAP acutely increased after L-NAME administration but not in HR (Table 17.1). Figure 17.1 shows the PO_2mv at rest and during contraction. Electrical stimulation was initiated at 60 s and terminated at 240 s. The PO_2mv decreased during the transition from baseline to contraction (Fig. 17.2). The two-way ANOVA revealed no interaction between time and L-NAME ($F = 3.02$, $P = 0.15$); however, a significant main effect of time ($F = 45.70$, $P < 0.01$) and L-NAME (CLP, $F = 10.69$, $P = 0.04$). The mean PO_2mv value during electrical stimulation was significantly lower in the L-NAME administration condition than in the non-administration (Fig. 17.2; $P < 0.01$). Figure 17.3 shows PO_2 profiles. There were no differences in time delay and baseline PO_2mv between the groups. The L-NAME administration decreased the Nadir PO_2mv and shortened the Tau ($P < 0.05$, respectively).

17.4 Discussion

In this study, we measured the oxygen dynamics in the diaphragm during the transition from rest to contraction and explored the contribution of NO. The main findings of this study are (1) the PO_2mv decreased during electrical stimuli-induced diaphragmatic contractions, (2) the NOS inhibitor L-NAME administration acutely decreased microvascular PO_2 during conditions but not at rest. These results suggest that NO has a crucial role in the diaphragm oxygen dynamics during contractions. To the best of our knowledge, this is the first study to show that NOS inhibition reduces PO_2mv during diaphragm contraction.

Poole et al. were the first to measure diaphragm PO_2mv in real time by using phosphorescence quenching technique (Poole et al. 1995). The PO_2mv value was lower in our study than in Poole's study (13 vs. 42 mmHg; this study vs. previous study; (Geer et al. 2002;

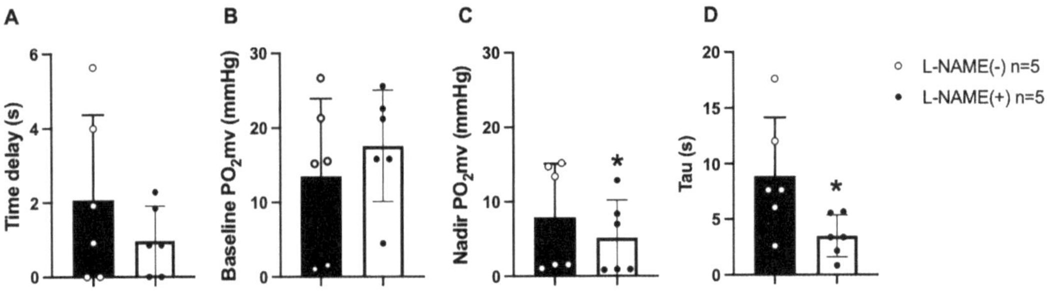

Fig. 17.3 PO_2 profiles. Open and closed circles indicate before (L-NAME (−)) and after (L-NAME (+)) the L-NAME administration, respectively. Time delay (A) and baseline PO_2mv (B) were not different between the groups. Nadir PO_2mv (C) and Tau (D) were significantly lower in the L-NAME (+) than in the L-NAME (−). Data were presented as mean ± standard deviation. * represents p-values less than L-NAME (−)

Poole et al. 1995)). There is a strong positive correlation between diaphragmatic PO_2mv and blood pressure (Poole et al. 1995). The blood pressure in this study was lower than in previous studies (Poole et al. 1995), which may have contributed to the low PO_2mv values. Even though the MAP increased after L-NAME administration (Table 17.1), PO_2mv at rest was not different between the groups (Fig. 17.3). The L-NAME inhibits NO production from endothelium, which results in increased vascular resistance and hypertension (Paulis et al. 2008). Consistent with the previous report (Geer et al. 2002; Poole et al. 2001), the PO_2mv dramatically decreased during the transition from baseline to contraction (Fig. 17.2). The PO_2mv is determined by the balance between O_2 supply to the capillaries and demand. The rapid fall of PO_2mv indicates the O_2 supply and demand imbalance. Kindig et al. reported the immediate and precipitous rise in red blood cell (RBC) velocity in capillaries compared with the prolonged increase of RBC flux, which may act to lower O_2 diffusing capacity (Kindig et al. 2002). Blood flow was reduced in contracting skeletal muscle following the administration of L-NAME, suggesting that the contribution of the NO pathway to the hyperemic blood flow responses (Hirai et al. 1995). Similar to previous studies, our study also suggests that the NO pathway may be involved in reactive hyperemia in the diaphragm during contraction. A previous study using electrical stimulation of rat diaphragm showed the MAP reduction during diaphragmatic contractions (117 ± 2 to 105 ± 3 mmHg; rest to 1 Hz contraction (Horn et al. 2024)). It has been demonstrated previously that phrenic artery blood flow is well regulated at arterial pressures of 90–120 mmHg (Hussain et al. 1988), and given that MAP was not different between groups, and in all instances above 90 mmHg, the lower MAP during contractions likely had little influence on the results herein. Even though the higher MAP during contraction, the Nadir PO_2mv was lower after the L-NAME administration than before the administration (Fig. 17.3), which indicates an obvious increased vascular resistance and limited blood flow to contracting diaphragm.

There are several limitations of this study. First, in order to strongly assert that the oxygen partial pressure measured in this study is NO dependent, it is necessary to directly measure diaphragm blood flow. When considering application to humans, it is necessary to capture the oxygen dynamics of the diaphragm in humans. Guenette et al. have already proposed a feasible method of measuring diaphragm blood flow by using near-infrared spectroscopy and the light-absorbing tracer indocyanine green dye. Our PO_2mv measurement followed the method of Geer et al. (2002). They exposed the diaphragm after laparotomy in mechanically ventilated rats and induced contractions by electrical stimulation. However, our resting PO_2mv values were lower than theirs. Several factors may explain this difference. First, the oxygen probe differed

(Oxyphor R2 vs. PdG4). Oxyphor R2 binds to albumin, while PdG4 dissolves freely in plasma. Second, spontaneous breathing may have affected our results. Although rats were anesthetized and mechanically ventilated, spontaneous breathing was not fully suppressed. We did not use muscle relaxants to eliminate diaphragm contractions. Since spontaneous breaths can strongly influence PO_2mv, these factors may partly explain the lower values in our study.

Based on the data presented, we conclude that NO is an important regulator of rat diaphragm oxygen dynamics.

Acknowledgments This work was supported by JSPS KAKENHI Grant Number and 24K14254.

References

Behnke BJ et al (2001) Dynamics of microvascular oxygen pressure across the rest-exercise transition in rat skeletal muscle. Respir Physiol 126(1):53–63

Chang HY, Chen CW, Hsiue TR (1997) Comparative effects of L-NOARG and L-NAME on basal blood flow and ACh-induced vasodilatation in rat diaphragmatic microcirculation. Br J Pharmacol 120(2):326–332

Colburn TD et al (2020) Transcapillary PO_2 gradients in contracting muscles across the fibre type and oxidative continuum. J Physiol 598(15):3187–3202

Delp MD, Evans MV, Duan C (1998) Effects of aging on cardiac output, regional blood flow, and body composition in Fischer-344 rats. J Appl Physiol (1985) 85(5):1813–1822

Esipova TV et al (2011) Two new "protected" oxyphors for biological oximetry: properties and application in tumor imaging. Anal Chem 83(22):8756–8765

Geer CM et al (2002) Dynamics of microvascular oxygen pressure in the rat diaphragm. J Appl Physiol (1985) 93(1):227–232

Hirai T, Zelis R, Musch TI (1995) Effects of nitric oxide synthase inhibition on the muscle blood flow response to exercise in rats with heart failure. Cardiovasc Res 30(3):469–476

Hirai DM et al (2018) Skeletal muscle microvascular and interstitial PO_2 from rest to contractions. J Physiol 596(5):869–883

Hirai DM et al (2019) Skeletal muscle interstitial PO_2 kinetics during recovery from contractions. J Appl Physiol (1985) 127(4):930–939

Hitosugi N et al (2024) The effect of sepsis and reactive oxygen species on skeletal muscle interstitial oxygen pressure during contractions. Microcirculation 31(1):e12833

Horn AG et al (2019) Impaired diaphragm resistance vessel vasodilation with prolonged mechanical ventilation. J Appl Physiol (1985) 127(2):423–431

Horn AG et al (2024) Effects of aging on diaphragm hyperemia and blood flow distribution in male and female Fischer 344 rats. Am J Physiol Regul Integr Comp Physiol 327(3):R328–R337

Hussain SN, Roussos C, Magder S (1988) Autoregulation of diaphragmatic blood flow in dogs. J Appl Physiol (1985) 64(1):329–336

Kindig CA, Richardson TE, Poole DC (2002) Skeletal muscle capillary hemodynamics from rest to contractions: implications for oxygen transfer. J Appl Physiol (1985) 92(6):2513–2520

Paulis L et al (2008) Regression of L-NAME-induced hypertension: the role of nitric oxide and endothelium-derived constricting factor. Hypertens Res 31(4):793–803

Poole DC, Wagner PD, Wilson DF (1995) Diaphragm microvascular plasma PO_2 measured in vivo. J Appl Physiol (1985) 79(6):2050–2057

Poole DC, Kindig CA, Behnke BJ (2001) Effects of emphysema on diaphragm microvascular oxygen pressure. Am J Respir Crit Care Med 163(5):1081–1086

Sieck GC et al (1998) Cross-bridge cycling kinetics, actomyosin ATPase activity and myosin heavy chain isoforms in skeletal and smooth respiratory muscles. Comp Biochem Physiol B Biochem Mol Biol 119(3):435–450

Takamizawa R et al (2024) Transcapillary PO(2) Gradients in contracting muscles of type I diabetic rats. Microcirculation 31(6):e12870

Wilson DF et al (2006) Oxygen pressures in the interstitial space and their relationship to those in the blood plasma in resting skeletal muscle. J Appl Physiol (1985) 101(6):1648–1656

Influence of Arterial Occlusion at Various Cuff Positions on Systemic Circulation Measured by Remote Photoplethymography (rPPG)

18

Leah DeVos, Gennadi Saiko, and Alexandre Douplik

Abstract

Background: Arterial occlusion is a ubiquitous medical procedure used in many clinical scenarios. However, no standard protocol exists for selecting the applied pressure or cuff position. As various pressures and positions may trigger different physiological responses, it is crucial to understand these peculiarities. The current work aims to investigate if there is any difference in the systemic response to the occlusion at various cuff positions (upper vs. lower arm). *Methods*: The hands of healthy volunteers (10 volunteers) were occluded by inflating the blood pressure cuff to 200 mmHg, and the cuff was placed on the upper or lower arm. The remote photoplethysmography (rPPG) measurements of control and experimental hands were taken. To assess systemic response, we have analyzed the behavior of AC (low, respiratory, and heart rate frequency) components in green and red channels during occlusion and reperfusion. *Results*: We have not found a statistically significant difference in the low, respiratory, or heart rate spectra between occlusion at 200 mmHg in either cuff position. *Conclusions*: Our preliminary results show that the systemic response is similar at both cuff positions. However, findings for both experimental and control hands in most cases have small or medium effect sizes. As such, to avoid a Type II error, further experimentation with proper power calculations and an increased sample size is required.

Keywords

Hemodynamics · Non-invasive monitoring · Blood perfusion · Vascular assessment · Cardiovascular physiology

L. DeVos (✉)
Department of Engineering, Toronto Metropolitan University, Toronto, ON, Canada
e-mail: leah.devos@torontomu.ca

G. Saiko
Department of Physics, Toronto Metropolitan University, Toronto, ON, Canada

A. Douplik
Department of Physics, Toronto Metropolitan University, Toronto, ON, Canada

iBest, Keenan Research Centre of the LKS Knowledge Institute, St. Michael's Hospital, Toronto, ON, Canada

18.1 Introduction

Arterial occlusion is a ubiquitous medical procedure used in many clinical scenarios. The most well-known use is blood pressure measurements using an auscultatory method, based initially on a stethoscope and a sphygmomanometer. Another important clinical use of arterial occlusion is to

S. C. Pias et al. (eds.), *Oxygen Transport to Tissue XLVI*, Advances in Experimental Medicine and Biology 1498, https://doi.org/10.1007/978-3-032-10389-5_18

detect endothelial dysfunction in critically ill patients (Joannides et al. 2006).

Typically, arterial occlusion is induced by placing a blood pressure cuff on the forearm or wrist and inflating it above systolic pressure (Lenders et al. 1991). The upper arm is often preferred as it effectively obstructs the brachial artery and is less impeded by tendons and bones. The wrist can also be used, offering increased patient comfort, although it may not always provide accurate measurements. This discrepancy is due to the anatomical differences and the impact on the underlying vessels. Research indicates that the cuff's location affects the brachial artery's flow-mediated dilation (FMD) response. For instance, upper arm occlusions tend to produce a greater and slower peak dilation than forearm occlusions, which results in a smaller and quicker peak response (Peretz et al. 2007). The relationship between cuff position and occlusion level can be influenced by factors like body mass index (BMI), making it necessary to balance the need for accurate vessel occlusion with patient comfort and pain tolerance. This is crucial in bedside tests for endothelial dysfunction where the artery is occluded for 3 min (Saldin 2019). Understanding differences at various cuff locations could provide valuable insights observed through local and systemic hemodynamic responses. Remote photoplethysmography (rPPG) has proven helpful for investigating skin microcirculation (Burton et al. 2023).

The primary signal in PPG is cardiac pulsation from left ventricle ejection during systole, causing blood vessel distension and tissue absorption changes. Blood waveforms show changes in five frequency bands linked to physiological processes. Average resting heart rates range from 60 to 100 bpm, or 0.83 to 1.67 Hz, extending to 50 bpm or 0.83 Hz for athletes (Doyen et al. 2019). Average respiration rates of 12–20 breaths per minute correspond to 0.20–0.33 Hz (Vital Signs: How to Check My Vitals at Home 2023). The PPG signal also includes oscillations in 0.01–0.02 Hz, 0.02–0.06 Hz, and 0.06–0.15 Hz ranges for endothelial-related metabolic, neurogenic, and myogenic activities (Li et al. 2006). The myogenic range includes Mayer waves at around 0.1 Hz, linked to a sympathetic baroreceptor response to hemodynamic disturbances (Julien 2006). The ability to capture Mayer waves using a smartphone camera was demonstrated by Burton et al. (2022).

Since occlusion and/or reperfusion can trigger hemodynamic disturbances, this study hypothesizes these disturbances depend on the position of occlusion and characterize this difference. We examined systemic physiological responses to 200 mmHg occlusion at various cuff positions (upper vs. lower arm) using rPPG to analyze skin microcirculation. The rPPG signals collected on experimental and control hands were analyzed to study systemic responses.

18.2 Methods

18.2.1 Data Collection

In the experimental setup, the subject sits with hands on a raised platform in a prone position. An iPhone 14 camera (Apple, Cupertino, CA, USA), mounted on a tripod above the hands, records the video. Two NEEWER LED panels (Shenzhen, China), each with 600 LEDs set to 4600 K and 100% intensity, illuminate the hands. A 1 cm white circle is placed next to the hands for color normalization.

During data collection, one hand is occluded and designated as the experimental hand, while the other serves as a control. A pressure cuff is applied around the wrist of the experimental hand. The procedure includes:

1. Measuring initial blood pressure from the experimental arm.
2. Begin recording a 60-fps video.
3. 1 min of no pressure (baseline interval).
4. Applying 200 mmHg pressure for 3 min (occlusion interval).
5. Releasing the pressure for 4 min (reperfusion interval).

6. Ending data collection and recording at the 8-min mark.
7. After a 5-min rest, the process is repeated with the opposite arm as the experimental hand and the pressure cuff are moved to the upper arm.

200 mmHg occlusion pressure was selected to create a standardized protocol, which ensures complete arterial occlusion in the experimental hand. The minimum required pressure for arterial occlusion is slightly above systolic pressure (typically around 150 mmHg). Standard practice is to add a 50 mmHg buffer to ensure consistent and effective occlusion across all participants.

Data were collected from 10 subjects (age 20–60, 4 males, 6 females) with no cardiovascular disease or extreme BMI values. The age range 20–60 was used as an inclusion criterion for participating in the research, but age was not collected for the study's data analysis. The Toronto Metropolitan Research Ethical Board approved the study.

18.2.2 Image Processing

The recorded video footage is processed and analyzed using MATLAB 2023b (Mathworks, Natick, MA, USA). The video is divided into three segments to improve processing time: no pressure, arterial occlusion, and pressure release (reperfusion). Each segment undergoes the following processing steps:

1. A Gaussian filter is applied to reduce noise and high-frequency components.
2. Three regions of interest (ROIs) are manually selected: one on the experimental hand, one on the control hand, and one around a white circle.
3. For each ROI, the pixels of each frame are averaged for each color channel (red, green, and blue), resulting in a single RGB value per frame. This creates time series signals for each color channel.

To normalize the time series for the experimental and control hands, specific equations are used,

$$e_{i,j} = 255\frac{E_{i,j}}{R_{i,j}} \quad (18.1)$$

$$c_{i,j} = 255\frac{C_{i,j}}{R_{i,j}} \quad (18.2)$$

with i representing the color channel (red, green, or blue), j representing the frame number, E the experimental hand time series, C the control hand time series, and R the white circle time series.

18.2.3 Data Analysis

This study focuses on the AC (alternating current) component of the rPPG signal, which corresponds to the pulsatile variations in light intensity caused by periodic blood volume changes during the cardiac cycle. These fluctuations are superimposed on the slower, non-pulsatile baseline (DC component) and are typically used to extract heart rate and other cardiovascular parameters. The analysis examines power amplitude variations within three frequency ranges (low-frequency (LF), respiratory rate (RR), and heart rate (HR)) over time during occlusion and reperfusion upon pressure release.

The red and green channel signals were first filtered using a bandpass filter (0.05–5 Hz) to isolate the LF, RR, and HR components. A continuous wavelet transform (CWT) was then applied to calculate the mean power. The power spectral density (PSD) was determined for each frequency range in both channels.

To quantify the percent change in power amplitude during occlusion and reperfusion compared to the baseline interval, the following equation was used:

$$\%\text{change} = \frac{\text{value}_i - \text{baseline}}{\text{baseline}} \times 100 \quad (18.3)$$

where i represents the segment number ($i = 2$–3), value_i is the PSD value of the frequency band of interest during occlusion (i=2) or reperfusion (i=3), and the baseline represents the PSD value before occlusion. This calculation was performed for both the red and green channels in the experimental and control hands for each segment.

18.2.4 Statistical Analysis

To evaluate regional differences in perfusion responses due to occlusion induced on the upper or lower arm, two-tailed t-tests were conducted to compare the percentage change in power spectral density (PSD) amplitude (baseline to occlusion and reperfusion states) from rPPG signals recorded when occlusion was induced from the upper and lower arm. These comparisons were performed for each frequency band of interest: LF, RR, and HR between the upper and lower arm occlusion protocol signals in the red and green channels.

The null hypothesis assumed no significant difference in the mean percentage change in PSD amplitude between the conditions being compared. The alternative hypothesis suggested a significant difference between these conditions. The practical significance (effect size) was assessed using Pearson's r-value. Statistical analyses were performed using MATLAB 2023b (MathWorks, Natick, MA, USA).

18.3 Results

We have analyzed the behavior of the AC components at low, respiratory, and heart rate spectral ranges in green and red channels during occlusion and reperfusion. The results are depicted in Figs. 18.1 and 18.2, respectively. Statistical (p-value) and practical (Pearson's r-value) significances are also summarized in Table 18.1 for the experimental hand and Table 18.2 for the control hand.

18.4 Discussion

We present a pilot study of microvasculature hemodynamics during occlusion/reperfusion captured by a smartphone camera. Data were collected from 10 healthy subjects.

For the experimental hand, the results indicate no statistically significant differences in low, respiratory, and heart rate frequency behaviors between upper and lower cuff positions during occlusion. Both positions show a drop in low, respiratory, and heart rate spectral amplitudes across color channels. During reperfusion, both positions show an increase in these amplitudes but no statistically significant differences between positions.

For the control hand, the results indicate mainly no statistically significant differences in low, respiratory, and heart rate frequency behaviors between upper and lower cuff positions during occlusion. Both positions show a drop in low, respiratory, and heart rate spectral amplitudes across color channels. During reperfusion, both positions show an increase in these amplitudes but mainly no statistically significant differences between positions. Both cuff positions elicit similar systemic and local responses in the experimental and control hands.

However, findings for both experimental and control hands in most cases have small or medium effect sizes. As such, to avoid a Type II error, further experimentation with proper power calculations and the increased sample size is required.

The work does have certain limitations. Firstly, the measurements were performed on just 10 participants. This was because it was a pilot study to determine the magnitude of the effect. Thus, more extensive studies are required to generalize the results. Secondly, a short interval (5 min) between the measurements on the left and right hands was taken. While quite a significant amount of time was allowed for baseline and reperfusion measurements (1 and 4 min, respectively), it may still impact the blood flow in the

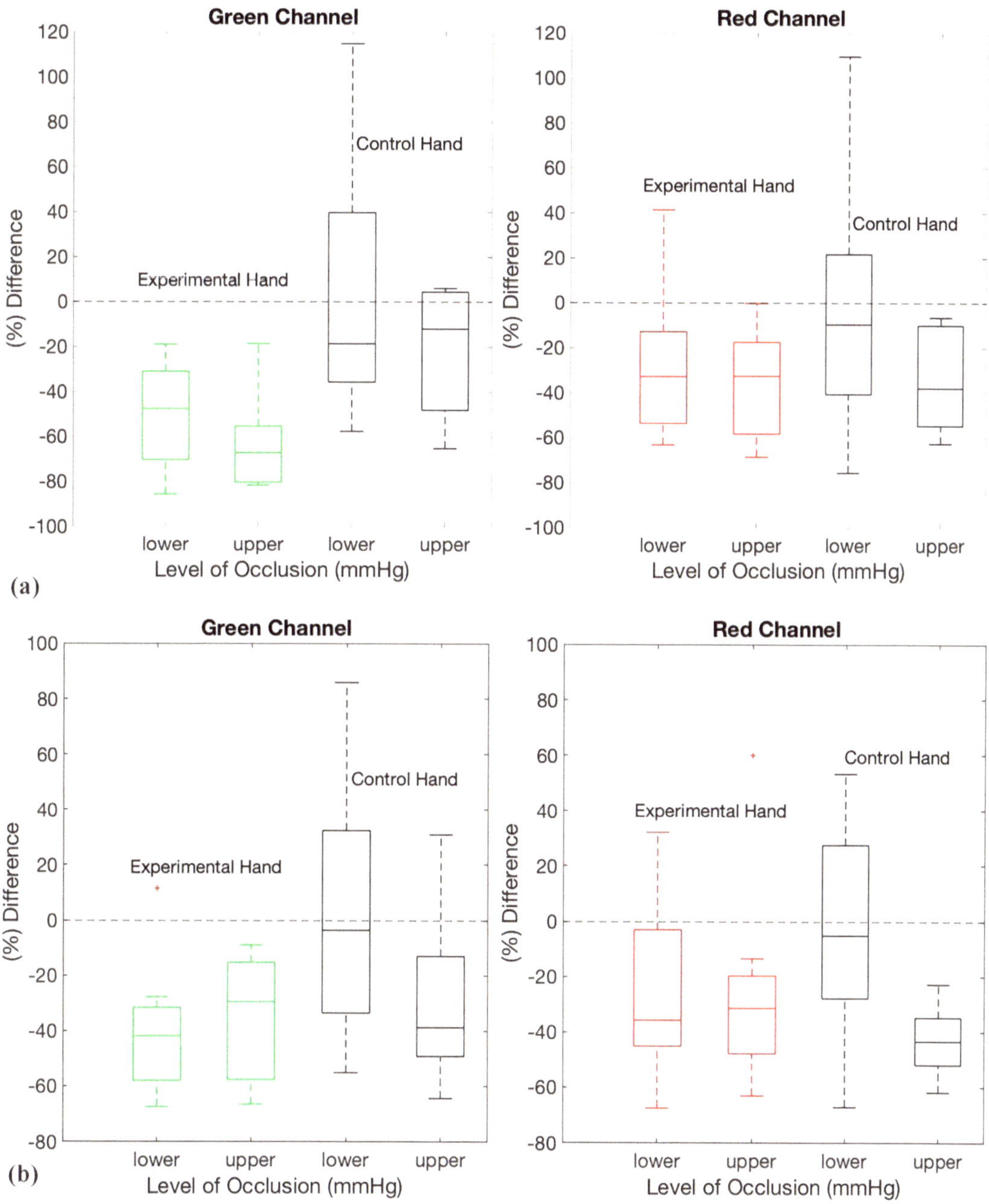

Fig. 18.1 The behavior of AC components in green (left panel) and red (right panel) channels during occlusion. The change has been calculated as a relative change in AC component between (**a**) 0.1 Hz, (**b**) 0.2–0.33 Hz, and (**c**) 0.83–1.67 Hz in segment 2—occlusion—compared to segment 1—baseline

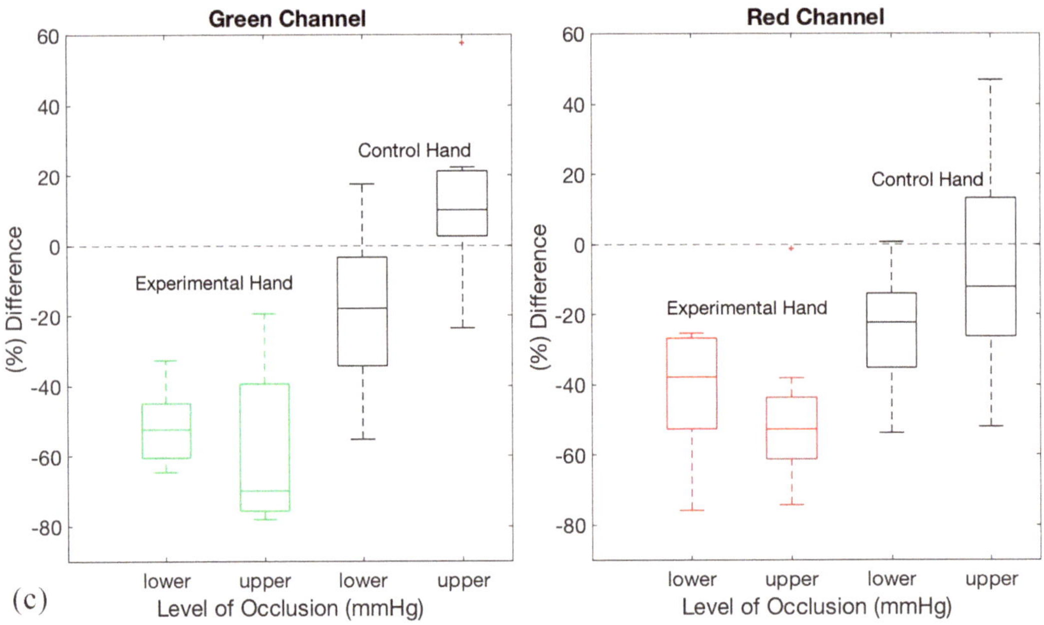

(c)

Fig. 18.1 (continued)

control (left) hand during the occlusion of the right hand. To mitigate this risk, more time (e.g., 10 min) must be allowed between future experiments. In future work, we plan to compare rPPG with contact PPG, which also measures microcirculation in the skin.

18.5 Conclusion

We have analyzed low, respiratory, and heart rate frequency components of remote photoplethysmography signals during arterial occlusion induced by a cuff pressure of 200 mmHg and reperfusion at upper and lower arm cuff positions. Our preliminary results show that the systemic response is similar during occlusion at both cuff positions. However, findings for both experimental and control hands in most cases have small or medium effect sizes. As such, to avoid a Type II error, proper power calculations and the increased sample size are required.

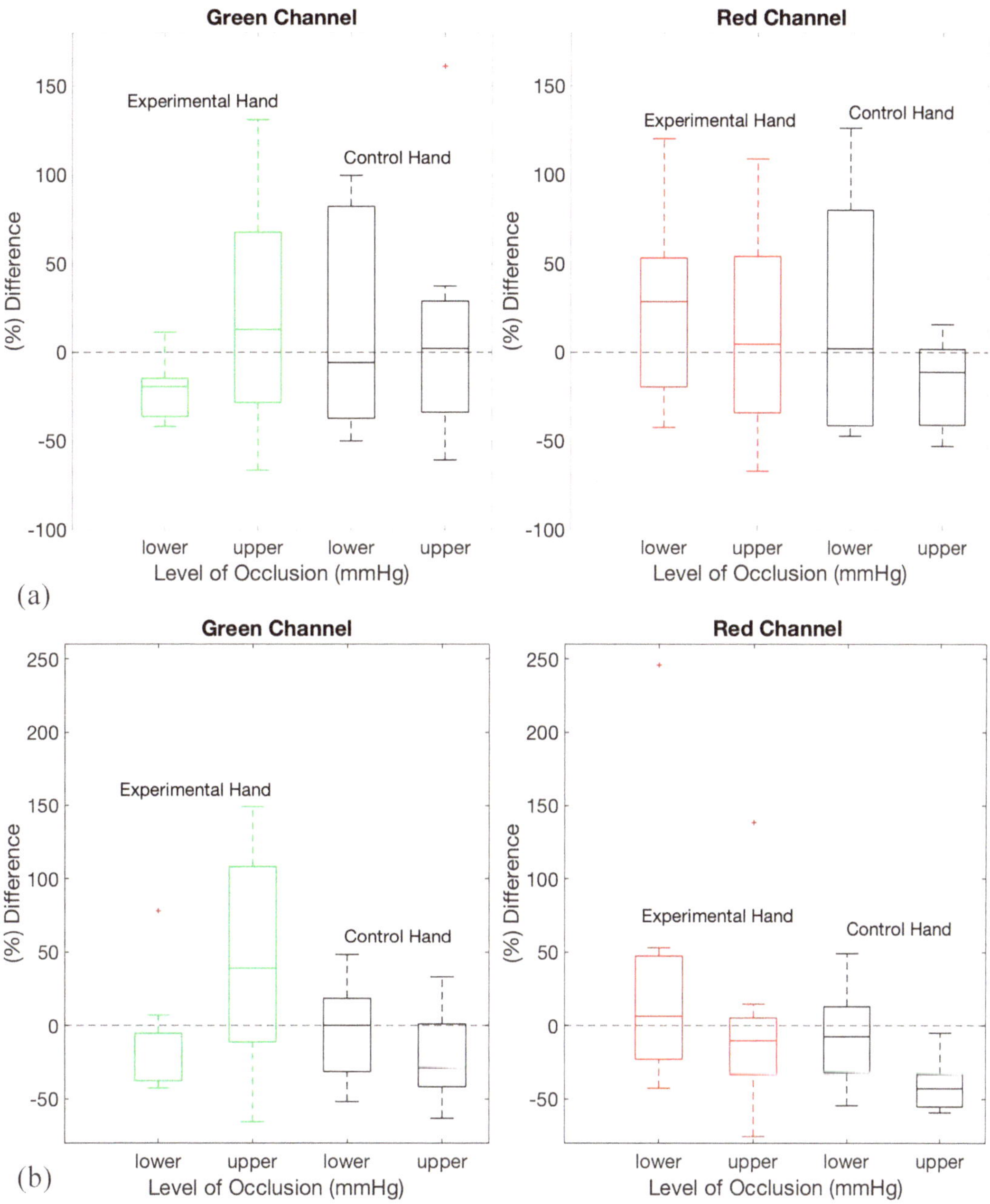

Fig. 18.2 The behavior of AC components in green (left panel) and red (right panel) channels during reperfusion. The change has been calculated as a relative change in AC component between (**a**) 0.1 Hz, (**b**) 0.2–0.33 Hz, and (**c**) 0.83–1.67 Hz in segment 3—reperfusion—compared to segment 1—baseline

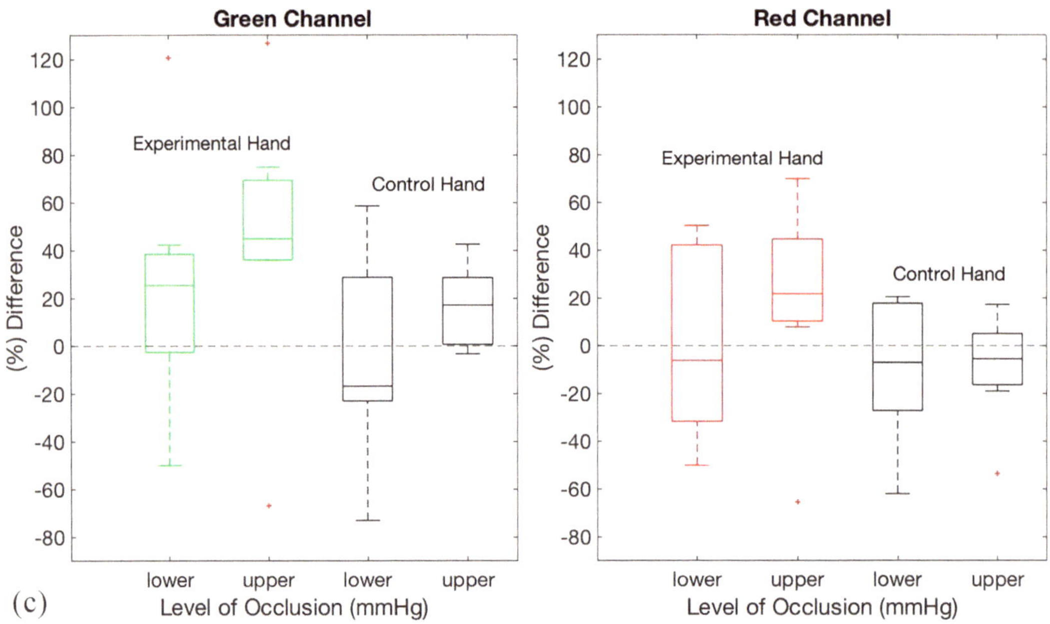

Fig. 18.2 (continued)

Table 18.1 Statistical (p-values, with significance noted by (ns), no statistical significance, when $p > 0.05$) and practical (Pearson's r-value) significance comparing lower and upper arm cuff positions on the experimental hand for low (LF), respiratory (RR), and heart rate (HR) spectral amplitude differences

Metric	p-value/r-value, LF	p-value/r-value, RR	p-value/r-value, HR
Occlusion green channel	0.11$^{(ns)}$/**0.59**	0.72$^{(ns)}$/−0.2	0.40$^{(ns)}$/0.46
Occlusion red channel	0.47$^{(ns)}$/**0.53**	0.99$^{(ns)}$/−0.17	0.45$^{(ns)}$/0.26
Perfusion green channel	0.10$^{(ns)}$/0.34	0.12$^{(ns)}$/−0.13	0.38$^{(ns)}$/0.3
Perfusion red channel	0.53$^{(ns)}$/0.44	0.44$^{(ns)}$/−0.17	0.34$^{(ns)}$/0.15

Large effect size in bold

Table 18.2 Statistical (p-values, with significance noted by (ns), no statistical significance, when $p > 0.05$) and practical (Pearson's r-value) significance comparing lower and upper arm cuff positions on the control hand for low (LF), respiratory (RR), and heart rate (HR) spectral amplitude differences

Metric	p-value/r-value, LF	p-value/r-value, RR	p-value/r-value, HR
Occlusion green channel	0.76$^{(ns)}$/−0.17	0.18$^{(ns)}$/−0.18	0.02$^{(*)}$/0.16
Occlusion red channel	0.11$^{(ns)}$/**0.61**	0.03$^{(*)}$/0.11	0.09$^{(ns)}$/**0.63**
Perfusion green channel	0.91$^{(ns)}$/−0.17	0.33$^{(ns)}$/0.03	0.22$^{(ns)}$/0.05
Perfusion red channel	0.18$^{(ns)}$/0.16	0.05$^{(*)}$/−0.11	0.96$^{(ns)}$/0.22

Large effect size in bold

References

Burton T, Saiko G, Douplik A (2022) Remote PPG imaging by a consumer-grade camera under rest and elevation-invoked physiological stress reveals mayer waves and venous outflow. In: Proceedings of the 15th international joint conference on biomedical engineering systems and technologies. Online Streaming: SCITEPRESS - Science and Technology Publications, pp 153–159. https://doi.org/10.5220/0010883100003123

Burton T, Saiko G, Cao M, Douplik A (2023) Remote photoplethysmography with consumer smartphone reveals temporal differences between glabrous and nonglabrous skin: pilot in vivo study. J Biophotonics 16(1):e202200187. https://doi.org/10.1002/jbio.202200187

Doyen B, Matelot D, Carré F (2019) Asymptomatic bradycardia amongst endurance athletes. Phys Sportsmed 47(3):249–252. https://doi.org/10.1080/00913847.2019.1568769

Joannides R, Bellien J, Thuillez C (2006) Clinical methods for the evaluation of endothelial function—a focus on resistance arteries. Fundam Clin Pharmacol 20(3):311–320. https://doi.org/10.1111/j.1472-8206.2006.00406.x

Julien C (2006) The enigma of Mayer waves: facts and models. Cardiovasc Res 70(1):12–21. https://doi.org/10.1016/j.cardiores.2005.11.008

Lenders J, Janssen GJ, Smits P, Thien T (1991) Role of the wrist cuff in forearm plethysmography. Clin Sci Lond Engl 1979 80(5):413–417. https://doi.org/10.1042/cs0800413

Li Z, Leung JY, Tam EW, Mak AF (2006) Wavelet analysis of skin blood oscillations in persons with spinal cord injury and able-bodied subjects. Arch Phys Med Rehabil 87(9):1207–1212. https://doi.org/10.1016/j.apmr.2006.05.025

Peretz A et al (2007) Flow mediated dilation of the brachial artery: an investigation of methods requiring further standardization. BMC Cardiovasc Disord 7(1):11. https://doi.org/10.1186/1471-2261-7-11

Saldin T (2019) Assessing endothelial dysfunction estimating the differences between 3 minute and 5 minute reactive hyperemia. Masters Theses. https://doi.org/10.15368/theses.2019.4

Vital Signs: How to Check My Vitals at Home. Cleveland Clinic. Accessed: Nov. 21, 2023. [Online]. Available: https://my.clevelandclinic.org/health/articles/10881-vital-signs

Physiological Effects of Photobiomodulation Therapy in the Treatment of Multiple Concussions

19

Fazle Kibria, Olga A. Bragina, Eduardo Colombari, Alexey Trofimov, Oxana Semyachkina-Glushkovskaya, Edwin M. Nemoto, Dmitriy N. Atochin, Satoshi Kashiwagi, Patric Stanton, and Denis E. Bragin

Abstract

Repeated concussion traumatic brain injury (TBI) results in long-term brain damage and cognitive dysfunctions, leading to neurodegenerative diseases. The brain clearance system plays a crucial role in TBI recovery and neurodegenerative disease amelioration by draining waste macromolecules from the brain. Pharmacological therapeutics have failed to demonstrate benefits in human TBI. Photobiomodulation (PBM) has gained interest in neuroscience and has been shown to improve brain drainage. Here, we evaluated the efficiency of PBM in the treatment of multiple concussions in mice and the augmentation of the brain clearance system. Three consecutive closed-head concussive TBIs were induced with a 1-h interval to the left hemisphere in C57BL/6 male mice. A near-infrared irradiation (1270 nm, 10 mW/cm^2) was used for PBM 4 h after the last TBI and the following 3 days twice a day. Laser speckle contrast imaging was used to assess cerebral blood flow (rCBF). In vivo 2-photon laser scanning microscopy assessed PBM effects on cerebral microcirculation, tissue oxygen supply (NADH), and meningeal lymphatics clearance. Brain compliance was evaluated by intracranial pressure waveform analysis. Neurological severity scores were

F. Kibria · O. A. Bragina
Lovelace Biomedical Research Institute, Albuquerque, NM, USA

E. Colombari
São Paulo State University, Araraquara, Brazil

A. Trofimov
Privolzhsky Research Medical University, Nizhny Novgorod, Russia

O. Semyachkina-Glushkovskaya
Saratov State University, Saratov, Russia

E. M. Nemoto
University of New Mexico School of Medicine, Albuquerque, NM, USA

D. N. Atochin
Boston VA Medical Center West Roxbury, Veterans Affairs Boston Healthcare System and Harvard Medical School, West Roxbury, MA, USA

S. Kashiwagi
Massachusetts General Hospital and Harvard Medical School, Charlestown, MA, USA

P. Stanton
New York Medical College, Valhalla, NY, USA

D. E. Bragin (✉)
Lovelace Biomedical Research Institute, Albuquerque, NM, USA

University of New Mexico School of Medicine, Albuquerque, NM, USA

New York Medical College, Valhalla, NY, USA
e-mail: dbragin@lovelacebiomedical.org

S. C. Pias et al. (eds.), *Oxygen Transport to Tissue XLVI*, Advances in Experimental Medicine and Biology 1498, https://doi.org/10.1007/978-3-032-10389-5_19

obtained at 0–3 days after TBI. Two-way ANOVA for multiple comparisons was used to test intergroup differences, with the statistical significance set at $p < 0.05$. Multiple concussions progressively impaired rCBF, cortical microcirculation, tissue oxygen supply, and brain drainage function ($p < 0.05$). Compared to the sham-treated group, PBM improved rCBF, microcirculation, tissue oxygenation, and the brain drainage system ($p < 0.05$). Neurological function was more preserved in the PBMT group than in sham-treated mice ($p < 0.05$). Our study demonstrated that PBMT can be used as an adjunct therapy even in the acute period of TBI.

Keywords

Traumatic brain injury · Concussion · Brain compliance · Photobiomodulation · Cerebral blood flow · Oxygen delivery to tissue · Meningeal lymphatics

19.1 Introduction

Severe or repeated mild traumatic brain injury (TBI) can result in chronic traumatic encephalopathy (CTE) that manifests in cognitive and mood dysfunctions and elevated risk for neurodegenerative diseases such as Alzheimer's and Parkinson's disease (McKee et al. 2009). Primary TBI is usually followed by secondary complications leading to prominent long-term neurologic disorders that can last for decades following the initial injury (Schwarzbold et al. 2008). Therefore, TBI is not a single event but the start of a chronic, progressive disease. A critical therapeutic strategy is to identify treatments that block the transition from primary to secondary injury in an attempt to prevent the progressive decline in brain function. Pharmacological therapeutics that typically target discrete pathways have failed to demonstrate clear benefits in clinical settings (Khellaf et al. 2019). Photobiomodulation (PBM) is a therapeutic modality that has gained interest in neuroscience applications, including acute TBI (Stevens et al. 2022).

Latest evidence suggests that recently discovered meningeal lymphatic vessels (MLVs) play a crucial role in maintaining brain homeostasis by draining macromolecules via both cerebrospinal fluid (CSF) and interstitial fluid (ISF) from the central nervous system (CNS) into the cervical lymph nodes (cLNs) (Louveau et al. 2015). An impaired cerebral lymphatic system is considered a risk factor for neuroinflammatory diseases, neurovascular diseases, and impaired recovery from brain injuries (Sun et al. 2018). In recent research, we demonstrated that PBM (1270 nm) effectively attenuates Alzheimer's disease and intraventricular hemorrhage via enhancement of the lymphatic clearance of blood products from the brain due to PBM-induced dilatation of the MLVs (Semyachkina-Glushkovskaya et al. 2021; Li et al. 2023). Here, we evaluated the efficiency of PBM in the treatment of multiple concussions in mice and the augmentation of the brain clearance system.

19.2 Methods

Twenty laboratory-acclimated male C57BL/6 mice (8 weeks of age, >25–30 g) were randomized into PBM and Sham groups. Mice received three consecutive modified closed-head concussive TBIs (Ren et al. 2013) to the left hemisphere at 1-h intervals. A near-infrared irradiation (1270 nm, 10 mW/cm^2) was used for PBM 4 h after the last TBI and twice daily for the following 3 days. Sham animals were subjected to the same procedures but without PBM. Laser speckle contrast imaging was used to evaluate regional cerebral blood flow (rCBF) at baseline, 1 h, and days 1 and 3 after TBI (Bragin et al. 2011). In vivo 2-photon laser scanning microscopy was used to assess PBM effects on cerebral microcirculation, tissue oxygen supply (inversely reflected by NADH autofluorescence), and meningeal lymphatics clearance at a baseline, 1 h, and days 1 and 3 after TBI, as we previously described (Semyachkina-Glushkovskaya et al. 2021; Li et al. 2023; Bragin et al. 2011). Brain compliance was evaluated by intracranial pressure waveform analysis (Fernandes et al. 2021). Neurological severity scores (Flierl et al. 2009) were obtained 0–3 days after TBI and quantified post-TBI impairment on a 0–10 scale. The score consists of 10 clinical parameters, including tasks on motor function, alertness, and physiological behavior. Failure on one of them added one point, so higher totals indicate more severe neurologic dysfunction (Flierl et al. 2009). Two-way ANOVA for multiple comparisons was used to test intergroup differences, with the statistical significance set at $p < 0.05$.

19.3 Results

Multiple concussions led to decreased regional CBF, cortical microcirculation, and tissue oxygen supply, as reflected by NADH autofluorescence increase (Fig. 19.1). One hour after the last TBI, rCBF decreased to 60 ± 9.2% from the baseline in both groups (Fig. 19.1a). On days 1 and 3 after TBI, rCBF remained reduced in the Sham group but progressively improved in the PBM group—72.3 ± 8.7% and 79.8 ± 8.1%, for days 1 and 3, respectively (Fig. 19.1a, $p < 0.05$ and $p < 0.01$ from the Sham). One hour after the last TBI, the number of perfused capillaries in the area of concussion reduced to 42 ± 6.9% in both groups,

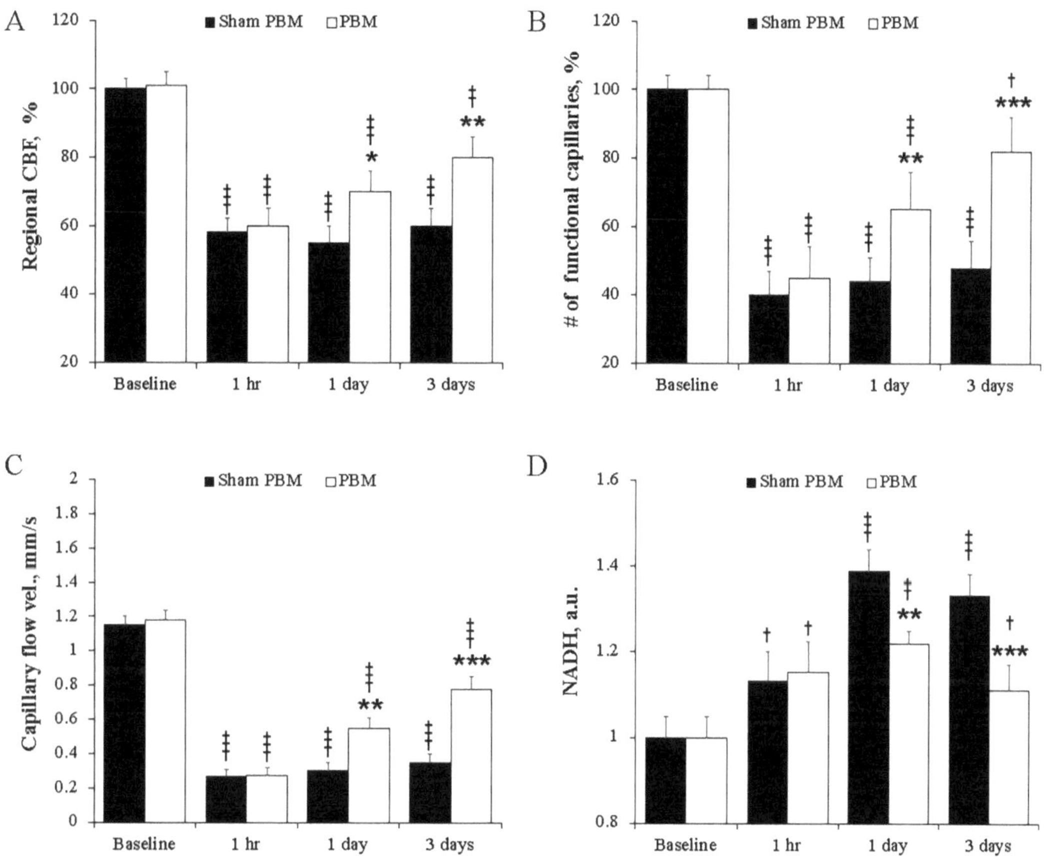

Fig. 19.1 After multiple TBIs: (**a**) regional CBF in the peri-lesion cortex dropped more significantly in mice without PBM; (**b**) the number of functional capillaries in the peri-lesion cortex was less in mice without PBM; (**c**) the capillary flow velocity in the peri-lesion cortex was less in mice without PBM; and (**d**) NADH autofluorescence in the peri-lesion cortex, inversely reflecting tissue oxygen supply, was higher in mice with sham PBM comparing to PBM treated mice. $N = 10$ per group; † $p < 0.05$, ‡ $p < 0.01$. ‡ $p < 0.001$ from baseline; and $^{*}p < 0.05$, $^{**}p < 0.01$. *** < 0.001 from sham PBM

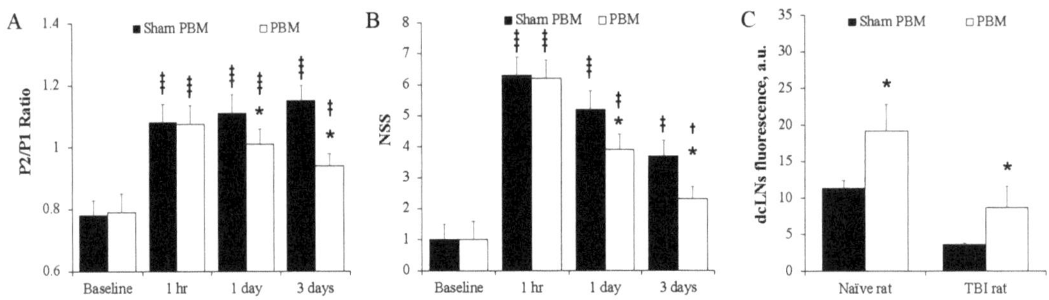

Fig. 19.2 After multiple TBIs: (**a**) P2/P1 ratio representing better-preserved brain compliance in the PBM group. (**b**) PBM improves neurologic function after multiple concussions. (**c**) Graph showing recovered meningeal lymphatics in the PBM group after multiple TBIs. $N = 10$ per group; † $p < 0.05$, ‡ $p < 0.01$. ‡ $p < 0.001$ from baseline; and * $p < 0.05$ from sham PBM

reflecting capillary stagnation (Fig. 19.1b). In the Sham group, the number of perfused capillaries remained low on the following days. In contrast, in the PBM group, the number of perfused capillaries significantly increased to 61.1 ± 7.7% ($p < 0.01$) and 83.3 ± 8.4% ($p < 0.001$) on days 1 and 3, from the Sham respectively (Fig. 19.1b). Similarly, the capillary flow velocity decreased in both groups from 1.1 ± 0.02 to 0.37 ± 0.07 mm/s 1 h after the last TBI (Fig. 19.1c). The capillary flow velocity remained the same in the Sham group while increased in the PBM group to 0.45 ± 0.01 and 0.78 ± 0.02 mm/s on days 1 and 3 (Fig. 19.1c, $p < 0.01$ and 0.001 from Sham, respectively). Cortical microcirculation impairment was associated with reduced tissue oxygen supply in both groups 1 h after the last TBI, as inversely reflected by the increase in NADH autofluorescence to 1.14 ± 0.07 a.u. (Fig. 19.1d). Prolonged microcirculation impairment in the Sham group led to a decrease in tissue oxygen supply as inversely reflected in an increase in NADH autofluorescence 1.39 ± 0.05 and 1.34 ± 0.06 a.u. on days 1 and 3 post-TBI (Fig. 19.1d). Improved microcirculation in the PBM group enhanced tissue oxygen supply reflected by a reduction in NADH autofluorescence to 1.20 ± 0.03 and 1.11 ± 0.06 normalized units at days 1 and 3 (Fig. 19.1d, $p < 0.01$ and 0.001 and 0.05, from Sham respectively).

Intracranial pressure wave analysis revealed a reduction in brain compliance 1 h after the last TBI as reflected by an increase in P2/P1 (percussion/tidal wave) pressure waves ratio to 1.08 ± 0.06 in both groups compared to 0.78 ± 0.05 at a baseline (Fig. 19.2a, $p < 0.05$). In the Sham group, brain compliance was further reduced on days 1 and 3, as suggested by an increase in the P2/P1 ratio to 1.11 ± 0.06 and 1.15 ± 0.05, respectively, while in the PBM group, the P2/P1 decreased to 1.01 ± 0.05 and 0.94 ± 0.04, respectively, reflecting better-preserved compliance (Fig. 19.2a, $p < 0.05$ from Sham). Similarly, the neurological severity score showed impairment of neurological functions 1 h after the last TBI in both groups and better-preserved neurologic function in the PBM group at days 1 and 3 post-TBI (Fig. 19.2b, $p < 0.05$ from Sham).

In another series of experiments, we evaluated the effects of multiple TBIs and PBM at 4 h after the last TBI on the meningeal lymphatics drainage into the deep cervical lymph nodes (dcLNs) (Fig. 19.2c). Evans blue dye (EBD) (5%) was injected into the cisterna magna, and its clearance in the dcLNs was quantified 1 h after injection by fluorescence microscopy as described previously (Semyachkina-Glushkovskaya et al. 2021; Li et al. 2023). As shown in Fig. 19.2c, the clearance of EBD was reduced to 3.6 ± 0.25 a.u. in those affected by multiple TBI compared to naïve mice (11.3 ± 1.1 a.u., $p < 0.05$), suggesting a reduced accumulation of the dye in dcLNs due to potential impairment of the meningeal lymphatic after multiconcussion TBI. One-hour PBM applied 4 h after TBI increased EBD fluorescence in dcLNs to 8.7 ± 2.9 a.u. (Fig. 19.2c, $p < 0.05$), reflecting a restoration of brain drainage after PBM.

19.4 Discussion

Although the precise mechanisms of PBM are still elusive and include an increase in antioxidants and energetics, a decrease in inflammation, protection against apoptosis, and modulated microglial activation, recent studies consistently report that PBM improves endothelial dysfunction via increasing bioavailable nitric oxide (NO) in a dose-dependent manner (Kashiwagi et al. 2023). We have shown that a short course of PBM, applied during the acute phase after multiple concussions, effectively improves cerebral microcirculation by dilating cerebral arterioles, likely through a nitric oxide-dependent mechanism. Similarly, the improvement in meningeal lymphatic function, as indicated by the faster movement of dye from the cisterna magna to the cervical lymph nodes, may also rely on nitric oxide. In previous studies using mouse models of Alzheimer's disease and intraventricular hemorrhage, we found that the effects of PBM could be blocked by inhibiting nitric oxide synthase, which may also apply to multiple concussions (Semyachkina-Glushkovskaya et al. 2021; Li et al. 2023). Brain drainage activation might play an essential role in blocking or at least inhibiting the development of secondary complications after TBI, thereby reducing sustained neuroinflammation, excitotoxicity, and oxidative stress—all major causatives of prominent long-term neurologic disorders.

19.5 Conclusion

Repetitive PBM applied in the acute period after multiple concussion TBI effectively restored cerebral microcirculation, tissue oxygen supply, and meningeal lymphatics function, improving brain compliance and neurological outcome. PBM can be used as an adjunct therapy, starting even in the acute period of TBI.

Acknowledgments This work was supported by the TU/LBRI Innovative Intramural Research Grant, NIH/NINDS 1R01NS112808, and NIH/NIA R01AG081809.

References

Bragin DE, Bush RC, Müller WS et al (2011) High intracranial pressure effects on cerebral cortical microvascular flow in rats. J Neurotrauma 28(5):775–785

Fernandes MV, Rosso Melo M et al (2021) Intracranial pressure during the development of renovascular hypertension. Hypertension 77(4):1311–1322

Flierl MA, Stahel PF, Beauchamp KM et al (2009) Mouse closed head injury model induced by a weight-drop device. Nat Protoc 4:1328–1337

Kashiwagi S, Morita A, Yokomizo S et al (2023) Photobiomodulation and nitric oxide signaling. Nitric Oxide 1(130):58–68

Khellaf A, Danyal, Khan Z et al (2019) Recent advances in traumatic brain injury. J Neurol 266:2878–2889. https://doi.org/10.1007/s00415-019-09541-4

Li D, Liu S, Yu T et al (2023) Photostimulation of brain lymphatics in male newborn and adult rodents for therapy of intraventricular hemorrhage. Nat Commun 14(1):6104

Louveau A, Smirnov I, Keyes TJ et al (2015) Structural and functional features of central nervous system lymphatic vessels. Nature 523(7560):337–341

McKee AC, Cantu RC, Nowinski CJ et al (2009) Chronic traumatic encephalopathy in athletes: progressive tauopathy after repetitive head injury. J Neuropathol Exp Neurol 68(7):709–735

Ren Z, Iliff JJ, Yang L et al (2013) 'Hit & Run' model of closed-skull traumatic brain injury (TBI) reveals complex patterns of post-traumatic AQP4 dysregulation. J Cereb Blood Flow Metab 33(6):834–845

Schwarzbold M, Diaz A, Martins ET et al (2008) Psychiatric disorders and traumatic brain injury. Neuropsychiatr Dis Treat 4:797–816

Semyachkina-Glushkovskaya O, Klimova M, Iskra T et al (2021) Transcranial photobiomodulation of clearance of beta-amyloid from the mouse brain: effects on the meningeal lymphatic drainage and blood oxygen saturation of the brain. Adv Exp Med Biol 1269:57–61

Stevens AR, Hadis M, Milward M et al (2022) Photobiomodulation in acute traumatic brain injury: a systematic review and meta-analysis. J Neurotrauma 40(3–4):210–227

Sun BL, Wang LH, Yang T et al (2018) Lymphatic drainage system of the brain: a novel target for intervention of neurological diseases. Prog Neurobiol 163:118–143

Photobiomodulation Increases Eye Movement Conjugation in Healthy Volunteers

20

Alex O. Trofimov, Elizabeth Kalinkina, Alexander Lazarev, Anastasia Kivenko, Edwin M. Nemoto, Olga A. Bragina, and Denis E. Bragin

Abstract

The aim of the study was to evaluate changes in eye movement conjugation in healthy volunteers in response to transcranial photobiomodulation (tPBM). *Materials and Methods.* 54 healthy volunteers (23 men, 31 women; median age was 23.5 years) were included in this retrospective, non-randomized study. Eye tracking (ET) was performed using the iPad app EyeTracker with a sampling rate of 20 Hz before and after 10 min tPBM (810 nm). The tracker calculated angular velocities in the vertical and horizontal planes (AV; vertical and horizontal). Data are shown as a median [interquartile range]. Pearson's correlation coefficients were used to assess agreement between computed variables. This coefficient value was designated as the vergence reactivity index (VRx), which reflects the eye movement conjugation. The differences were determined using the Wilcoxon T-criterion. The level of significance was $p < 0.05$. *Results.* Vertical and horizontal VRx after tPBM were significantly higher than before the procedure: 0.878 [0.843; 0.902] vs. 0.783 [0.702; 0.829], $p < 0.001$, respectively and 0.875 [0.834; 0.912] vs. 0.927 [0.921; 0.93], $p < 0.001$, respectively. *Conclusion.* Transcranial PBM causes a significant increase in vertical and horizontal vergence indices, which indicates an improvement in eye movement conjugation in healthy volunteers.

Keywords

Photobiomodulation · Eye tracking · Eye movements conjugation

20.1 Introduction

Currently, photobiomodulation (PBM) is understood as a method of exposing tissues and organs

A. O. Trofimov (✉)
Department of Neurological Diseases, Privolzhsky Research Medical University, Nizhny Novgorod, Russia

Department of Physiology, Saratov State Researh University, Saratov, Russia

E. Kalinkina · A. Lazarev · A. Kivenko
Department of Neurological Diseases, Privolzhsky Research Medical University, Nizhny Novgorod, Russia

E. M. Nemoto
Department of Neurology, University of New Mexico School of Medicine, Albuquerque, NM, USA

O. A. Bragina
Lovelace Biomedical Research Institute, Albuquerque, NM, USA

D. E. Bragin
Department of Neurology, University of New Mexico School of Medicine, Albuquerque, NM, USA

Lovelace Biomedical Research Institute, Albuquerque, NM, USA

S. C. Pias et al. (eds.), *Oxygen Transport to Tissue XLVI*, Advances in Experimental Medicine and Biology 1498, https://doi.org/10.1007/978-3-032-10389-5_20

to red or infrared light (wavelength 600–1100 nm) at low energy (1–20 J/cm^2) to improve function (Lapchak and Boitano 2016). Numerous experimental studies have demonstrated the safety and optimal parameters of PBM both in healthy volunteers and in patients with various pathologies, such as traumatic brain injury, ischemic stroke, Alzheimer's disease, Parkinson's disease, and psychological disorders such as depression and anxiety, as well as age-related cognitive decline (Zivin et al. 2014).

However, while the mechanisms of action on cellular structures are generally clear and associated with the excitation of mitochondrial cytochrome-C-oxidase (CCO) and copper-dependent photodissociation of nitric oxide (NO), the effects of PBM on subcortical structures and pathways remain poorly understood (Nairuz and Lee 2024). Existing methods for this assessment are still too expensive and cumbersome (functional magnetic resonance imaging [fMRI] and magnetic encephalography [MEG]). However, the introduction of compact devices like eye trackers into research can significantly expand our understanding of the systemic effects of transcranial PBM (tPBM), particularly in assessing visual perception. The aim of the study was to evaluate the changes in eye movement conjugation in healthy volunteers in response to tPBM.

20.2 Methods

20.2.1 Design and Study Population

This retrospective, non-randomized, observational, single-cohort, single-center study complied with Helsinki Declaration standards and was approved by the Local Ethics Committee. It included 54 healthy volunteers from September 2023 to April 2024 (23 men, 31 women; median age was 23.5 years, min 22, max 24; all right-handed).

Informed consents were obtained from all subjects. All study participants were free of any neurological (cognitive impairment, bipolar disorder, depression, cerebrovascular disease or traumatic brain injury history, alcohol or drug addiction disorder, etc.) or ophthalmological disease (including myopia, hypermetropia, astigmatism, strabismus). All subjects had a visual acuity of 1.0.

20.2.2 Eye Tracking Protocol

The eye tracking procedure took place in the presence of a researcher (AT), who controlled the position of the subject's face on the tracker screen during the study.

After calibration, a round marker with a diameter of 3 mm changed color from white to red with a frequency of 1 Hz and moved randomly across the screen. The position of the center of the pupils was estimated using the iPad app EyeTracker (BVG Software Group LLC, San Francisco, CA, USA) with a sampling rate of 20 Hz. In total, 450 frames were collected during the eye tracking procedure. During post-processing, frames with incorrectly recognized pupils, blinking, etc. were identified and removed. Finally, the 300 best consecutive eye tracking frames were analyzed. The total duration of the procedure was 45–50 s.

Subsequently, the tracker calculated angular velocities in the vertical and horizontal planes (AV; vertical and horizontal). As a result, the following values were obtained separately for each eye (in radiant per second): left vertical velocity (LVV), right vertical velocity (RVV), left horizontal velocity (LHV), and right horizontal velocity (RHV).

Time-averaged LVV values versus RVV and LHV versus RHV were calculated in a linear moving Pearson correlation between 30 consecutive 1.5-s averages of the corresponding angular velocities of the right and left eyes. The coefficient value was designated as the vergence reactivity index (VRx), which actually reflects the subject's eye movement conjugation in a time window lasting about 1 s.

20.2.3 Transcranial Photobiomodulation Protocol

The transcranial PBM parameters were: 810 nm laser wavelength, number of NIRS optodes—10, summary output power—6 W, an irradiance—300 mW/cm^2, fluence rate—103.2 J/cm^2 (total fluence 2064 J per procedure), light spot area of single optode—2 cm^2, the total size of the light delivery—20 cm^2 in both fronto-temporal areas, duration—10 min.

20.2.4 Statistical Analysis

Data are shown as a median [interquartile range]. Pearson's correlation coefficients were used to assess agreement between computed variables. The differences were determined using the Wilcoxon T-criterion. The level of significance was $p < 0.05$. The software package Statistica 12 (TIBCO Inc., Palo Alto, CA, USA) was used for data analysis.

20.3 Results

tPBM led to a significant increase in both VRx. Vertical VRx after tPBM were significantly higher than before the procedure: 0.878 [0.843; 0.902] vs. 0.783 [0.702; 0.829], $p < 0.001$, respectively (Fig. 20.1).

At the same time, horizontal VRx before tPBM were significantly lower than after tPBM; 0.875 [0.834; 0.912] vs. 0.927 [0.921; 0.93], $p < 0.001$, respectively (Fig. 20.2). We did not observe any adverse effects either during or after the tPBM procedure.

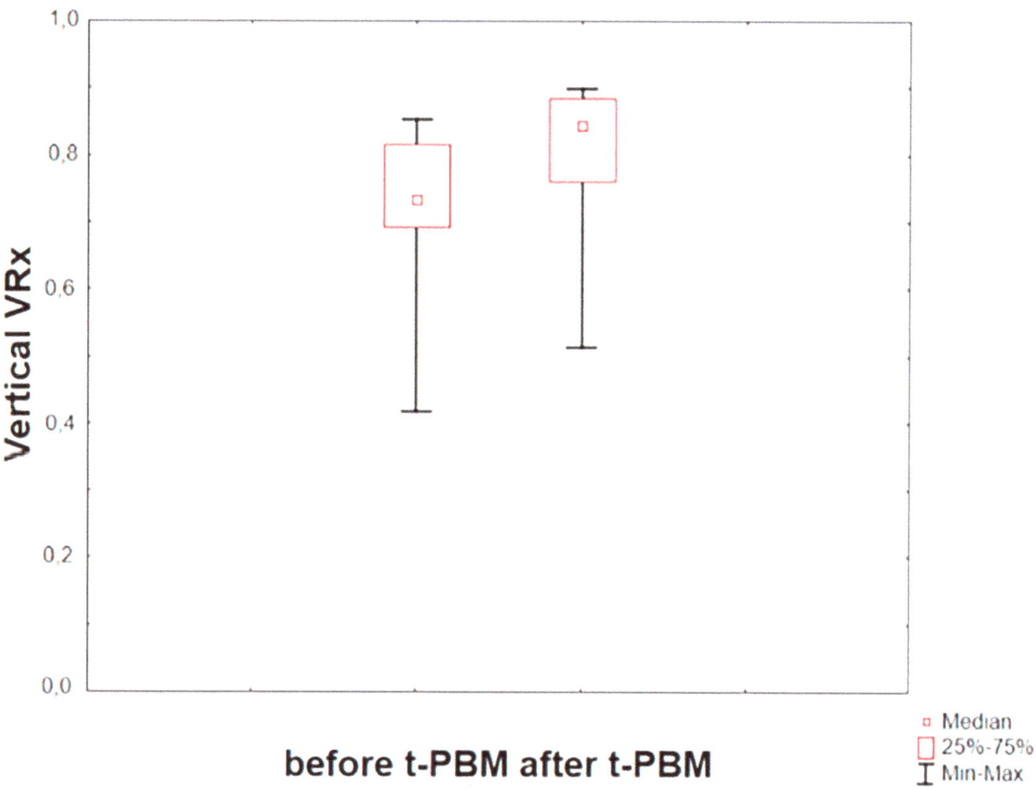

Fig. 20.1 Comparison of vertical VRx before and after tPBM

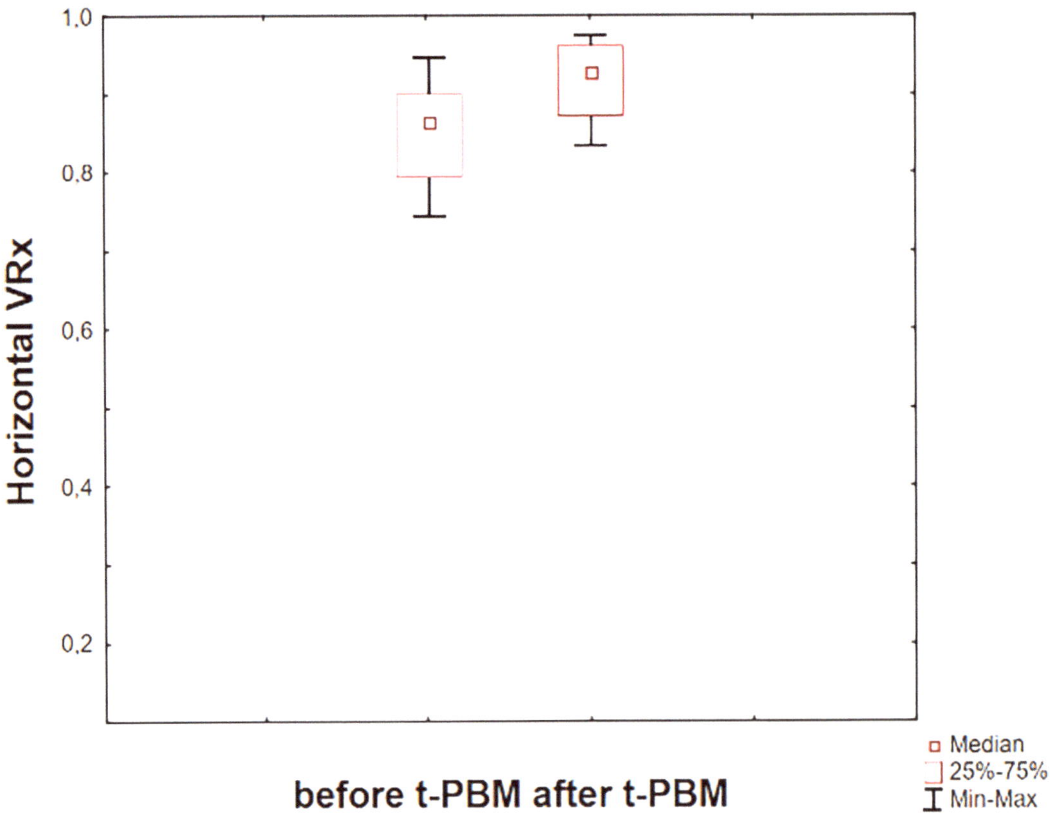

Fig. 20.2 Comparison of horizontal VRx before and after tPBM

20.4 Discussion

We found an improvement in the conjugation of eye movements in response to tPBM in healthy volunteers. Both vergence indices significantly increased after tPBM. We believe there are several reasons for this effect. As previously shown, NIR increases CCO-related metabolism and NO-related cerebral vasodilation, leading to reactive cerebral hyperemia development and tissue oxygenation increase in the tPBM area, which might improve neuronal function and connectivity (Wang et al. 2016).

Although the depth of penetration of NIR in the brain is limited to several centimeters (Tedford et al. 2015), it appears quite enough since, according to some data, the structure involved in the regulation of oculomotor synergies is the dorsomedial prefrontal cortex (dmPFC), which is located superficially and in the tPBM zone (Abubshait et al. 2023). Recent studies have revealed its involvement in controlling saccades, which confirms our assumptions. In addition, the dmPFC is part of the default mode network (DMN), the operation of which remains poorly defined (Fateh et al. 2023). This confirms previous suggestion that tPBM may have more significant effects than just metabolic and hemodynamic effects (Lemos et al. 2016). Thus, the assumption is being discussed that PBM has a remodulating effect on the functioning of neural networks in the underlying areas of the brain, which is confirmed by data on changes in bioelectrical activity according to electroencephalography and magnetoencephalography. However, despite the promising prospects of experimental results, the use of PBM in the clinic is still sporadic and requires further research.

20.5 Conclusion

Transcranial PBM causes a significant increase in vertical and horizontal vergence indices, which indicates an improvement in eye movement conjugation in healthy volunteers.

Acknowledgments DB was supported by NIH R01 NS112808. The study was funded by RSF 24-45-00010

References

Abubshait A, Kompatsiari K, Cardellicchio P et al (2023) Modulatory effects of communicative gaze on attentional orienting are driven by dorsomedial prefrontal cortex but not right temporoparietal junction. J Cogn Neurosci 35(10):1670–1680. https://doi.org/10.1162/jocn_a_02032

Fateh AA, Huang W, Hassan M, Zhuang Y et al (2023) Default mode network connectivity and social dysfunction in children with attention deficit/hyperactivity disorder. Int J Clin Health Psychol: IJCHP 23(4): 100393. https://doi.org/10.1016/j.ijchp.2023.100393

Lapchak PA, Boitano PD (2016) Transcranial near-infrared laser therapy for stroke: how to recover from futility in the NEST-3 clinical trial. Acta Neurochir Suppl 121:7–12

Lemos J, Pereira D, Almendra L et al (2016) Distinct functional properties of the vertical and horizontal saccadic network in health and Parkinson's disease: an eye-tracking and fMRI study. Brain Res 1648 (Pt A):469–484. https://doi.org/10.1016/j.brainres.2016.07.037

Nairuz T, Sangwoo-Cho, Lee JH (2024) Photobiomodulation therapy on brain: pioneering an innovative approach to revolutionize cognitive dynamics. Cells 13(11):966. https://doi.org/10.3390/cells13110966

Tedford CE, DeLapp S, Jacques S, Anders J (2015) Quantitative analysis of transcranial and intraparenchymal light penetration in human cadaver brain tissue. Lasers Surg Med 47(4):312–322. https://doi.org/10.1002/lsm.22343

Wang X, Tian F, Soni SS, Gonzalez-Lima F, Liu H (2016) Interplay between up-regulation of cytochrome-c-oxidase and hemoglobin oxygenation induced by near-infrared laser. Sci Rep 6:30540. https://doi.org/10.1038/srep30540

Zivin JA, Sehra R, Shoshoo A et al (2014) NeuroThera (R) Efficacy and Safety Trial-3 (NEST-3): a double-blind, randomized, sham-controlled, parallel group, multicenter, pivotal study to assess the safety and efficacy of transcranial laser therapy with the NeuroThera (R) Laser System for the treatment of acute ischemic stroke within 24 h of stroke onset. Int J Stroke 9:950–955

Part VI

Brain Functional Dynamics

21 Brain-to-Brain Synchronization When Sharing Positive Emotions Based on Memories: An fNIRS Hyperscanning Study

Piroska Herpai, Sabino Guglielmini, Christoph Amstad, Julia Buff, Felix Scholkmann, and Martin Wolf

Abstract

Background: Interpersonal neural synchrony (i.e., brain-to-brain synchronization, BBS) during collaborative tasks is an emerging area of research.

The *aim* was to investigate whether sharing positive emotions through storytelling among colleagues enhances neural synchrony levels measured by functional near-infrared spectroscopy (fNIRS).

Subjects and Methods: Fifteen pairs of co-workers (age: 39.5 ± 12.2 years, range: 21–60 years) from different professional backgrounds participated in the study. Each pair participated in three measurement sessions, with simultaneous fNIRS recording (left/right temporal-parietal and prefrontal cortex) during three different tasks: puzzle completion, sharing positive memories, and a verbal reasoning task. Participants were divided into three groups based on the frequency of practicing positive memory sharing between sessions. BBS was assessed using wavelet transform coherence (frequency range: 0.015–0.15 Hz). Linear mixed model analysis with bootstrapping was employed to examine the effects of "session" and "group" on BBS, with "pair ID" treated as a random effect.

Piroska Herpai and Sabino Guglielmini shared first authorship

P. Herpai · S. Guglielmini (✉)
Biomedical Optics Research Laboratory, Department of Neonatology, University Hospital Zurich, University of Zurich, Zurich, Switzerland

Family Larsson-Rosenquist Foundation Center for Neurodevelopment, Growth and Nutrition of the Newborn, Department of Neonatology, University Hospital Zurich, University of Zurich, Zurich, Switzerland
e-mail: Sabino.Guglielmini@usz.ch

C. Amstad
Obwaldner Kantonalbank, Sarnen, Switzerland

J. Buff
Biomedical Optics Research Laboratory, Department of Neonatology, University Hospital Zurich, University of Zurich, Zurich, Switzerland

F. Scholkmann
Biomedical Optics Research Laboratory, Department of Neonatology, University Hospital Zurich, University of Zurich, Zurich, Switzerland

Institute of Complementary and Integrative Medicine, University of Bern, Bern, Switzerland, Neuroscience Center Zurich, University of Zurich and ETH Zurich, Zurich, Switzerland

Institute of Complementary and Integrative Medicine, University of Bern, Bern, Switzerland, Neuroscience Center Zurich, University of Zurich and ETH Zurich, Zurich, Switzerland

M. Wolf
Biomedical Optics Research Laboratory, Department of Neonatology, University Hospital Zurich, University of Zurich, Zurich, Switzerland

Institute of Complementary and Integrative Medicine, University of Bern, Bern, Switzerland, Neuroscience Center Zurich, University of Zurich and ETH Zurich, Zurich, Switzerland

S. C. Pias et al. (eds.), *Oxygen Transport to Tissue XLVI*, Advances in Experimental Medicine and Biology 1498, https://doi.org/10.1007/978-3-032-10389-5_21

Results: Significant increases in [O_2Hb] BBS in the left temporal-parietal region were observed between session 1 and session 3 (0.004–0.02) and between session 1 and session 2 (0.01–0.06). Groups practicing more frequently showed higher BBS levels (0.005–0.6 in [O_2Hb]; 0.009–0.06 in [tHb]).

Discussion: In our fNIRS hyperscanning study, we found that continuous sharing of personal experiences with positive emotions significantly enhances neural synchrony among co-workers, particularly in the left temporal-parietal region. More frequent practice led to higher BBS levels, suggesting the importance of regular collaborative activities, such as team-building exercises, in workplace settings. While bootstrapping was applied to improve statistical robustness, the small sample size remains a limitation, and further studies with larger sample size are needed to validate these findings.

Keywords

Functional near-infrared spectroscopy hyperscanning · fNIRS hyperscanning · Brain-to-brain synchronization · Positive emotions · Team collaboration

21.1 Introduction

In the workplace, where individuals spend a significant portion of their time, the promotion of positive social interactions is essential to create a pleasing environment. Positive social interactions are linked to improved team well-being and performance, while negative ones, such as ostracism, can inhibit work success and personal relationships (Ferris et al. 2008). Effective teamwork, characterized by dynamic information sharing and coordination, is essential for task execution and overall team performance (Schmutz et al. 2019). Engaging in social activities in the workplace has been found to improve employee well-being and reduce work-family conflict (Ilies et al. 2007).

Interpersonal neural synchrony, also known as brain-to-brain synchronization (BBS), during collaborative tasks is an emerging topic of research aimed at understanding the social interactions and teamwork (Réveillé et al. 2024). BBS refers to the phenomenon whereby the brain activity of two or more individuals becomes synchronized during social interaction, i.e., a harmonization of neural processes that reflects better communication and coordination (Xu et al. 2023). Recent advances in neuroimaging, such as functional near-infrared spectroscopy (fNIRS) hyperscanning (Scholkmann et al. 2013b), have enabled researchers to explore BBS in real-time, offering insights into how naturalistic social interactions impact neural synchronization (Guglielmini et al. 2022).

This pilot study aimed to investigate the potential of sharing positive emotions through storytelling among colleagues as a means to enhance neural synchrony levels. Specifically, we hypothesized that regular sharing of positive experiences will lead to greater BBS, particularly in brain regions associated with social cognition and emotional processing. Understanding how effective teamwork and social bonding can provide insights into strategies to improve workplace dynamics and facilitate a more supportive and collaborative environment.

21.2 Material and Methods

The study involved 15 pairs of co-workers (age: 39.5 ± 12.2 years, range: 21–60 years) from different professional backgrounds. All participants provided informed consent prior to participation in the study.

The study was conducted over three sessions, each lasting 43 min and spaced 1 month apart. During each session, pairs engaged in three cooperative tasks while fNIRS signals were recorded. In the first task, participants completed a puzzle of varying complexity for 5 min. The second task was designed to elicit positive emotions. Each participant chose two or three cards from a bundle of 81 cards with different activity symbols, which took 1 min. They then shared a positive past memory related to the symbols for 4 min, while the other participant maintained eye contact. The roles were alternated between those of the storyteller and listener. The third task involved a continuous exchange of thoughts based on a fictive case study, requiring participants to discuss personal values and solve logic-based problems through verbal interactions for 5 min.

fNIRS signals, including changes in the concentrations of O_2Hb, HHb, and total hemoglobin (i.e., [O_2Hb], [HHb], and [tHb]), were measured on the left and right prefrontal cortex (LP, RP) and the left and right temporo-parietal regions (LTP, RTP) during the three tasks and the baseline periods. During the baselines, participants were instructed to keep their eyes closed for 8 min before the first task and for 5 min after each task. The experimental paradigm and the layout of the optodes with the regions of interest (ROIs) are displayed in Fig. 21.1a and b.

During the intervals between sessions, pairs were instructed to perform the second task for 10 min each week during their lunch breaks at work (Fig. 21.1c). Participants were divided into three groups based on their practice intervals. Group 1 practiced only during the first interval, group 2 solely during the second interval, and group 3 practiced during both intervals. Additionally, participants were categorized based on their reported work relationship dynamics, but this factor was not further analyzed as no significant differences were found.

fNIRS data were preprocessed using the pipeline previously described by Guglielmini et al. (2022). Channels were averaged according to the four ROIs, resulting in a single channel in [O_2Hb], [HHb], and [tHb] for each region per participant.

BBS was assessed using the wavelet transform coherence (WTC) method, whereby the data were filtered over the frequency band 0.015–0.15 Hz in order to obtain a coherence signal over the experimental time for each pair, each fNIRS signal, and ROI.

The coherence signal was divided into three segments, corresponding to the three conditions and the baselines. The mean of the last 3 min of the first baseline was subtracted from each condition. A linear mixed model analysis with bootstrapping was employed to examine the effects of "session" and "group" on BBS, with "pair ID" included as a random effect to account for repeated measures within pairs. Baseline-corrected BBS in each ROI and the fNIRS signal was compared by splitting the analysis into two parts: comparing session 1 with session 3 to evaluate changes in BBS from the beginning to after 3 months of practice and comparing session 1 with session 2 to assess the increase in BBS after 1 month of sharing positive emotion exercises using the cards. 95% confidence intervals (CIs) were obtained using the bootstrapping method with 1000 samples and the percentile method for CI estimation.

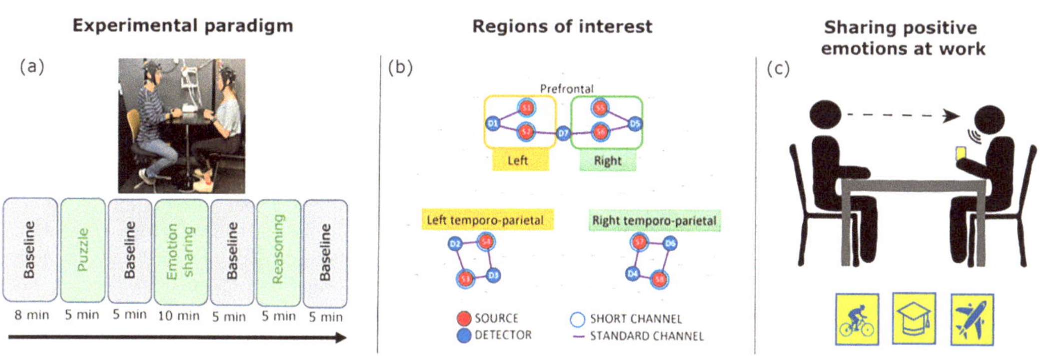

Fig. 21.1 Visualizations of the experimental paradigm (**a**), regions of interest (**b**), sharing positive emotions through storytelling at work (**c**)

21.3 Results

The BBS analysis revealed a significant increase in neural synchrony in the LTP region. The results are visualized in Fig. 21.2, where each fNIRS signal is displayed separately for the four ROIs. Statistically significant effects, indicated by confidence intervals that do not include 0, are marked with an asterisk.

Comparing session 1 with session 3, we observed a significant increase in [O_2Hb] BBS in the LTP region (Fig. 21.2a), with values ranging from 0.004 to 0.02. Similarly, when comparing session 1 with session 2, an increase in

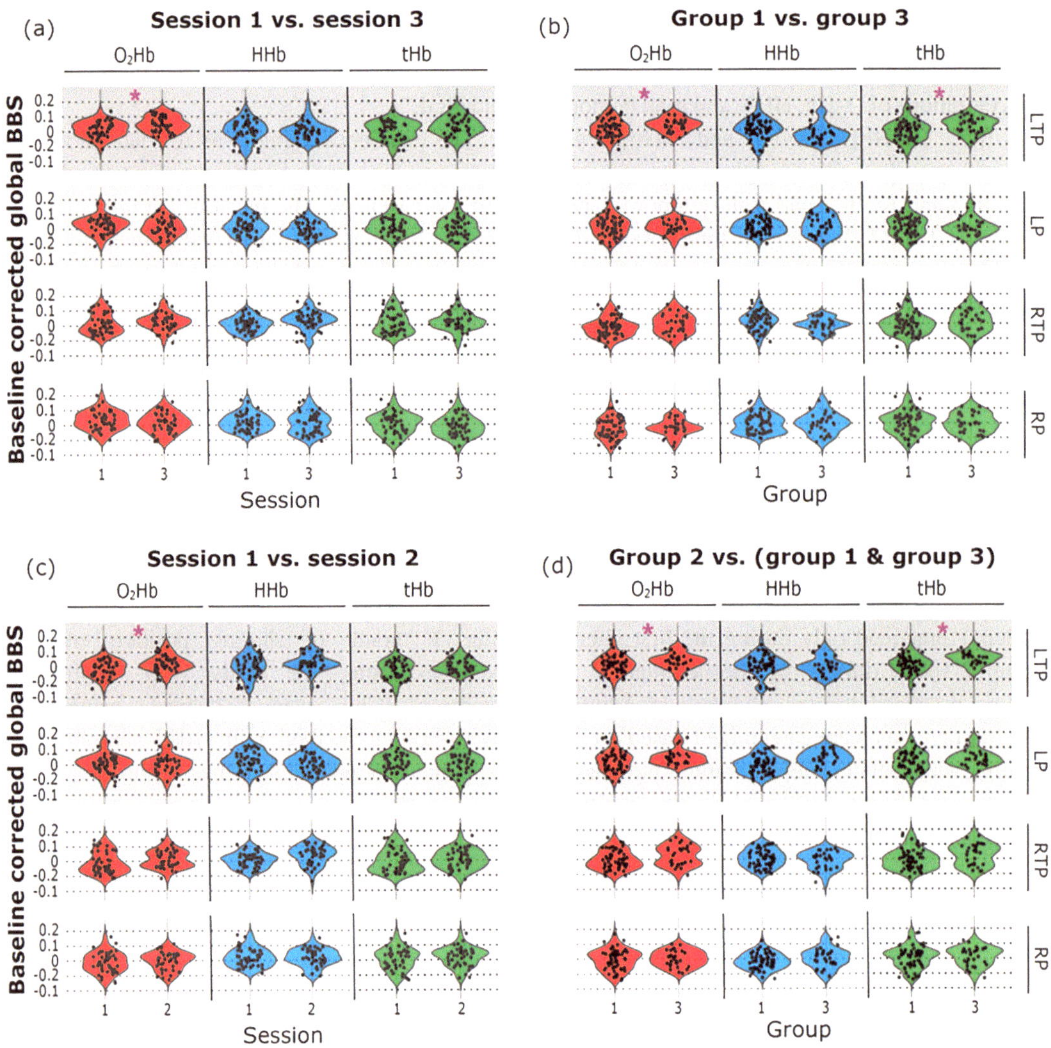

Fig. 21.2 Global brain-to-brain synchronization (BBS) of the 15 co-worker pairs during the three cooperative tasks corrected with the mean coherence of the last 3 min of the first baseline in [O_2Hb], [HHb], and [tHb] displayed separately for the four ROIs (left and right prefrontal cortex (LP, RP) and the left and right temporo-parietal regions (LTP, RTP)) (comparison between session 1 and session 3, (**a**), and between session 1 and session 3, (**c**)). (**b**) Comparison of the global BBS of the group who practiced only during the first interval (group 1) and during both intervals (group 3). (**d**) Comparison of the global BBS of the group that still has not practiced at work (group 2) and the two groups that practiced during the first interval (group 1 and group 3). Asterisks in magenta indicate statistically significant differences

[O_2Hb] BBS was observed (Fig. 21.2c), with values ranging from 0.01 to 0.06. These findings suggest that the collaborative activities and sharing of positive emotions had a significant impact on enhancing neural synchrony over time.

Additionally, the extent of practice significantly influenced BBS levels in the LTP region. The group of pairs who practiced more frequently showed a statistically significant higher BBS level. Specifically, group 3, which practiced during both intervals, exhibited significantly higher BBS compared to group 1 (Fig. 21.2b), with [O_2Hb] levels ranging from 0.01 to 0.5 and [tHb] levels from 0.009 to 0.06. Furthermore, when comparing session 1 with session 2, both group 1 and group 3 exhibited higher BBS compared to group 2 Fig. 21.2d, with [O_2Hb] levels ranging from 0.005 to 0.6 and [tHb] levels from 0.009 to 0.06.

21.4 Discussion, Conclusions, and Outlook

The findings of this pilot study indicate that increased practice and engagement in shared emotional exercises, using the cards, resulted in enhanced neural synchronization between participants, particularly in the LTP region. The observed increases in BBS indicate that the continuous sharing of personal experiences with positive emotions is a crucial factor in the development of neural synchrony among co workers. Our results demonstrate that the more frequently pairs engaged in these positive emotional sharing exercises, the higher their BBS levels were. This was particularly evident in group 3, which engaged in these exercises during both intervals and exhibited significantly greater BBS than groups 1 and 2. These findings are consistent with previous research indicating that positive social interactions can enhance neural coupling and facilitate more effective communication and cooperation within teams (Kinreich et al. 2017).

With regard to the physiological interpretation of the BBS values that were measured with fNIRS hyperscanning, it should be borne in mind that our basic assumption was that the BBS values are based directly and primarily on neuronal activity. However, it is important to note that deviations from this assumption (i.e., the influence of physiological factors that are not caused by neurovascular coupling) should not be completely ruled out (Scholkmann et al. 2013a, 2022; Tachtsidis and Scholkmann 2016).

Bootstrapping was applied to enhance statistical robustness, but the small sample size remains a limitation. While results indicate strong effects of positive memory sharing on neural synchrony, larger studies are needed to confirm and expand these findings.

The insights from our study have practical implications for workplace dynamics and team-building strategies. Organizations may derive benefit from the implementation of regular interaction tasks and positive emotional sharing in their routines. Team-building exercises that promote positive social interactions and emotional sharing could enhance neural synchrony, thereby improving team cohesion, communication, and overall job satisfaction. Further research could build upon the findings of this pilot study by examining the long-term effects of regular positive emotional sharing on neural synchrony and team performance.

Acknowledgments We thank Prof. Jochen Menges of the University of Zurich for his kind and insightful support, as well as all the participants for their involvement.

We gratefully acknowledge funding by UZH Candoc/Postdoc Grant 2024 (grant number FK-24-050).

References

Ferris DL, Brown DJ, Berry JW, Lian H (2008) The development and validation of the workplace ostracism scale. J Appl Psychol 93:1348–1366

Guglielmini S, Bopp G, Marcar VL, Scholkmann F, Wolf M (2022) Systemic physiology augmented functional near-infrared spectroscopy hyperscanning: a first evaluation investigating entrainment of spontaneous activity of brain and body physiology between subjects. Neurophotonics 9:026601

Ilies R, Schwind KM, Wagner DT, Johnson MD, Derue DS, Ilgen DR (2007) When can employees have a family life? The effects of daily workload and affect on work-family conflict and social behaviors at home. J Appl Psychol 92:1368–1379

Kinreich S, Djalovski A, Kraus L, Louzoun Y, Feldman R (2017) Brain-to-brain synchrony during naturalistic social interactions. Sci Rep 7:17060

Réveillé C, Vergotte G, Perrey S, Bosselut G (2024) Using interbrain synchrony to study teamwork: a systematic review and meta-analysis. Neurosci Biobehav Rev 159:105593

Schmutz JB, Meier LL, Manser T (2019) How effective is teamwork really? The relationship between teamwork and performance in healthcare teams: a systematic review and meta-analysis. BMJ Open 9:e028280

Scholkmann F, Gerber U, Wolf M, Wolf U (2013a) End-tidal CO_2: an important parameter for a correct interpretation in functional brain studies using speech tasks. Neuroimage 66:71–79

Scholkmann F, Holper L, Wolf U, Wolf M (2013b) A new methodical approach in neuroscience: assessing inter-personal brain coupling using functional near-infrared imaging (fNIRI) hyperscanning. Front Hum Neurosci 7:813

Scholkmann F, Tachtsidis I, Wolf M, Wolf U (2022) Systemic physiology augmented functional near-infrared spectroscopy: a powerful approach to study the embodied human brain. Neurophotonics 9:030801

Tachtsidis I, Scholkmann F (2016) False positives and false negatives in functional near-infrared spectroscopy: issues, challenges, and the way forward. Neurophotonics 3:031405

Xu M, Morimoto S, Hoshino E, Suzuki K, Minagawa Y (2023) Two-in-one system and behavior-specific brain synchrony during goal-free cooperative creation: an analytical approach combining automated behavioral classification and the event-related generalized linear model. Neurophotonics 10:013511

22 Unlocking Unique Physiological Reactions: Novel Insights into Four Cerebral and Systemic Physiological Responses to Colored Light

Hamoon Zohdi, Sarusan Jegatheeswaran, Felix Scholkmann, and Ursula Wolf

Abstract

Background: For several years, our research group has been systematically investigating how changes in cerebral hemodynamics as well as systemic physiology can be induced by colored light exposure (CLE). We have observed that these physiological changes can differ between subjects for exactly the same CLE stimulus. In the present case series, we report specific physiological reactivity patterns from four subjects. *Methods:* A total of 36 subjects participated in the study. These subjects were exposed to blue and red light while performing a 2-back task. For the present analysis, we selected four subjects as representative examples of distinct response patterns. We simultaneously measured changes in cerebral and systemic physiological parameters using the systemic physiology augmented functional near-infrared spectroscopy (SPA-fNIRS) method. The SPA-fNIRS method included oxygenated hemoglobin ([O_2Hb]) and deoxygenated hemoglobin ([HHb]) in the prefrontal cortex, mean arterial pressure (MAP), end-tidal carbon dioxide ($P_{ET}CO_2$), skin conductance level (SCL), and respiration rate (RR). *Results:* We identified four distinct patterns of physiological responses. Pattern 1: For all parameters except MAP, the subject's response to blue and red light remained consistent. Pattern 2: Similar to the previous case, there was no significant contrast between red and blue with respect to changes in cerebral hemodynamics. Similar patterns were also detected for $P_{ET}CO_2$ and MAP, whereas completely different patterns were found for SCL. Surprisingly, an inverse correlation was found for RR. Pattern 3: Changes in most parameters exhibited mirrored effects. Consistent trends were observed for SCL and RR. Pattern 4: Mirrored patterns were observed for [O_2Hb] and [HHb], the same patterns for $P_{ET}CO_2$ and RR, and divergent trends for MAP and SCL. *Conclusions:* In our analysis of SPA-fNIRS measurements, we identified four subjects with specific physiological reactivity patterns to a combined CLE and cognitive task. Individual physiological reactivity or physiological constitution should

Hamoon Zohdi and Sarusan Jegatheeswaran shared the first authorship.

H. Zohdi (✉) · S. Jegatheeswaran · U. Wolf
Institute of Complementary and Integrative Medicine, University of Bern, Bern, Switzerland
e-mail: hamoon.zohdi@unibe.ch

F. Scholkmann
Institute of Complementary and Integrative Medicine, University of Bern, Bern, Switzerland

Biomedical Optics Research Laboratory, Department of Neonatology, University Hospital Zurich, University of Zurich, Zurich, Switzerland

Neuroscience Center Zurich, University of Zurich and ETH Zurich, Zurich, Switzerland

S. C. Pias et al. (eds.), *Oxygen Transport to Tissue XLVI*, Advances in Experimental Medicine and Biology 1498, https://doi.org/10.1007/978-3-032-10389-5_22

be taken into account when analyzing physiological data on CLE to enable a more adequate interpretation of the results.

Keywords

Colored light exposure · Blue and red light · SPA-fNIRS · Four distinct patterns · Individual physiological reactivity · Cerebral hemodynamics · Systemic physiology

22.1 Introduction

Light has a profound effect on human physiology and behavior, extending beyond mere visibility to affect circadian rhythms, mood, and cognitive functions (Vandewalle et al. 2009; Vetter et al. 2022; Yuan et al. 2021). In particular, colored light has been demonstrated to evoke a range of physiological responses. However, emerging evidence suggests that these responses are not uniform across individuals (Peeters et al. 2020; Spitschan and Santhi 2022; Zohdi et al. 2021a). In our previous studies (Zohdi et al. 2021a, b, 2022; Scholkmann et al. 2022a), employing systemic physiology augmented functional near-infrared spectroscopy (SPA-fNIRS) (Scholkmann et al. 2022b), we have also observed that physiological changes can be different between subjects for exactly the same-colored light exposure (CLE) stimulus, highlighting the importance of detecting such individual differences in cerebral hemodynamic and systemic physiological responses.

Understanding these individual differences is crucial for a number of reasons. First, it enables personalized lighting solutions that can optimize well-being and performance in various settings, from workplaces to healthcare environments. Second, it provides insights into the mechanisms underlying light perception and its broader impacts on human health. By studying how different individuals react to colored light, researchers and clinicians can develop tailored interventions for issues ranging from sleep disorders to mood regulation, advancing both scientific knowledge and practical applications in lighting design.

This chapter explores the variability in individual responses to colored light, examining the cerebral and systemic physiological dimensions that influence these differences. In this case series, we report specific physiological reactivity patterns from four subjects.

22.2 Subjects and Methods

Thirty-six healthy right-handed subjects were exposed to blue and red light on two different days in a randomized crossover study design. During the CLE, subjects were asked to perform a 2-back task at specified intervals according to our previously published experimental protocol (Zohdi et al. 2024). In a 2-back task, a series of consonants was presented auditorily, and the subjects had to respond as quickly as possible by pressing a key each time the current letter was identical to the one presented two positions back in the sequence. The subjects were initially exposed to 7 min of darkness, followed by three sessions of the CLE/2-back task, each lasting 2 min, with a 30-s rest interval between the tasks. The subjects were seated in an upright position with their feet on the floor, facing a white wall. The light exposure was provided by six stage lights (Rush PAR 2 RGBW Zoom, Martin by HARMAN, Arhus, Denmark) with an illuminance of 120 lux.

Changes in the concentration of oxygenated hemoglobin ([O_2Hb]) and deoxygenated hemoglobin ([HHb]) were measured over the prefrontal cortex (PFC) using frequency-domain near-infrared spectroscopy (Imagent, ISS Inc., Champaign, IL, USA). Systemic physiological parameters, including mean arterial pressure (MAP), end-tidal carbon dioxide ($P_{ET}CO_2$), skin conductance level (SCL), and respiration rate (RR), were measured concurrently.

Data analysis, signal processing, and statistical analysis were performed using MATLAB (R2023b, MathWorks, Inc., Natick, MA, USA). Data were manually inspected for quality, completeness, and reliability. For the present analysis, we selected four subjects as representative examples of distinct response patterns. All signals, except SCL, were low-pass filtered using a robust second-degree polynomial moving average with a window length of 1 min. SCL data were processed using the Ledalab toolbox (Benedek and Kaernbach 2010). All signals were normalized to the last 5 min of the baseline period (darkness) by subtracting the average baseline values, allowing us to assess changes relative to the resting state before exposure. Finally, the three sessions of task (CLE/2-back) periods for each color exposure were extracted and averaged.

Pearson's correlation coefficients were calculated to examine the relationship between changes in cerebral and systemic physiological parameters during the CLE for blue versus red at the individual level for each case. The statistical significance level was set to 0.05 in all comparisons ($p < 0.05$).

22.3 Results

Figures 22.1 and 22.2 show the time series of changes in cerebral and systemic physiological parameters during the CLE for blue and red at the individual level for each case. Subjects exhibited unique response patterns: These included consistent responses regardless of light color (case 1), minimal differences between light colors (case 2), mirrored effects across parameters (case 3), and divergent trends (case 4). To more effectively examine the relationship between changes in all parameters for blue versus red, scatter plots were created and correlation coefficients were calculated for all parameters in each case (Fig. 22.3).

Pattern 1: With the exception of MAP, the subject's response to blue and red light remained consistent across all other variables: O_2Hb: $r = 0.94$, $p < 0.001$; HHb: $r = 0.89$, $p < 0.001$; $P_{ET}CO_2$: $r = 0.79$, $p = 0.002$; MAP: $r = 0.10$, $p = 0.766$; SCL: $r = 0.95$, $p < 0.001$; RR: $r = 0.88$, $p < 0.001$. Pattern 2: Similar to the previous case, there was no notable contrast between red and blue with respect to changes in cerebral hemodynamics (O_2Hb: $r = 0.95$, $p < 0.001$; HHb: $r = 0.88$, $p < 0.001$). Similar patterns were also detected for $P_{ET}CO_2$ ($r = 0.90$, $p < 0.001$) and MAP ($r = 0.73$, $p = 0.007$), whereas completely different patterns were found for SCL ($r = -0.16$, $p = 0.62$). Surprisingly, an inverse correlation was noted in RR ($r = -0.70$, $p = 0.011$). Pattern 3: Although not all parameters exhibit significant correlations, the majority of parameters showed mirroring effects (Fig. 22.2), which changes increasing with blue light exposure and decreasing with red light, and vice versa (O_2Hb: $r = -0.012$, $p = 0.971$; HHb: $r = -0.88$, $p < 0.001$; $P_{ET}CO_2$: $r = -0.52$, $p = 0.082$; MAP: $r = -0.41$, $p = 0.180$). Consistent trends were observed for SCL and RR (SCL: $r = 0.74$, $p = 0.006$; RR: $r = 0.73$, $p = 0.008$). Pattern 4: Mirror-like patterns were evident for O_2Hb and HHb (O_2Hb: $r = -0.72$, $p = 0.008$; HHb: $r = -0.71$, $p = 0.009$), with the same patterns observed for $P_{ET}CO_2$ ($r = 0.99$, $p < 0.001$). In contrast, divergent trends were evident for MAP ($r = -0.17$, $p = 0.592$) and SCL ($r = 0.34$, $p = 0.276$). The same patterns and mirror-like patterns were observed at the beginning and the end of the task, respectively, for RR.

The accuracy levels of the 2-back task performance for subjects 1, 2, 3, and 4 were 95%, 93%, 95%, and 84%, respectively.

22.4 Discussion and Conclusions

In this study, we report the cerebral and systemic physiological responses of four subjects to colored light exposure, namely red and blue light. Our study identified four distinct patterns of physiological reactivity in response to these two colored lights. A number of factors, including the individual physiological constitution, the emotional state, and environmental variables, may contribute to differential reactions to colored light (Zohdi et al. 2021a, 2020; Chellappa et al. 2014; Chellappa 2021; Lok et al. 2018).

Genetic differences have an influence on the individual variability of the physiological reaction to light. Variations in genes related to the

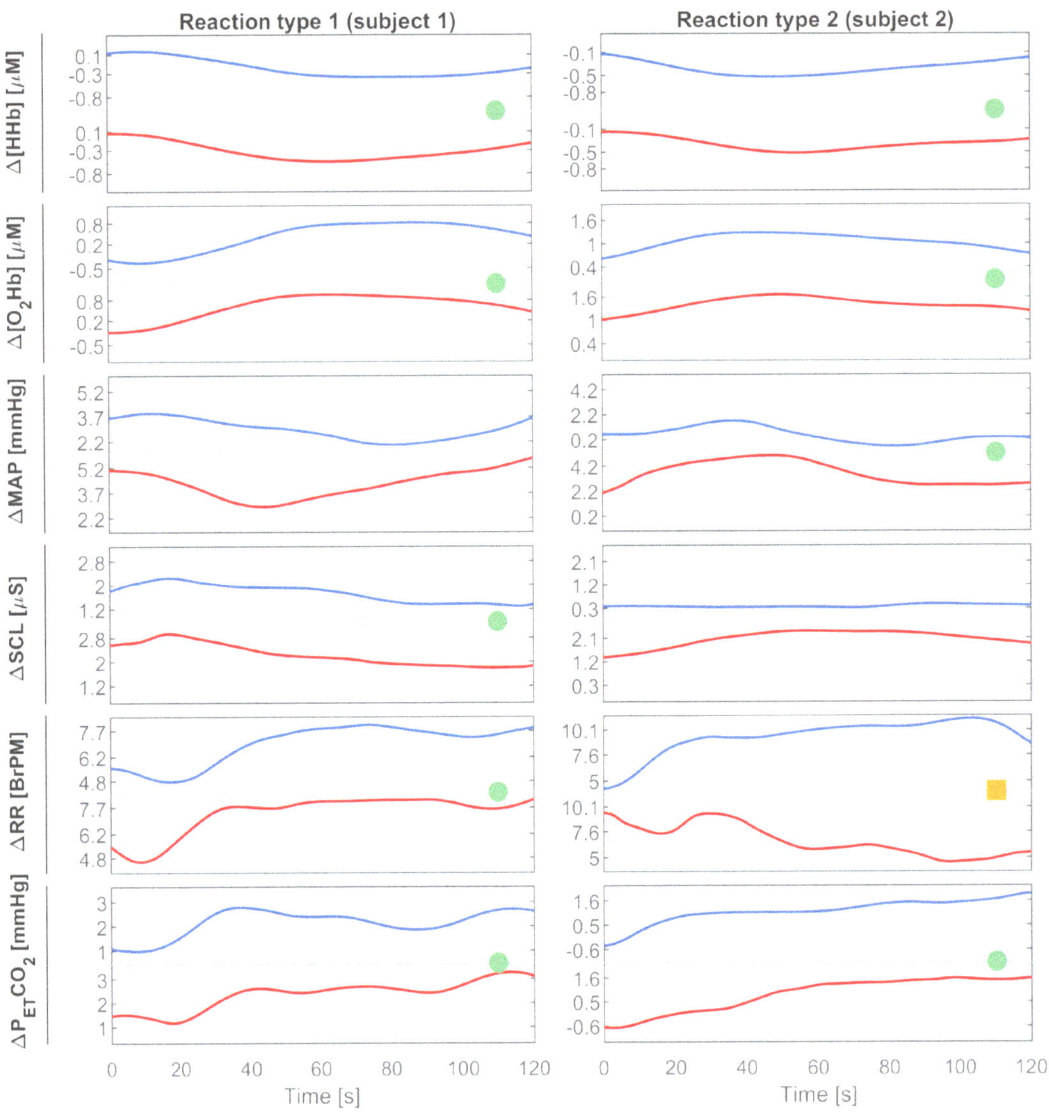

Fig. 22.1 Changes in cerebral and systemic physiological parameters during the CLE for blue and red light exposure at the individual level for subjects 1 and 2. Green circles: similar patterns between red and blue light exposures. An orange squares: mirror-like (inverse) patterns between red and blue light exposures

circadian clock can influence how individuals respond to light exposure, affecting their sleep patterns, mood, and cognitive functions (Chellappa 2021; Partonen 2012; Charrier et al. 2017). Moreover, the emotional state of each individual, prior experiences with light, and personal preferences can shape how one reacts to light exposure, influencing both physiological and psychological outcomes (Zohdi et al. 2021a; Roenneberg et al. 2004). Finally, the environment and certain situational variables in which light exposure occurs can significantly affect individual responses. Factors such as seasonal changes, time of day, temperature, humidity, and the context in which the light is experienced can modulate how light influences cerebral and systemic physiological processes (Zohdi et al. 2021a, 2020).

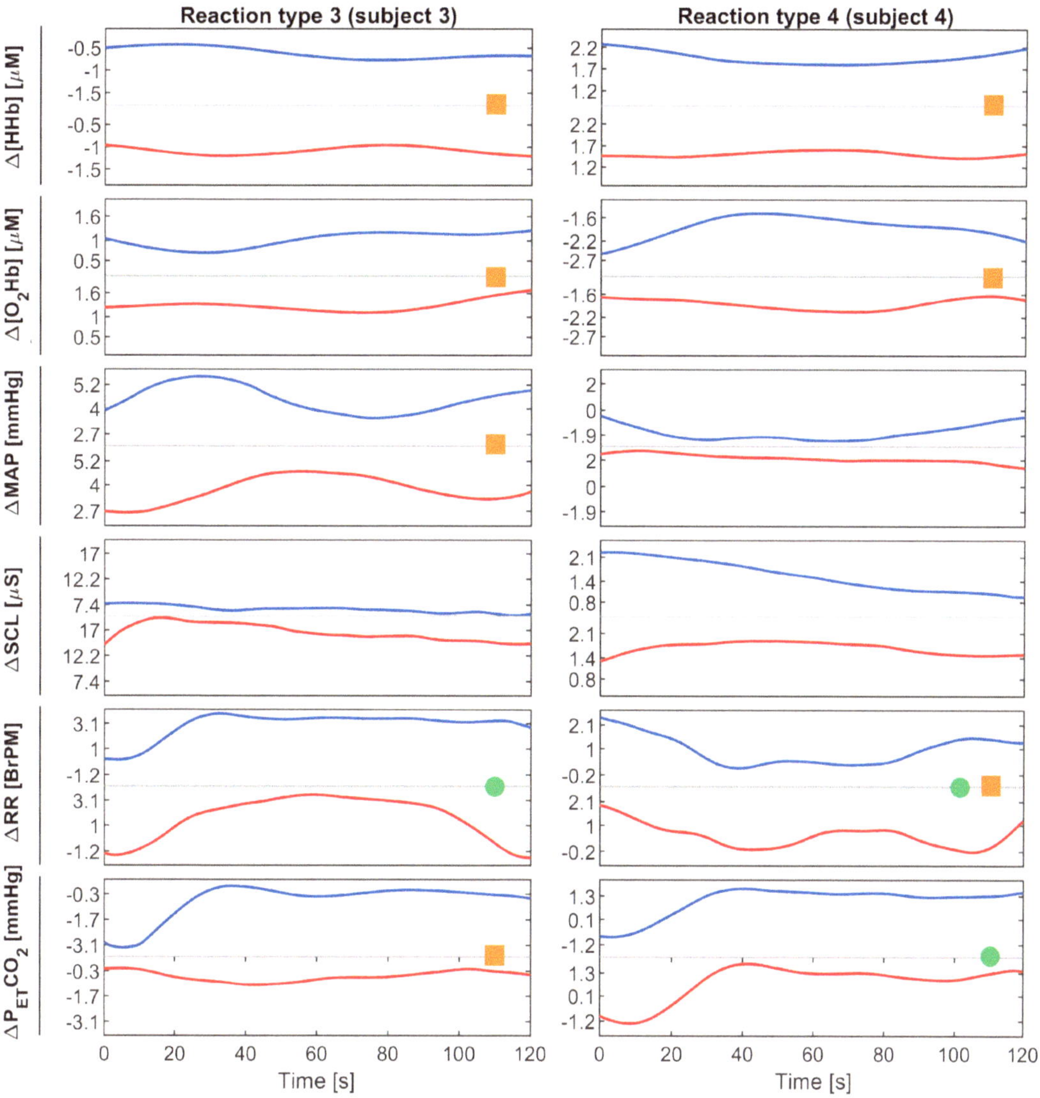

Fig. 22.2 Changes in cerebral and systemic physiological parameters during the CLE for blue and red light exposure at the individual level for subjects 3 and 4. Green circles: similar patterns between red and blue light exposures. Orange squares: mirror-like (inverse) patterns between red and blue light exposures

In our analysis, we identified four subjects with specific physiological reactivity patterns to CLE. Our findings are important for a better understanding of physiological changes after CLE and to show that the individual physiological reactivity or physiological constitution must be taken into account.

Our findings are also of significance for fNIRS neuroimaging studies in general since they highlight that the often-accompanying systemic physiological changes can be different (in magnitude and sign) in individuals despite taking part in the same experimental paradigm. Since systemic physiological changes can impact fNIRS measurements (Scholkmann et al. 2022b; Tachtsidis and Scholkmann 2016; Caldwell et al. 2016), this insight is important to be considered in

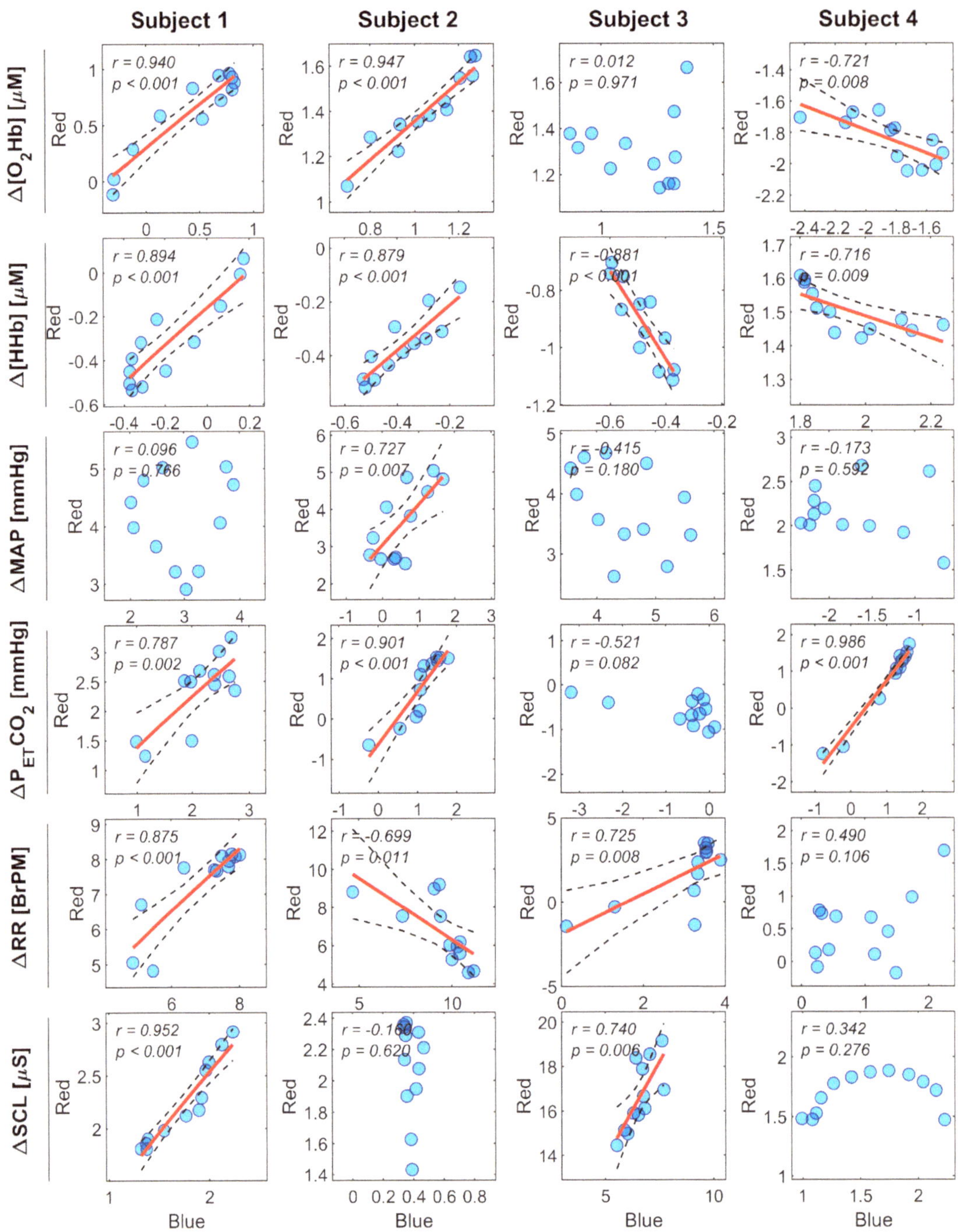

Fig. 22.3 Scatter plots displaying changes in cerebral and systemic physiological parameters during the CLE for blue vs. red at the individual level for each case. The linear fit is presented for pairs with a significant correlation (red lines). The areas between dashed lines show 95% confidence intervals

planning the study and interpreting the study results.

To address the issue of individual differences in physiological reactivity, future research should adopt a personalized approach to studying the effects of colored light. This could involve genetic profiling, as well as emotional, mood, and chronotype assessments while controlling external factors (i.e., environmental conditions) that might influence the results. By considering these approaches, researchers can develop more effective and individualized lighting strategies that cater to the unique needs of each person, thereby enhancing well-being and performance as well as optimizing lighting environments for health and productivity.

References

Benedek M, Kaernbach C (2010) A continuous measure of phasic electrodermal activity. J Neurosci Methods 190: 80–91

Caldwell M, Scholkmann F, Wolf U et al (2016) Modelling confounding effects from extracerebral contamination and systemic factors on functional near-infrared spectroscopy. NeuroImage 143:91–105

Charrier A, Olliac B, Roubertoux P, Tordjman S (2017) Clock genes and altered sleep–wake rhythms: their role in the development of psychiatric disorders. Int J Mol Sci 18:938

Chellappa SL (2021) Individual differences in light sensitivity affect sleep and circadian rhythms. Sleep 44: zsaa214

Chellappa SL, Viola AU, Schmidt C et al (2014) Light modulation of human sleep depends on a polymorphism in the clock gene Period3. Behav Brain Res 271:23–29

Lok R, Woelders T, Gordijn MCM et al (2018) White light during daytime does not improve alertness in well-rested individuals. J Biol Rhythm 33:637–648

Partonen T (2012) Clock gene variants in mood and anxiety disorders. J Neural Transm 119:1133–1145

Peeters ST, Smolders KCHJ, de Kort YAW (2020) What you set is (not) what you get: how a light intervention in the field translates to personal light exposure. Build Environ 185:107288

Roenneberg T, Kuehnle T, Pramstaller PP et al (2004) A marker for the end of adolescence. Curr Biol 14: R1038–R1039

Scholkmann F, Zohdi H, Wolf U (2022a) The role of systemic physiology in individual hemodynamic responses measured on the head due to long-term stimulation involving colored light exposure and a cognitive task: an SPA-fNIRS study. Brain Sci 12:597

Scholkmann F, Tachtsidis I, Wolf M, Wolf U (2022b) Systemic physiology augmented functional near-infrared spectroscopy: a powerful approach to study the embodied human brain. Neurophotonics 9:030801

Spitschan M, Santhi N (2022) Individual differences and diversity in human physiological responses to light. EBioMedicine 75:103640

Tachtsidis I, Scholkmann F (2016) False positives and false negatives in functional near-infrared spectroscopy: issues, challenges, and the way forward. Neurophotonics 3:031405

Vandewalle G, Maquet P, Dijk DJ (2009) Light as a modulator of cognitive brain function. Trends Cogn Sci 13:429–438

Vetter C, Pattison PM, Houser K et al (2022) A review of human physiological responses to light: implications for the development of integrative lighting solutions. Leukos 18:387–414

Yuan Y, Li G, Ren H, Chen W (2021) Effect of light on cognitive function during a stroop task using functional near-infrared spectroscopy. Phenomics 1:54–61

Zohdi H, Scholkmann F, Wolf U (2020) Frontal cerebral oxygenation asymmetry: Intersubject variability and dependence on systemic physiology, season, and time of day. Neurophotonics 7:025006

Zohdi H, Scholkmann F, Wolf U (2021a) Individual differences in hemodynamic responses measured on the head due to a long-term stimulation involving colored light exposure and a cognitive task: a SPA-fNIRS study. Brain Sci 11:54

Zohdi H, Scholkmann F, Wolf U (2021b) Long-term blue light exposure changes frontal and occipital cerebral hemodynamics: not all subjects react the same. Adv Exp Med Biol 1269:217–222

Zohdi H, Natale L, Scholkmann F, Wolf U (2022) Intersubject variability in cerebrovascular hemodynamics and systemic physiology during a verbal fluency task under colored light exposure: clustering of subjects by unsupervised machine learning. Brain Sci 12:1449

Zohdi H, Märki J, Scholkmann F, Wolf U (2024) Cerebral, systemic physiological and behavioral responses to colored light exposure during a cognitive task: a SPA-fNIRS study. Behav Brain Res 462:114884

Long-Term GABAergic Modulation of Brain Tissue Oxygenation and Electrophysiological Activity Following Neonatal Anesthesia Exposure

23

E. D. Doubovikov, N. A. Serdyukova, and D. P. Aksenov

Abstract

Millions of children are exposed to general anesthesia annually, raising concerns about its impact on neuronal development, subsequent behavioral impairments, and the conditions under which these functional deficiencies can be detected instrumentally. In this study, we investigated the delayed consequences of administering sevoflurane, a widely used anesthetic, to newborn rabbits to mimic repeated clinical anesthesia exposure. When these rabbits reached adolescence/adulthood, we implanted electrodes into their cerebral cortex to record electrophysiology and tissue oxygen. Tissue oxygen recording was chosen to show translational changes in hemodynamics, which can be detected by functional MRI. Despite observing no differences in baseline oxygen between the group exposed to neonatal anesthesia and the control group, subsequent administration of GABA antagonists in small concentrations near the oxygen electrodes revealed significant findings. While the local administration of the GABA antagonist picrotoxin resulted in a noticeable shift in oxygen oscillations to lower frequencies in both groups, the rabbits exposed to neonatal anesthesia also generated large and sustained response curves associated with epileptic activity. This study suggests the specific effects of neonatal anesthesia in inducing a state of susceptibility to epileptic activity, which can be triggered by specific and mild challenges (e.g., the application of picrotoxin in small concentrations). We generalize these findings to suggest that this mechanism can explain the inconsistency of clinical manifestations in populations exposed to neonatal anesthesia, as the anesthesia alone may not be sufficient for clinical symptoms to appear; an additional trigger or challenge is required.

Keywords

Neonatal anesthesia · GABAergic system · Latent epilepsy · Brain tissue oxygen · Picrotoxin

E. D. Doubovikov
Department of Radiology, NorthShore University HealthSystem, Evanston, IL, USA

N. A. Serdyukova
Department of Pediatrics, NorthShore University HealthSystem, Evanston, IL, USA

D. P. Aksenov (✉)
Department of Radiology, NorthShore University HealthSystem, Evanston, IL, USA

Department of Anesthesiology, NorthShore University HealthSystem, Evanston, IL, USA

University of Chicago, Pritzker School of Medicine, Chicago, IL, USA

Department of Biomedical Engineering, Northwestern University, Evanston, IL, USA
e-mail: daksenov@northshore.org

S. C. Pias et al. (eds.), *Oxygen Transport to Tissue XLVI*, Advances in Experimental Medicine and Biology 1498, https://doi.org/10.1007/978-3-032-10389-5_23

23.1 Introduction

Concerns about the long-term behavioral effects of neonatal anesthesia are growing (Aksenov et al. 2023a; Borzage and Peterson 2024; Robinson et al. 2022). Research has shown that neonatal anesthesia can impact learning in later age, with effects varying depending on the cumulative dose of general anesthetics such as sevoflurane, isoflurane, and propofol. Human studies present a variable picture, with some showing behavioral deficiency and others not (Aksenov 2024). Moreover, it remains unclear how neuronal or neurovascular markers of these functional deficiencies can be detected instrumentally. Previous studies (Aksenov et al. 2020; Taghon et al. 2015; Neudecker et al. 2023) have shown that functional MRI (fMRI) has inconsistently detected these effects, and the reasons for this variability remain unclear.

We propose a plausible explanation for this inconsistency. Numerous studies have demonstrated that neonatal anesthesia induces apoptosis, which selectively affects certain neurons, particularly subpopulations of GABAergic interneurons (Gascoigne et al. 2021; Zhou et al. 2021). As the number of these interneurons decreases, those remaining attempt to compensate by forming additional synapses. This compensation is effective only up to a certain limit. Consequently, children exposed to neonatal anesthesia may appear functionally identical to non-exposed children under normal conditions. However, under stress, overstimulation, excessive workload, and other challenges, this compensation fails, and the differences between exposed and non-exposed children become apparent.

We hypothesize that if interneurons are affected by neonatal anesthesia, and if functional deficiency of interneurons is a cause of epileptic activity (Mueller-Buehl et al. 2023), then neonatal anesthesia may result later in a state of "hidden" (clinically and electrophysiologically silent) epilepsy. We term it "hidden" because it can only be revealed by specific challenges. To expose the functional deficiency of the inhibitory system, we employed local injections of the GABA-antagonist picrotoxin (PTX) at a concentration that does not induce epileptic activity in the control group (Aksenov et al. 2023b).

Since it is known that fMRI provides better volume resolution compared to EEG, and we have shown that brain tissue oxygen (PO_2) and fMRI follow each other (Aksenov et al. 2023c), we simultaneously recorded local field potentials as a marker of neuronal postsynaptic activity and PO_2 fluctuations as a marker of hemodynamics.

23.2 Methods

Animal preparation. Dutch-belted rabbits ($N = 10$) were used in accordance with the National Institutes of Health guidelines and NorthShore University HealthSystem Research Institute Institutional Animal Care and Use Committee-approved protocol. Infant rabbits underwent sevoflurane exposures ($N = 5$) on days 8, 11, and 14 for 2 h each following a previously described approach (Aksenov et al. 2020). At the age of 3 months, animals were implanted with restraining headbolts and a microdrive containing electrodes and an injection cannula (Aksenov et al. 2019). After 1 week of recovery from surgery, each subject was habituated for 3–5 days. The electrophysiological signals from the microwires were fed through a miniature preamplifier to a multi-channel differential amplifier system (Neuralynx, Inc., Bozeman, MT, USA). The signals were amplified, band-pass filtered (0.3–3 kHz for single units and 1–150 Hz for local field potentials (LFP)), and digitized (32 kHz/channel) using a Neuralynx data acquisition system. Single units were monitored to ensure that the volume effect was minimized (Aksenov et al. 2014).

For oxygen recording, microwires were polarized to −0.7 V with respect to a reference electrode (located between the skull and dura), and the current was measured with a Keithley model 614 electrometer (Keithley Instruments, Cleveland, OH, USA). The output voltage from the electrometer was low-pass filtered at 30 Hz, amplified, and digitized at 2 Hz. LFP were recorded simultaneously with PO_2 recordings.

Epileptic events were recorded for rabbits and were determined from LFP measurements. Picrotoxin (50 μM) was dissolved in artificial cerebrospinal fluid (ACSF) for injection. The concentration of picrotoxin (PTX) was based on our previous studies which showed this concentration does not cause epileptic activity (Aksenov et al. 2023b).

23.3 Results

For rabbits exposed to anesthesia as neonates, a small concentration of picrotoxin (PTX) induces focal seizures, as highlighted in Fig. 23.1. Figure 23.1a and b shows the drastic change in the LFP trace of a neonatal anesthesia-exposed rabbit after the injection of PTX. Figure 23.1a displays a representative 2-min LFP trace before injection, and Fig. 23.1b displays a 5-min LFP trace after injection, containing two major epileptic events. These major epileptic events are characterized by highly frequent large deviations in LFP from the baseline, followed by a refractory period with minimal LFP fluctuations. The increased magnitude of LFP fluctuations is evident in the LFP spectrograms (Fig. 23.1c and d). Figure 23.1c shows the LFP spectrogram for a rabbit not exposed to neonatal anesthesia. After PTX injection at 15 min, the effect on the power

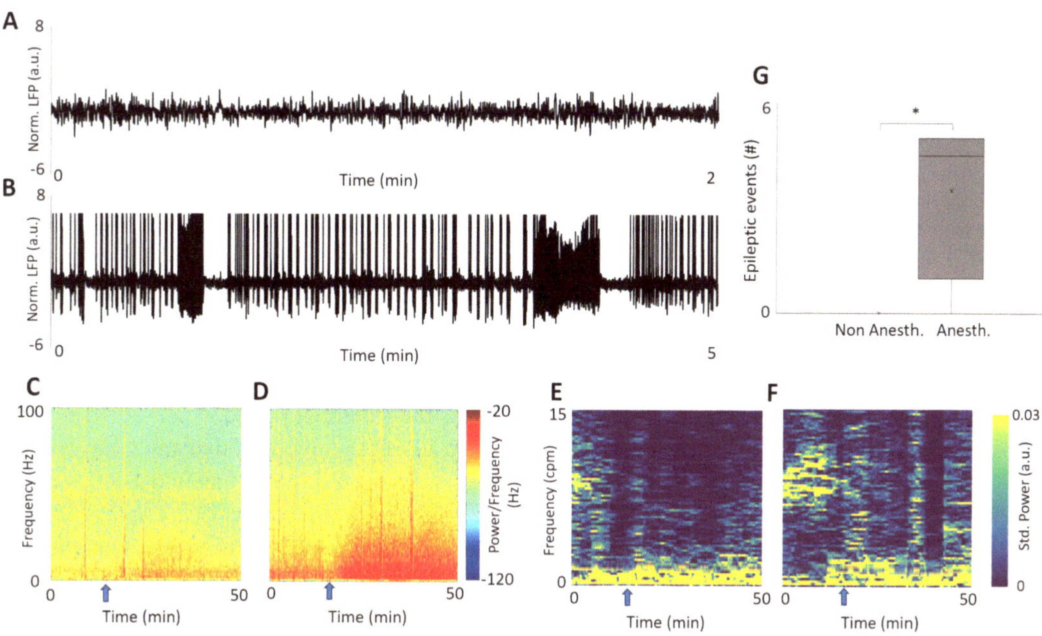

Fig. 23.1 Injections of picrotoxin induce epileptic activity in rabbits exposed to neonatal anesthesia. From a rabbit that was exposed to anesthesia as a neonate, (**a**) shows an illustrative 2-min time window of normalized LFP prior to the injection of PTX and (**b**) shows a 5-min window of normalized LFP after injection of PTX, which contains two epileptic events. LFP spectrograms for neonates (**c**) not exposed to neonatal anesthesia and (**d**) exposed to neonatal anesthesia suggest the substantial change in the power of the lower frequency bands shortly after the timing of the PTX injection. (**e**) A PO_2 spectrogram of a rabbit not exposed to neonatal anesthesia demonstrates the abolition of the resting state higher frequency component after the injection of PTX. (**f**) A PO_2 spectrogram of a rabbit exposed to neonatal anesthesia shows an initial decrease in the power of the higher frequency component after PTX injection; however, during major epileptic events, there is a sudden increase in the power across most frequency bands, with a subsequent abolition across all frequencies above 1 cpm. (**g**) shows a significant increase ($p < 0.05$) in the occurrence of epilepsy between the rabbits not exposed to neonatal anesthesia and the rabbits exposed to neonatal anesthesia. Blue arrows in (**c–f**) denote the time of PTX injection. Examples are representative of the group means reported above

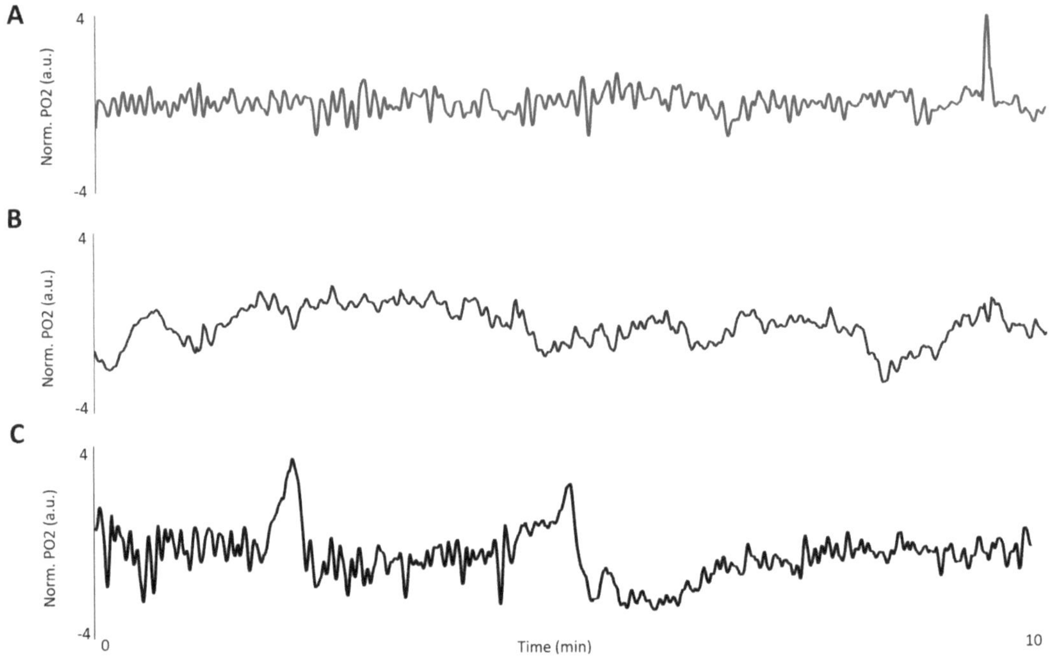

Fig. 23.2 PO_2 reveals different levels of functional deficiency in the inhibitory system. (**a**) displays 10 min of PO_2 prior to the injection of PTX, highlighting the dominant 10 cpm higher frequency component typical for the resting state. (**b**) displays 10 min of PO_2 after PTX injection for a rabbit not exposed to anesthesia as a neonate, where slower frequencies have higher power over higher frequencies. (**c**) displays 10 min of PO_2 after PTX injection for a rabbit exposed to neonatal anesthesia, which has large fluctuations in PO_2 corresponding to epileptic events

of LFP frequencies is minimal. In contrast, Fig. 23.1d shows the LFP spectrogram for a rabbit exposed to neonatal anesthesia, revealing a substantial increase in the power of all frequencies below 40 Hz, consistent with an epileptic state.

Figure 23.1e and f compare the PO_2 spectrograms between a rabbit not exposed to neonatal anesthesia and a rabbit exposed to neonatal anesthesia, respectively. In both rabbits, the higher frequency (approximately 10 cpm) of PO_2 oscillations is substantially reduced after the injection of PTX. However, during epileptic events, the rabbits exposed to neonatal anesthesia (Fig. 23.1f) exhibit increases in the power of all PO_2 frequencies up to 15 cpm, corresponding to major epileptic events. The exposed group showed 3.5 ± 2.38 events per 30-min recording (3/5), compared with 0 events in controls (unpaired t-test, $p < 0.05$). Focal seizures were observed in 0/5 control rabbits and 3/5 anesthesia-exposed rabbits. All anesthesia-exposed rabbits had epileptiform discharges across the recording period.

To illustrate how hemodynamics can be used to gauge various levels of functional deficiency in the inhibitory system, Fig. 23.2 displays illustrative PO_2 traces from our experiments. Figure 23.2a shows the typical PO_2 fluctuations during the resting state across 10 min (prior to the injection of PTX). Notably, there is a prominent high-frequency component at around 10 cpm. Figure 23.2b shows a 10-min PO_2 trace after PTX injection from the same rabbit in Fig. 23.2a, which was not exposed to neonatal anesthesia. The higher frequency fluctuations of PO_2 are smaller compared to the lower frequency fluctuations, suggesting a neurovascular decoupling resulting in reduced frequencies. Figure 23.2c shows a 10-min PO_2 trace after PTX injection from a different rabbit that was exposed to neonatal anesthesia. Major epileptic events are reflected by the extended and significant fluctuations in PO_2, along with a preceding increase in higher frequency oscillations.

23.4 Discussion

Our results indicate the presence of "hidden" epileptic activity in adults previously exposed to neonatal anesthesia. We used the GABA antagonist picrotoxin at a small concentration (negligible for healthy subjects at the level of single neurons (Aksenov et al. 2023b)) to mimic a minor functional deficiency of the inhibitory system. The rabbits exposed to neonatal anesthesia experienced focal seizures and epileptic activity, which could be visualized using recordings of both electrophysiology and brain tissue oxygen. This suggests that any conditions, diseases, or states that induce a minor functional deficiency of the inhibitory system can trigger epileptic-like activity in subjects exposed to neonatal anesthesia. All animals underwent identical implantation surgery, so the differential response cannot be attributed to acute wound-healing effects.

Neonatal anesthesia produces initial damage, but the brain can compensate for it under normal conditions—that is why we do not see epileptic activity in the subjects exposed to neonatal anesthesia without picrotoxin. However, when a challenge (small concentration of picrotoxin in rabbits or, presumably, stress, overstimulation, or additional damage in humans) occurs, this compensation can fail, leading to the manifestation of epileptic activity or increased excitation. This can explain the inconsistency of clinical manifestations in populations exposed to neonatal anesthesia, as the anesthesia alone may not be sufficient for clinical symptoms to appear; an additional trigger or challenge is required. Moreover, different clinical manifestations can appear based on specific brain regions involved, depending on the nature of secondary damage or challenge.

This phenomenon concerns not only clinical epilepsy but also other conditions related to increased excitability in specific brain regions, such as autism and attention-deficit/hyperactivity disorder (ADHD) (Rubenstein and Merzenich 2003; Bogdanska-Chomczyk et al. 2024). These conditions have also been inconsistently reported in patients previously exposed to neonatal anesthesia. Additionally, post-traumatic conditions can follow the same logic.

Recordings of brain tissue oxygen illustrate the utility of hemodynamic signals in diagnosing this epileptic-like activity, even if it is highly localized, clinically silent, and located in the deep structures of the brain.

Acknowledgments This work was supported by R01GM112715 (National Institute of General Medical Sciences).

References

Aksenov DP (2024) Early exposure to general anaesthesia: considerations for age-related vulnerability and behavioural outcomes. Lancet Respir Med 12(9):662–663

Aksenov DP et al (2014) Volume effect of localized injection in functional MRI and electrophysiology. Magn Reson Med 72(4):1170–1175

Aksenov DP et al (2019) Role of the inhibitory system in shaping the BOLD fMRI response. NeuroImage 201: 116034

Aksenov DP et al (2020) Effects of neonatal isoflurane anesthesia exposure on learning-specific and sensory systems in adults. Sci Rep 10(1):13832

Aksenov DP et al (2023a) Preclinical and clinical paediatric anaesthesia research. Lancet Child Adolesc Health 7(1):6–8

Aksenov DP et al (2023b) Functional deficiency of interneurons and negative BOLD fMRI response. Cells 12(5)

Aksenov DP et al (2023c) Brain tissue oxygen and BOLD fMRI under different levels of neuronal activity. Adv Exp Med Biol 1438:3–8

Bogdanska-Chomczyk E et al (2024) Parvalbumin interneuron deficiency in the prefrontal and motor cortices of spontaneously hypertensive rats: an attention-deficit hyperactivity disorder animal model insight. Front Psych 15:1359237

Borzage MT, Peterson BS (2024) A scoping review of the mechanisms underlying developmental anesthetic neurotoxicity. Anesth Analg

Gascoigne DA, Serdyukova NA, Aksenov DP (2021) Early development of the GABAergic system and the associated risks of neonatal anesthesia. Int J Mol Sci 22(23)

Mueller-Buehl C et al (2023) Regulation of the E/I-balance by the neural matrisome. Front Mol Neurosci 16: 1102334

Neudecker V et al (2023) Early-in-life isoflurane exposure alters resting-state functional connectivity in juvenile non-human primates. Br J Anaesth 131(6):1030–1042

Robinson EJ, Lyne TC, Blaise BJ (2022) Safety of general anaesthetics on the developing brain: are we there yet? BJA Open 2:100012

Rubenstein JL, Merzenich MM (2003) Model of autism: increased ratio of excitation/inhibition in key neural systems. Genes Brain Behav 2(5):255–267

Taghon TA et al (2015) A comparison of functional magnetic resonance imaging findings in children with and without a history of early exposure to general anesthesia. Paediatr Anaesth 25(3):239–246

Zhou H et al (2021) Behavioural impairments after exposure of neonatal mice to propofol are accompanied by reductions in neuronal activity in cortical circuitry. Br J Anaesth 126(6):1141–1156

24 Unravelling the Connection Between Reduced Mobility and Mild Cognitive Impairment in Chronic Pain Patients Through Cluster Analysis

Masamichi Moriya, Masaki Kitahara, Kenji Karako, and Kaoru Sakatani

Abstract

Cognitive impairment has been closely associated with systemic metabolic disorders, such as oxygen and energy metabolism disorders. We used deep learning to estimate the risk of mild cognitive impairment (MCI) in outpatients with pain (pain group) and those without pain (non-pain control group) from general blood test data reflecting systemic metabolic disorders. Univariate analysis was performed on blood test data from both groups to calculate the estimated Mini-Mental State Examination (MMSE) scores. Additionally, principal component analysis was conducted on data obtained from eight assessment batteries, comprising 12 items commonly used by Japanese pain treatment institutions. Cluster analysis was also performed using Ward's method on seven components with cumulative proportions exceeding 90%. Our results showed that patients suffering from pain had significantly lower estimated MMSE scores than controls, despite no significant age differences. Patients suffering from pain had significantly higher white blood cell, triglyceride, glucose, and potassium values and significantly lower red blood cell, hemoglobin (Hgb), and uric acid values. Notably, Hgb values were significantly lower only in men, with no significant differences in women. Cluster analysis of the assessment battery data revealed five distinct clusters (Euclidean distance: 10.14%). The average MMSE score of the cluster with an extremely low value for the first component was 27.1, whereas the scores for the other clusters fell below the MCI cutoff. The first component primarily reflects mobility, suggesting that patients suffering from pain with reduced mobility are at increased risk of mild dementia. This finding indicates that impaired mobility due to pain can promote systemic metabolic disorders, subsequently increasing the risk of MCI.

Keywords

Chronic pain · Deep learning · Physiotherapy · Metabolic disorders · Mini-Mental State Examination

M. Moriya (✉)
Department of Physiology, Showa University School of Medicine, Shinagawa, Japan

Department of Anesthesiology, Yokohama City University Medical Center, Kanagawa, Japan

Department of Human and Engineered Environmental Studies Graduate School of Frontier Sciences, The University of Tokyo, Chiba, Japan

M. Kitahara
Department of Anesthesiology, Yokohama City University Medical Center, Kanagawa, Japan

K. Karako · K. Sakatani
Department of Human and Engineered Environmental Studies Graduate School of Frontier Sciences, The University of Tokyo, Chiba, Japan

S. C. Pias et al. (eds.), *Oxygen Transport to Tissue XLVI*, Advances in Experimental Medicine and Biology 1498, https://doi.org/10.1007/978-3-032-10389-5_24

24.1 Introduction

Cognitive impairment has been closely associated with systemic metabolic disorders, such as oxygen and energy metabolism disorders (Komuro et al. 2020; Mathew et al. 1976). The brain is highly dependent on oxygen supply, with evidence showing that fluctuations in oxygen levels easily affect its function (Chen et al. 2020). In particular, studies have revealed that aging and hypoxia can induce the production of reactive oxygen species and that dysfunction of oxygen metabolism can be a risk factor for Alzheimer's disease (Guo et al. 2023; Guzy et al. 2005). Although age-related systemic metabolic disorders have been widely recognized to cause cognitive impairment, reports have found that patients suffering from chronic pain experience a greater degree of age-related cognitive decline compared to the general elderly population (Rouch et al. 2021; Whitlock et al. 2017).

Chronic pain has been shown to not only cause physical and psychological suffering but also significantly impact cognitive functions, such as processing speed, psychomotor speed, memory, learning, attention, and executive function (Higgins et al. 2018; Phelps et al. 2021). Reduced mobility in chronic pain may accelerate cognitive decline via neurovascular and metabolic dysfunction. The underlying mechanisms are thought to be structural and biochemical changes in the hippocampal extracellular matrix (Tajerian et al. 2018) and dysfunction of the locus coeruleus–norepinephrine system (Cao et al. 2019), although these mechanisms have yet to be fully understood.

The current study aimed to investigate risk factors for cognitive dysfunction in patients suffering from chronic pain. We developed a method to estimate mild cognitive impairment (MCI) using deep learning from general blood test data that reflects systemic metabolic disorders (Sakatani et al. 2020), which we used to clarify whether patients visiting pain clinics were at risk for MCI (Moriya et al. 2024). However, basic analysis of blood data reflecting systemic metabolic disorders and detailed analysis of data from the chronic pain research group of the Ministry of Health, Labor and Welfare in Japan (assessment batteries) remain unresolved. Therefore, we compared blood data obtained from outpatients with and without chronic pain who visited a pain clinic and comprehensively analyzed data from eight assessment batteries comprising 12 items.

24.2 Methods

This study targeted 114 patients who visited a multidisciplinary outpatient pain clinic (patient group) and 252 individuals who visited a dementia prevention clinic for routine cognitive screening (control group). The control group consisted of patients without a history of chronic pain or clinical signs of cognitive impairment based on medical records. MMSE scores were not directly measured in either group; instead, estimated MMSE scores were calculated using a validated DNN model based on age and blood test data. First, blood test results for both groups were compared using univariate analysis. Items included WBC, RBC, hemoglobin, hematocrit, PLT, total protein, A/G ratio, albumin, uric acid, BUN, creatinine, sodium, potassium, chloride, total cholesterol, triglycerides, AST (GOT), ALT (GPT), γ-GTP, and glucose. The MMSE estimation model was based on a previously published DNN trained on age and blood data from 202 patients with systemic metabolic disorders (Sakatani et al. 2020). It was validated in three groups: stroke patients, healthy individuals, and those undergoing health checkups. The original study reported a significant correlation between predicted and actual MMSE scores ($r = 0.66$). In the present study, we applied this pre-trained, validated model without further training.

In addition, to identify subgroups of patients with chronic pain who shared similar symptom profiles, principal component analysis was performed on the results of the 108 patients who underwent eight assessment batteries (12 items) commonly used by pain treatment institutions in Japan and had no missing data. Afterward, cluster analysis was performed using Ward's method.

The questionnaire items included the Brief Pain Inventory (BPI), the Pain Disability Assessment Scale (PDAS), the Hospital Anxiety and Depression Scale (HADS), Pain Catastrophizing Scale (PCS), EuroQol-5D (EQ-5D), Pain Self-Efficacy Questionnaire (PSEQ), Athens Insomnia Scale (AIS), and the 25-Question Geriatric Locomotive Function Scale (Locomo).

This study was approved by the Life Science and Medical Research Ethics Committee (F230400058). As an observational study using anonymized records, informed consent was waived. However, information was disclosed and participants were allowed to opt out.

24.3 Results

We determined the presence of MCI using the DNN to estimate MMSE scores and assess cognitive dysfunction risk in patients with chronic pain. Estimated MMSE scores were significantly lower in patients than in controls, despite no age difference. Blood tests showed elevated WBC, TG, GLU, and K were significantly higher, but RBC, Hgb, and UA were significantly lower. Hgb values were significantly lower only in men (Table 24.1). These results show no abnormal blood data but indicate a trend toward anemia and obesity, which may contribute to cognitive decline.

Principal component analysis was performed to clearly verify the characteristics of the multidimensional questionnaire data. The first component (mobility-related) loaded highly on EQ-5D, Locomo, and PDAS; the second component (pain and emotional status) on BPI and HADS; and the third component (sleep quality) on Athenae. Seven components accounting for >90% cumulative variance were then used for hierarchical

Table 24.1 Comparison of two groups of general blood tests

		Pain patients $n = 114$ (M/F: 48/66)	Normal adults $n = 252$ (M/F: 138/151)	p
Age, yr		57.0 (49.8–67.0)	55.0 (50.0–61.0)	0.222
Estimated MMSE, P		26.9 (25.5–27.9)	27.2 (26.2–28.0)	0.043
WBC count (10^3/μL)		**5420.0**	**4930.0**	**0.0001**
RBC count (10^6/ μL)		**449.0**	**464.0**	**0.007**
Hemoglobin (g/dL)	**Male**	**14.3**	**15.0**	**0.007**
	Femele	13.2	13.2	0.747
Hematocrit (%)		41.4	41.4	0.629
Platelet count (10^3/μL)		24.3	23.9	0.863
Total protcin (g/dL)		7.2	7.2	0.381
A/G		1.6	1.6	0.467
Albumin (mg/dL)		4.5	4.5	0.080
UA	**Male**	**5.0**	**6.0**	**0.011**
	Femele	**4.1**	**4.7**	**0.010**
BUN (mg/dL)		14.0	13.9	0.383
Creatinine (mg/dL)		0.7	0.7	0.779
Na		142.0	141.0	0.143
K		**4.2**	**4.1**	**0.007**
Cl		105.0	105.0	0.558
Cholesterol (mg/dL)		206.5	211.0	0.661
Triglyceride (mg/dL)		**108.0**	**81.0**	**0.0001**
AST		22.0	21.0	0.380
ALT		19.0	20.0	0.747
γ-GT		22.0	24.0	0.267
GLU		**98.0**	**96.0**	**0.036**

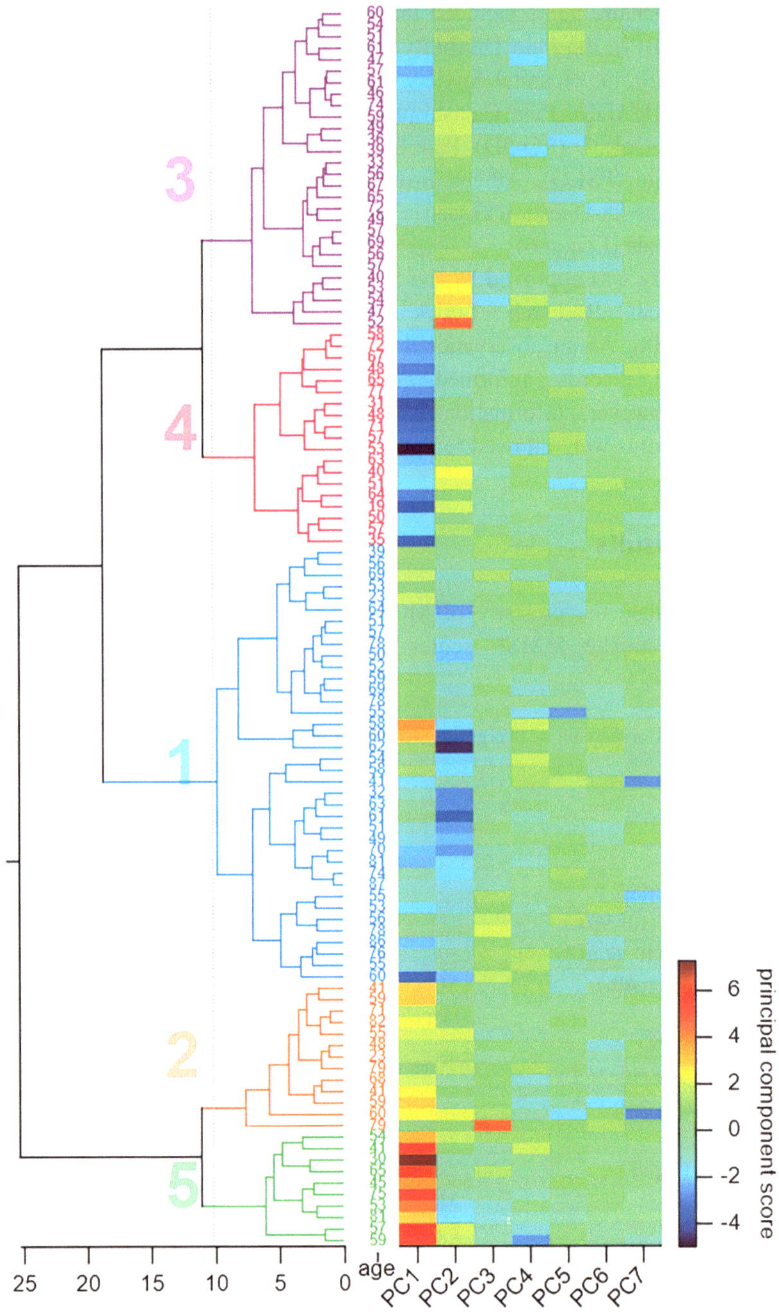

Fig. 24.1 Heatmap and dendrogram of participant clustering based on standardized clinical features

clustering (Ward's method), revealing five clusters (Euclidean distance: 10.14%) (Fig. 24.1).

To clarify the characteristics of each cluster, we compared MMSE, mobility scores, and metabolic blood markers (Hgb, TG, and GLU). As shown in Fig. 24.2, all clusters except Cluster 4 fell below the MCI threshold. Notably, age did not differ significantly across clusters. Figure 24.3 further illustrates the divergence in mobility-related questionnaires, highlighting favorable mobility scores in Cluster 4, with modest variation also observed in blood-based metabolic

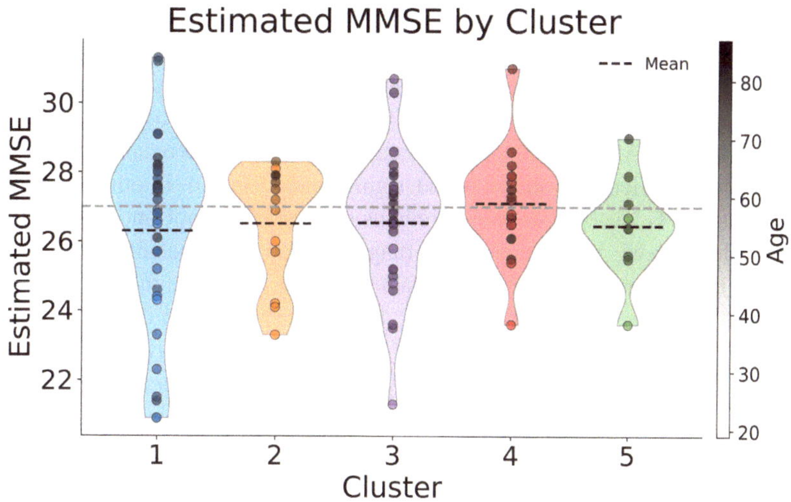

Fig. 24.2 Cluster-wise estimated MMSE distribution with age gradient and mean indicators

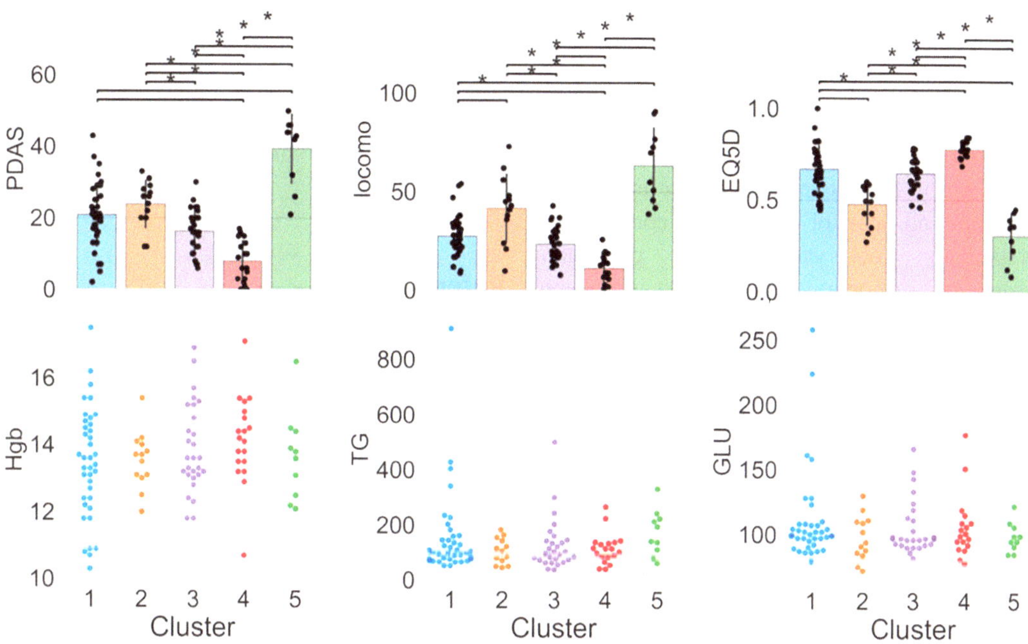

Fig. 24.3 Cluster-wise comparisons of questionnaire and blood parameters

markers. These results indicate that regardless of age, people with pain who have mobility problems may be at high risk for MCI.

24.4 Discussion

The current study showed that patients suffering from pain have a high risk for MCI and are prone to anemia and obesity, with studies showing that anemia is a risk factor for cognitive dysfunction (Wang et al. 2023; Hong et al. 2013). Moreover, the current study found that Hgb values differed only among men suffering from pain. Anemia is usually defined using the World Health Organization criteria (Hgb concentration of <13 and <12 g/dL for men and women, respectively). Notably, our results for both men and women were within the normal range. Obesity has also been reported to be a risk factor for cognitive dysfunction, with particular attention having been placed on TGs (Farr et al. 2008; Banks et al. 2018). Although the present study found that those suffering from pain had significantly increased TG levels, supporting previous studies, their levels were within the normal range similar to Hgb. One can assume that these factors do not affect cognitive dysfunction alone, but rather interact with each other in a complex manner (Zhang et al. 2022). Reduced mobility may contribute to cognitive decline through neurovascular dysfunction, neuroinflammation, or metabolic disruption, including abnormal TG or Hgb. Hgb and TG may serve as early indicators for identifying cognitive risk in chronic pain patients. This study does not attempt to diagnose MCI but instead uses estimated MMSE scores as a proxy for screening-level cognitive function. A clinical diagnosis of MCI requires a comprehensive assessment that was beyond the scope of this study. Using Hgb as an indicator for anemia and TG as an indicator for obesity may contribute to preliminary observation and preventive treatments for cognitive dysfunction in patients suffering from pain.

Utilizing the results of these assessment batteries to classify patients into several clusters may help physicians understand the complex characteristics of chronic pain and plan appropriate treatment strategies. We used principal component analysis to clarify the existence of three main components: degree of mobility, degree of pain, anxiety/depression, and degree of sleep.

Figure 24.1 shows cluster analysis results. For the first component, blue indicates high mobility; red, low. Clusters other than cluster 4 showed MMSE scores below the MCI cutoff, suggesting mobility's importance in cognitive risk. The second component reflected pain and anxiety; cluster 1 included patients with mixed mobility and high emotional distress, implying that both mobility and psychological factors may affect cognition.

Several reports have shown that spending a substantial amount of time sitting is associated with cognitive decline (Huang et al. 2022; Raichlen et al. 2023). Interestingly, the current study suggested that the same results may hold true for the mobility of patients suffering from chronic pain. The effects of the decline in mobility on cognitive function may be attributed to fear-avoidance beliefs (Vlaeyen and Linton 2000). In other words, it is possible that patients did not move due to pain, that patients had become unable to move due to lack of movement, and that the inability to move promoted cognitive function decline. Chronic pain and obesity share the behavioral feature of physical inactivity and the common neuropathological feature of inactivation or dysfunction of the mesolimbic dopamine system (Senba and Kami 2017). Our findings suggest a link between elevated TG and reduced mobility. However, chronic pain is influenced by medications, comorbidities, and activity levels, which may have confounded our results. Given the small, single-center sample, further research is needed. Still, these findings may inform cluster-specific physiotherapy approaches, as in the SQRP study (Bäckryd et al. 2018).

References

Bäckryd E, Persson EB, Larsson AI et al (2018) Chronic pain patients can be classified into four groups: clustering-based discriminant analysis of psychometric data from 4665 patients referred to a multidisciplinary pain centre (a SQRP study). PLoS One 13(2): e0192623

Banks WA, Farr SA, Salameh TS et al (2018) Triglycerides cross the blood-brain barrier and induce central leptin and insulin receptor resistance. Int J Obes 42(3):391–397

Cao S, Fisher DW, Yu T et al (2019) The link between chronic pain and Alzheimer's disease. Review. J Neuroinflammation 16(1):204

Chen PS, Chiu WT, Hsu PL et al (2020) Pathophysiological implications of hypoxia in human diseases. J Biomed Sci 27(1):63

Farr SA, Yamada KA, Butterfield DA et al (2008) Obesity and hypertriglyceridemia produce cognitive impairment. Endocrinology 149(5):2628–2636

Guo Y, Guan T, Shafiq K et al (2023) Mitochondrial dysfunction in aging. Ageing Res Rev 88:101955

Guzy RD, Hoyos B, Robin E et al (2005) Mitochondrial complex III is required for hypoxia-induced ROS production and cellular oxygen sensing. Cell Metab 1(6): 401–408

Higgins DM, Martin AM, Baker DG et al (2018) The relationship between chronic pain and neurocognitive function: a systematic review. Clin J Pain 34(3): 262–275

Hong CH, Falvey C, Harris TB et al (2013) Anemia and risk of dementia in older adults: findings from the Health ABC study. Neurology 81(6):528–533

Huang SY, Li YZ, Zhang YR et al (2022) Sleep, physical activity, sedentary behavior, and risk of incident dementia: a prospective cohort study of 431,924 UK biobank participants. Mol Psychiatry 27(10): 4343–4354

Komuro Y, Oyama K, Hu L et al (2020) Relationship between cognitive dysfunction and systemic metabolic disorders in elderly: dementia might be a systematic disease. Adv Exp Med Biol 1232:91–97

Mathew NT, Meyer JS, Achari AN et al (1976) Hyperlipidemic neuropathy and dementia. Eur Neurol 14(5):370–382

Moriya M, Hu L, Sakatani K et al (2024) Estimation of cognitive impairment in chronic pain patients and characteristics of estimated mild cognitive impairment. Front Neurol 15:1344190

Phelps CE, Navratilova E, Porreca F (2021) Cognition in the chronic pain experience: preclinical insights. Trends Cogn Sci 25(5):365–376

Raichlen DA, Aslan DH, Sayre MK et al (2023) Sedentary behavior and incident dementia among older adults. JAMA 330(10):934–940

Rouch I, Edjolo A, Laurent B et al (2021) Association between chronic pain and long-term cognitive decline in a population-based cohort of elderly participants. Pain 162(2):552–560

Sakatani K, Oyama K, Hu L (2020) Deep learning-based screening test for cognitive impairment using basic blood test data for health examination. Front Neurol 14(11):588140

Senba E, Kami K (2017) A new aspect of chronic pain as a lifestyle-related disease. Neurobiol Pain 17(1):6–15

Tajerian M, Hung V, Nguyen H et al (2018) The hippocampal extracellular matrix regulates pain and memory after injury. Mol Psychiatry 23(12):2302–2313

Vlaeyen JWS, Linton SJ (2000) Fear-avoidance and its consequences in chronic musculoskeletal pain: a state of the art. Pain 85(3):317–332

Wang J, Wang C, Li X et al (2023) Association of Anemia with cognitive function and dementia among older adults: the role of inflammation. J Alzheimer's Dis 96(1):125–134

Whitlock EL, Diaz-Ramirez LG, Glymour MM et al (2017) Association between persistent pain and memory decline and dementia in a longitudinal cohort of elders. JAMA Intern Med 177(8):1146–1153

Zhang YR, Xu W, Zhang W et al (2022) Modifiable risk factors for incident dementia and cognitive impairment: an umbrella review of evidence. J Affect Disord 314:160–167

Part VII

Cerebral Hemodynamics

Time-Dependent Efficacy of Drag-Reducing Polymers in the Treatment of Permanent Middle Cerebral Artery Occlusion Injury in Rats

25

Olga A. Bragina, Jun Wu, Zefeng Peng, Edwin M. Nemoto, Michael Chopp, Alex O. Trofimov, Afshin A. Divani, Fawaz Al-Mufti, and Denis E. Bragin

Abstract

Current treatments for ischemic stroke are not focused on microvascular cerebral blood flow (mvCBF), which actually delivers oxygen to tissue and is significantly impaired during a stroke. We previously showed that drag-reducing polymers (DRPs, 2 μg/ml), injected just after permanent middle cerebral artery occlusion (pMCAO) in rats, effectively restored mvCBF and tissue oxygenation. The aim of this work was to assess the efficiency of DRPs administered at various time points post-onset. DRPs (2 μg/ml) or saline was i.v. injected 0.5, 3, or 6 h after pMCAO induction in rats ($n = 10$/group). Laser speckle contrast imaging (LSCI) was used to evaluate regional cerebral blood flow in the parietal cortex and estimate the ischemic area. In vivo 2-photon laser scanning microscopy (2PLSM) was used to assess the effects on mvCBF and tissue hypoxia (NADH). Two-way ANOVA for multiple comparisons was used to test intergroup differences with the statistical significance level set at $p < 0.05$. PMCAO progressively decreased cortical mvCBF, causing tissue hypoxia ($p < 0.05$). Intravenous administration of DRPs at 0.5, 3, and 6 h post-pMCAO effectively improved cerebral microcirculation and oxygen delivery to tissue in a time-dependent manner compared to the saline-treated group ($p < 0.05$). DRPs efficacy reduced with the time of application from 0.5 to 6 h after pMCAO but remained significant compared to the saline-treated group ($p < 0.05$). DRPs augment collateral microcirculatory flow through leptomeningeal and pial

O. A. Bragina
Lovelace Biomedical Research Institute, Albuquerque, NM, USA

J. Wu · Z. Peng
Department of Neurosurgery, Central South University, Changsha, Hunan, China

E. M. Nemoto · A. A. Divani
Department of Neurology, University of New Mexico School of Medicine, Albuquerque, NM, USA

M. Chopp
Department of Neurology, Henry Ford Hospital, Detroit, MI, USA

Department of Physics, Oakland University, Rochester, MI, USA

A. O. Trofimov
Privolzhsky Research Medical University, Nizhny Novgorod, Russia

F. Al-Mufti
New York Medical College, Valhalla, NY, USA

Westchester Medical Center, Valhalla, NY, USA

D. E. Bragin (✉)
Lovelace Biomedical Research Institute, Albuquerque, NM, USA

Department of Neurology, University of New Mexico School of Medicine, Albuquerque, NM, USA

New York Medical College, Valhalla, NY, USA
e-mail: dbragin@lovelacebiomedical.org

S. C. Pias et al. (eds.), *Oxygen Transport to Tissue XLVI*, Advances in Experimental Medicine and Biology 1498, https://doi.org/10.1007/978-3-032-10389-5_25

anastomoses that interconnect the middle, anterior, and posterior cerebral artery watershed territories. We demonstrated that DRPs can be used as a new effective adjunct therapy for ischemic stroke, even without reperfusion and after delayed administration following the stroke onset.

Keywords

Ischemic stroke · Cerebral blood flow · Drag-reducing polymers · Collateral flow · Penumbra

25.1 Introduction

Acute ischemic stroke (AIS) is a leading cause of death and disability, accounting for 75 million new cases annually (Pu et al. 2023). Although two therapies—recombinant tissue plasminogen activator (thrombolysis) and endovascular arterial thrombectomy—are approved for AIS treatment, only 20–25% of patients are eligible for these interventions (Henninger and Fisher 2016). Thus, up to 75% of AIS still urgently need treatment to suspend the evolution of the ischemic penumbra. Another issue is that current treatments do not specifically target microvascular cerebral blood flow (mvCBF), which actually delivers oxygen to tissue and is significantly impaired during a stroke. We previously showed that drag-reducing polymers (DRPs, 2 μg/ml), injected i.v. just after permanent middle cerebral artery occlusion (pMCAO) in rats, effectively restored mvCBF and tissue oxygenation (Bragin et al. 2017). However, it is unlikely that treatment can be administered immediately after stroke onset. Therefore, the aim of this work was to assess the efficiency of DRPs administered at various time points post-ischemia in a rat model of pMCAO.

25.2 Methods

Most of the procedures used in these studies were previously described (Bragin et al. 2017; Bragin et al. 2016). Forty laboratory-acclimated male Sprague-Dawley rats weighing 250–300 grams were randomly assigned to four experimental groups. During the study, the rats were maintained on isoflurane (1%), nitrous oxide (69%), and oxygen (30%) via face-mask anesthesia. A pMCAO model of ischemic stroke was induced by advancing a 2.5-cm 4–0 nylon filament through the external into the internal carotid artery, as previously reported (Bragin et al. 2017; Foley et al. 2010). DRPs were intravenously injected with a final blood concentration of 2 μg/ml at 0.5, 3, and 6 h after pMCAO (Group I-III, respectively, $n = 10$/group). The control pMCAO group was injected with the same volume of saline (Group IV). Cerebral blood flow and tissue oxygen supply were monitored at baseline, just after, and 7 h after pMCAO induction. Laser speckle contrast imaging (LSCI) was used to evaluate regional cerebral blood flow in the parietal cortex using an RFLSI III System (RWD Life Science, Shenzhen, China). In vivo 2-photon laser scanning microscopy (2PLSM) using a Prairie View Ultima System (Bruker, Billerica, MA) was utilized to monitor cerebral microcirculation (i.v. fluorescein isothiocyanate dextran, FITC), and mitochondrial respiration and brain tissue oxygen supply (nicotinamide adenine dinucleotide autofluorescence, NADH) in the penumbra area of the parietal cortex. For imaging, the rat's head was secured in a stereotactic frame, the skull was exposed, and the ipsilateral parietal bone was thinned to ~15 μm with a hand-held drill to achieve optical clarity and then overlaid with cyanoacrylate glue and a cover glass using a modified protocol from Shih et al. (2012). For DRPs preparation, polyethylene oxide (PEO, MW ~4000 kDa, Sigma-Aldrich) was dissolved in saline to 0.1% (1000 ppm) concentration and dialyzed overnight using a 100 kDa cutoff membrane (Bragin et al. 2017; Bragin et al. 2016). Before injection, the PEO solution was diluted with saline to 50 ppm and slowly rocked for ~2 h. The solution was sterilized using a 0.22 μm filter and i.v. injected over 2 min to achieve a final blood concentration of 2 μg/ml. The volume of DRP solution (ml) to be injected was determined by: ([rat weight (grams) × 0.07]/

50 ppm) × 2. Data analysis was done by GraphPad Prism. Differences between groups and from baseline were determined using a two-way analysis of variance for multiple comparisons and post hoc testing. Data were presented as a Mean ± SEM, and the significance level was $p < 0.05$.

25.3 Results

Permanent MCAO progressively decreased cortical mvCBF, leading to a dramatic reduction in brain tissue oxygen supply and mitochondrial respiration as reflected by NADH autofluorescence increase ($p < 0.05$ from baseline). In the saline group, by the end of the monitoring period (7 h post-pMCAO), the capillary flow velocity reduced to 0.29 ± 0.04 from 0.96 ± 0.05 mm/s at a baseline, while the number of perfused capillaries reduced from 224.3 ± 34.2 to 98.4 ± 28.7 per 0.075 mm^3 (Fig. 25.1a and b; $p < 0.05$). This was associated with a reduction in tissue oxygen supply as reflected by NADH autofluorescence increase to 1.59 ± 0.9 normalized units (Fig. 25.1c; $p < 0.05$). LSCI showed that in the saline group, the ischemic area reached 4.87 ± 0.9 sq. mm by the end of the monitoring period (Fig. 25.2a).

DRPs (injected at 0.5, 3, and 6 h after pMCAO) preserved capillary flow velocities that were 0.78 ± 0.06, 0.71 ± 0.05, and 0.59 ± 0.05 mm/s, respectively (Fig. 25.1a, $p < 0.001$, 0.01, and 0.05 from the saline group, respectively). The increase in capillary blood flow velocity restored circulation in collapsed capillaries (178.9 ± 42.4, 156.1 ± 41.1 and 148.6 ± 38.4, for DRPs injected 0.5, 3, and 6 h after pMCAO with $p < 0.05$, 0.01 and 0.001, respectively (Fig. 25.1b). Enhanced mvCBV preserved tissue oxygen supply reflected by lower NADH autofluorescence—1.19 ± 0.8, 1.22 ± 0.9 and 1.31 ± 1.1 normalized units comparing to saline group (Fig. 25.1c, $p < 0.001$, 0.01, and 0.05, respectively). LSCI demonstrated that DRPs injection at 0.5, 3, and 6 h after pMCAO substantially reduced the ischemic area in the parietal cortex—1.96 ± 0.4, 2.6 ± 0.5, and 3.1 ± 0.5 sq. mm, with $p < 0.001$, 0.01, and 0.05, respectively (Fig. 25.2a).

LSCI revealed that DRP injection increased collateral flow, depending on the time of application after the pMCAO. The watershed territory was 1.51 ± 0.2, 1.43 ± 0.3, and 1.29 ± 0.3 normalized units between ACA and MCA for DRPs injected at 0.5, 3, and 5 h after pMCAO (Fig. 25.2b, $p < 0.001$, 0.01, and 0.05, compared to saline control, respectively). Similarly, the watershed territory was 1.58 ± 0.3, 1.41 ± 0.4,

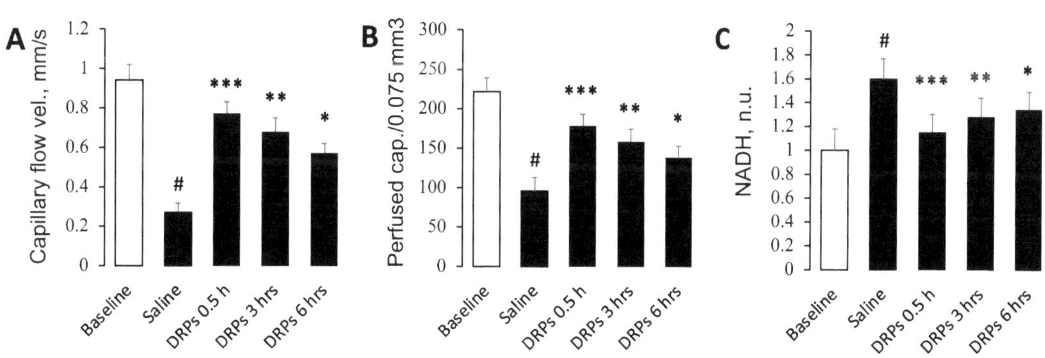

Fig. 25.1 DRPs injection enhances cerebral microvascular flow, attenuates capillary stasis, and improves oxygen delivery in the cerebral cortex after pMCAO. (**a**) pMCAO reduced microvascular flow velocity, while DRPs injected at 0.5, 3, or 6 h after pMCAO partially restored it. (**b**) pMCAO caused progressive capillary stasis; DRPs injected at 0.5, 3, or 6 h after pMCAO attenuated capillary thrombosis. (**c**) pMCAO led to tissue hypoxia, reflected by increased NADH autofluorescence. Improved by DRPs, microvascular perfusion attenuated tissue hypoxia, reflected by a decrease in NADH. (* and # = $p < 0.05$, ** = $p < 0.01$, *** = $p < 0.001$, $N = 10$)

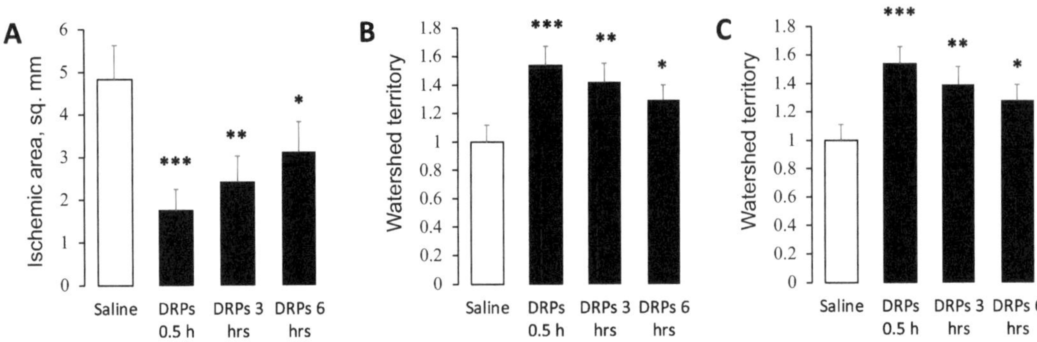

Fig. 25.2 LSCI demonstrated that DRPs enhanced collateral flow, reducing the ischemic area in the cortex. (**a**) Ischemic area in the cortex reduced after DRP injection; (**b**) DRPs injected 0.5, 3, or 6 h after pMCAO increased perfusion in the watershed territory between the middle cerebral artery (MCA) and the anterior cerebral artery (ACA); (**c**) DRPs injected at 0.5, 3, or 6 h after pMCAO increased perfusion between the MCA and the posterior cerebral artery (PCA) collaterals area. (* = $p < 0.05$, ** = $p < 0.01$, *** = $p < 0.001$, $N = 10$)

and 1.34 ± 0.3 between PCA and MCA in DRPs injected at 0.5, 3, and 5 h after pMCAO (Fig. 25.2c, $p < 0.001$, 0.01, and 0.05, compared to saline control, respectively).

25.4 Discussion

Timely restoration of CBF remains the only effective treatment for AIS. Currently, it is achieved either by thrombolysis using recombinant tissue plasminogen activator or by surgical endovascular arterial thrombectomy. CBF-enhanced delay of the evolution of the ischemic penumbra and core is a promising intervention (Yi et al. 2021), and collateral circulation plays an important role (Liebeskind 2003). However, neither of the approved therapies specifically targets microvascular cerebral blood flow (mvCBF), which actually delivers oxygen to tissue and is significantly impaired during a stroke. The New Paradigm for Neuroprotection Clinical Trials in Acute Ischemic Stroke suggests using neuroprotectants to suspend the evolution of the penumbra and core before imaging (Nemoto 2023). However, pharmacological treatments suggested so far have failed to prove clinical efficiency as they require intact tissue or receptor sensitivity. For example, phosphodiesterase inhibitors and endothelin-1 antagonists increase CBF but require intact vascular sensitivity and tissue viability, likely impaired in AIS (Biose et al. 2023). Pharmacologic arterial vasodilation could shunt blood flow from ischemic regions in a reverse steal "Robinhood syndrome," and is associated with arterial hypotension that might worsen stroke severity (Alexandrov et al. 2009). As we demonstrated, DRPs delay the evolution of the ischemic penumbra and core by enhancing the perfusion of compromised tissue poststroke, which is associated with the enhancement of collateral flow. DRPs reach microvessels distal to the ligation via collateral flow through leptomeningeal and pial anastomoses that interconnect the middle, anterior, and posterior cerebral arteries. Unlike pharmacologic vasodilators, DRPs enhance blood flow by altering the physical dynamics of blood (Kameneva 2012) independent of tissue viability or vascular reactivity. DRPs reduce flow separations at arterial bifurcations and vortices, narrowing the vascular wall's cell-free layer and increasing vascular wall shear stress (Kameneva 2012). These mechanisms, validated in vitro and in vivo, collectively elevate precapillary pressure, enhance capillary perfusion and oxygen delivery, and improve tissue survival (Bragin et al. 2017; Kameneva 2012). Therefore, DRPs can be used as a new effective adjunct therapy for ischemic stroke even without reperfusion and after delayed administration following the stroke onset.

25.5 Conclusion

Intravenous DRP application restores cerebral microcirculation, thereby reducing tissue hypoxia. The efficacy decreased with the time of application from 0.5 to 6 h after pMCAO but remained significant compared to the saline-treated group. One mechanism of the DRPs' effects is enhancing blood flow through leptomeningeal or pial collaterals between MCA and ACA or PCA watershed territories after pMCAO, leading to improved mvCBF and tissue oxygenation in the ischemic areas distal to the ligation. DRPs can be used as a new effective adjunct therapy for ischemic stroke, even without reperfusion and after delayed administration following the stroke onset.

Acknowledgments This work was supported by NIH NINDS R21NS091600 and 1R01NS112808.

References

Alexandrov AV, Nguyen HT, Rubiera M, Alexandrov AW, Zhao L, Heliopoulos I, Robinson A, Dewolfe J, Tsivgoulis G (2009) Prevalence and risk factors associated with reversed Robin Hood syndrome in acute ischemic stroke. Stroke 40(8):2738–2742. https://doi.org/10.1161/STROKEAHA.109.547950. Epub 2009 May 21

Biose IJ, Oremosu J, Bhatnagar S, Bix GJ (2023) Promising cerebral blood flow enhancers in acute schemic stroke. Transl Stroke Res 14(6):863–889. https://doi.org/10.1007/s12975-022-01100-w. Erratum in: Transl Stroke Res. 2022

Bragin DE, Peng Z, Bragina OA, Statom GL, Kameneva MV, Nemoto EM. Improvement of impaired cerebral microcirculation using rheological modulation by drag-reducing polymers. Adv Exp Med Biol 2016;923:239–244. doi: https://doi.org/10.1007/978-3-319-38810-6_32

Bragin DE, Kameneva MV, Bragina OA, Thomson S, Statom GL, Lara DA, Yang Y, Nemoto EM (2017) Rheological effects of drag-reducing polymers improve cerebral blood flow and oxygenation after traumatic brain injury in rats. J Cereb Blood Flow Metab 37(3):762–775. https://doi.org/10.1177/0271678X16684153. Epub 2016 Dec 22

Foley LM, Hitchens TK, Barbe B, Zhang F, Ho C, Rao GR, Nemoto EM (2010) Quantitative temporal profiles of penumbra and infarction during permanent middle cerebral artery occlusion in rats. Transl Stroke Res 1(3):220–229. https://doi.org/10.1007/s12975-010-0032-6

Henninger N, Fisher M (2016) Extending the time window for endovascular and pharmacological reperfusion. Transl Stroke Res 7(4):284–293. https://doi.org/10.1007/s12975-015-0444-4. Epub 2016 Jan 7

Kameneva MV (2012) Microrheological effects of drag reducing polymers in vitro and in vivo. Int J Eng Sci 59:168–183

Liebeskind DS (2003) Collateral circulation. Stroke 34(9): 2279–2284. https://doi.org/10.1161/01.STR.0000086465.41263.06. Epub 2003 Jul 24

Nemoto EM (2023) A new paradigm for neuroprotection clinical trials for acute ischemic stroke. Transl Stroke Res 14(6):829–832. https://doi.org/10.1007/s12975-022-01091-8

Pu L, Wang L, Zhang R, Zhao T, Jiang Y, Han L. Projected global trends in ischemic stroke incidence, deaths and disability-adjusted life years from 2020 to 2030. Stroke. 2023;54(5):1330–9. https://doi.org/10.1161/STROKEAHA.122.040073. Epub 2023 Apr 24. Erratum in: Stroke. 2024;55(1):e23. 10.1161/STR.0000000000000452.

Shih AY, Mateo C, Drew PJ, Tsai PS, Kleinfeld D (2012) A polished and reinforced thinned-skull window for long-term imaging of the mouse brain. J Vis Exp 61: 3742. https://doi.org/10.3791/3742

Yi Y, Liu Z, Wang M et al (2021) Penumbra in acute ischemic stroke. Curr Neurovasc Res 18(5):572–585

Prefrontal Cortex Oxygenation Dynamics During Walking in Patients with Acute Stroke

26

Tomoya Takahashi, Kentaro Iwata, Kanji Yamada, Yukihiro Maekawa, Yuta Inagaki, Natsuki Takahara, Tomoki Hirai, Yuto Sasaki, Nobuo Kohara, and Atsuhiro Tsubaki

Abstract

This study aimed to elucidate the cerebral oxygenation dynamics during walking in patients with acute stroke. Participants were required to walk for 70 s while measurements of oxy-hemoglobin (O_2Hb), deoxy-hemoglobin (HHb), and total hemoglobin (THb) in the prefrontal cortex (PFC) were obtained using near-infrared spectroscopy. The Friedman test and Wilcoxon signed-rank tests with Bonferroni correction were applied to determine the oxygenation dynamics within the PFC during the task, making comparison to the pre-walking baseline at 10 s prior to walking onset. Analysis of 50 acute stroke patients (post-stroke onset: 8.7 [3.3] days) revealed significant main effects for O_2Hb, HHb, and THb ($p < 0.001$ for all). Post hoc analyses further revealed that O_2Hb significantly decreased at 40 s post walking onset, while HHb significantly increased at 45 s, and THb significantly decreased from 35 s ($p < 0.05$). These findings indicate that ischemia and vascular constriction in the PFC may occur during walking in the acute phase of stroke. However, the lack of a control group and the sole focus on the PFC limit the generalizability of the study findings.

Keywords

Acute stroke · Prefrontal cortex · Infrared spectroscopy · Oxygenation dynamics · Walking · Rehabilitation

26.1 Introduction

Stroke is characterized by an interruption of blood supply to the brain or hemorrhage within the brain tissue, which often results in impaired brain function. Approximately half of first-time stroke patients develop residual walking disabilities post-stroke (Louie et al. 2022). Rehabilitation is essential for mitigating post-stroke sequelae, and recent studies have underscored the benefits of

T. Takahashi · N. Takahara · T. Hirai
Department of Rehabilitation, Kobe City Medical Center General Hospital, Kobe, Japan

Graduate School, Niigata University of Health and Welfare, Niigata, Japan

K. Iwata · K. Yamada · Y. Maekawa · Y. Inagaki · N. Kohara
Department of Rehabilitation, Kobe City Medical Center General Hospital, Kobe, Japan

Y. Sasaki
Department of Rehabilitation, Nagaoka Red Cross Hospital, Nagaoka, Japan

A. Tsubaki (✉)
Graduate School, Niigata University of Health and Welfare, Niigata, Japan

Institute for Human Movement and Medical Sciences, Niigata University of Health and Welfare, Niigata, Japan
e-mail: tsubaki@nuhw.ac.jp

S. C. Pias et al. (eds.), *Oxygen Transport to Tissue XLVI*, Advances in Experimental Medicine and Biology 1498, https://doi.org/10.1007/978-3-032-10389-5_26

early intervention (Liu et al. 2022). However, in patients with acute stroke, impaired cerebral autoregulation due to vasoparalysis may reduce cerebral blood flow during rehabilitation, potentially impacting therapy effectiveness (Claassen et al. 2021).

Near-infrared spectroscopy (NIRS) is used to evaluate changes in blood flow and oxygen metabolism in living tissues (Herold et al. 2018). NIRS enables the measurement of oxy-hemoglobin (O_2Hb), deoxy-hemoglobin (HHb), and total hemoglobin (THb), allowing for the interpretation of tissue oxygenation dynamics based on fluctuations in these parameters (Hillman 2014). A study examining oxygenation dynamics in the prefrontal cortex (PFC) using NIRS in both patients with acute stroke and healthy individuals found that lowering the head-of-bed angle increased O_2Hb and THb levels while decreasing HHb levels in healthy subjects. However, these changes were not observed in patients with acute stroke (Truijen et al. 2018). This indicates the presence of characteristic oxygenation dynamics during movement in the brains of patients with acute stroke. Furthermore, studies on patients with chronic stroke have indicated that the O_2Hb in the PFC increases during 30 s of walking (Lim et al. 2022). However, in patients with acute stroke who are believed to have impaired cerebral autoregulation, the same walking task may elicit different oxygenation dynamics, with these changes potentially persisting beyond 30 s.

Therefore, this study aimed to investigate cerebral oxygenation dynamics from the onset of walking in patients with acute stroke.

26.2 Materials and Methods

26.2.1 Participants

The study cohort comprised patients with acute stroke admitted to the Stroke Care Unit at the Kobe City Medical Center General Hospital in Kobe, Japan, between May 2023 and March 2024, who presented with acute stroke symptoms. Stroke events included cerebral infarction, intracerebral hemorrhage, and subarachnoid hemorrhage, with the acute phase defined as within 14 days of stroke onset. All participants received written and verbal explanations of the study and provided informed consent. Ethical approval was obtained from the Institutional Review Board of Kobe City Medical Center General Hospital (approval number: zn230408). Inclusion criteria were first-ever stroke patients aged ≥ 20 years who were ambulatory without assistance before stroke onset. Exclusion criteria included cognitive impairments, other neurological disorders, uncontrolled hypertension, recent fractures, joint replacements within the past 6 months, and inability to comprehend instructions in Japanese. Data collected included age, sex, body mass index, clinical stroke characteristics, National Institutes of Health Stroke Scale (NIHSS), days post-stroke onset, type of gait aid, 6-min walking distance, maximum walking speed, functional ambulatory categories (FAC), and Mini-Mental State Examination.

26.2.2 Protocol and Equipment

The measurements were conducted on a rectangular course (16 m long, 7 m wide, with a total perimeter of 46 m) equipped with handrails. The task involved walking for approximately 70 s at a comfortable pace. This duration was chosen to balance the need to observe both initial and sustained cerebral oxygenation responses during walking. Rest periods of approximately 10 s of quiet standing were provided before and after the task. Baseline measurements were taken only after the participants had maintained a quiet standing posture for at least 30 s and visual confirmation was made that NIRS-measured parameters were stable.

Measurements of O_2Hb, HHb, and THb (μmol/L) in the PFC during walking and rest were obtained using a wearable continuous wave-NIRS device (Brain Activity Monitor Hb133, Astem, Kawasaki City, Japan). NIRS optode locations were determined using the International 10–20 system at Fp1 and Fp2, and the region of interest was set as PFC. The interchannel distances were 30 and 35 mm, corresponding to the distance between the source and detector. Near-infrared light wavelengths were set at 760 and 850 nm, and the sampling rate was set at 10 Hz. Using a video camera, the walking ability of each patient was assessed using walking speed (m/s), cadence (steps/s), and stride length (m).

26.2.3 Near-Infrared Spectroscopy and Task Performance Analysis

As data obtained from the NIRS may include spike-like artifacts caused by heartbeat, respiration, walking, and speech (which can induce physical movement of the NIRS device), we applied a moving average filter with a window size of 11 points to attenuate high-frequency fluctuations, including potential motion-related artifacts. While this approach helps reduce the impact of minor fluctuations, it may not completely eliminate motion artifacts, particularly those caused by large movements. Therefore, we utilized the built-in triaxial accelerometer of the NIRS device to identify periods of excessive motion, and visually inspected the corresponding O_2Hb and HHb signals. Segments with apparent motion artifacts, confirmed through accelerometer data and visual inspection of NIRS parameters, were excluded from the analysis. Further, given that the acquired NIRS signals may contain systemic hemodynamic fluctuations, including skin blood flow and other extracerebral influences, we employed a signal separation method based on the hemodynamic modalities proposed by Yamada et al. (2012) (Yamada et al. 2012). This method distinguishes between functional (cortical) and systemic components based on the characteristic correlations between changes in O_2Hb and HHb; specifically, a negative correlation (i.e., O_2Hb increases as HHb decreases) is typically associated with functional cortical activation, whereas a positive correlation is indicative of systemic physiological changes. This procedure was applied to further mitigate the effects of systemic and motion-related confounding components on cortical oxygenation signals. The analysis interval for the data obtained from the NIRS was set to 10 s before the start of the walking task (i.e., −10 s) and continued for a total of 90 s, including the task start at 0 s, task end at 70 s, and a 10-s rest period. The obtained data were all computed as changes from the baseline at −10 s, and the mean values were calculated by averaging the values over the preceding 5-s period. Walking speed, cadence, and stride length were calculated as the average values from the start to the end of the walking task.

26.2.4 Statistical Analysis

To investigate temporal variations during the task, we initially considered using a repeated-measures one-way ANOVA. However, the Shapiro–Wilk test revealed that the assumption of normality was violated for several time points. Although Mauchly's test confirmed that the assumption of sphericity was met, we opted for the Friedman test—a non-parametric alternative to repeated-measures ANOVA—due to the lack of normality. Wilcoxon signed-rank tests were conducted for post hoc comparisons using the −10 s baseline as the reference point, while Bonferroni correction was applied to adjust for multiple comparisons. A p-value <0.05 was deemed statistically significant.

26.3 Results

Data analysis was conducted for 50 patients who met the inclusion criteria and provided informed consent (Table 26.1). The Friedman test revealed significant time effects for O_2Hb, HHb, and THb during the walking task ($p < 0.001$ for all). Post hoc Wilcoxon signed-rank tests (with Bonferroni correction) indicated that O_2Hb significantly decreased from 40 s after the onset of walking, while HHb significantly increased from 45 s, and THb significantly decreased from 35 s ($p < 0.05$, see Fig. 26.1). The walking speed during the task was 1.0 (0.2) m/s, cadence was 1.8 (0.2) steps/s, and stride length was 0.5 (0.1) m (mean [standard deviation]).

Table 26.1 Patients' characteristics

	$n = 50$
Age, years	63.7 (12.8)
Sex (male/female)	37/13
Body Mass Index, km/m^2	23.7 (4.0)
Lesion type (INF/ICH/SAH)	37/7/6
Lesion depth (cortical/subcortical/mixed)	3/36/11
Lesion side (left/right/bilateral)	15/25/10
NIHSS on admission	5.1 (7.0)
NIHSS at discharge	0.7 (1.2)
Days post-stroke onset, days	8.7 (3.3)
Gait aids (none/walking stick/4-wheeled walker)	43/2/5
Maximum walking speed, m/s	1.5 (0.4)
6-min walking distance, m	414.4 (92.4)
Functional Ambulation Category (4/5)	15/35
Mini-Mental State Examination	28.3 (1.9)

INF cerebral infarction, *ICH* intracerebral hemorrhage, *SAH* subarachnoid hemorrhage, *NIHSS* National Institutes of Health Stroke Scale (mean [standard deviation])

26.4 Discussion

This study investigated the cerebral oxygenation dynamics in PFC during walking in ambulatory patients with acute stroke using NIRS. Our findings revealed a pattern in which both O_2Hb and THb levels decreased, whereas HHb levels increased approximately 30 s after the initiation of walking.

In patients with acute stroke, cerebral tissue and/or cerebral blood vessel ischemia due to impaired cerebral autoregulation may occur during early rehabilitation, even in patients with mild stroke severity. Prior reviews on healthy individuals have indicated an increase in O_2Hb during walking, whereas our results indicate a decrease in these parameters in acute stroke patients (Bishnoi et al. 2021). This contrast suggests that the changes observed in the present study may be specific to stroke pathology. Hyperventilation-induced hypocapnia, which can lead to cerebral vasoconstriction and reduced oxygen delivery, may possibly account for the observed changes in PFC oxygenation (Meng and Gelb 2015; Zhang et al. 2019). Additionally, fluctuations in blood pressure during walking may also influence cerebral autoregulation, potentially influencing oxygenation dynamics (Nogueira et al. 2022). The observed decrease in O_2Hb and THb levels, accompanied by an increase in HHb as measured by NIRS, is commonly interpreted as indicative of vascular constriction and tissue ischemia (Hillman 2014). Previous studies have reported an association between stroke severity, as measured by the NIHSS, and impaired cerebral autoregulation (Nogueira et al. 2022). The patients included in our study had a mean NIHSS score of 5.1 at onset, indicating mild stroke severity (Salinet et al. 2019). Furthermore, all patients demonstrated an FAC score of four or greater at the time of measurement, indicating functional independence in indoor walking. These findings underscore the importance of monitoring cerebral oxygenation dynamics during early rehabilitation, even in patients with mild stroke severity.

Our study had several limitations. First, the absence of a control group of healthy individuals limited our ability to attribute the observed changes solely to post-stroke conditions. Second, our examination focused exclusively on PFC and did not assess oxygenation dynamics in other cerebral regions. Although it is plausible that similar dynamics occur in other regions of the brain (Lim et al. 2022), our study design did not allow for the investigation of these potential changes beyond the PFC. Third, the measurement of oxygenation dynamics in the PFC in this study was conducted only during the 70 s of walking, and therefore, subsequent changes could not be captured. Future studies with extended measurement periods may provide insights into how oxygen dynamics evolve over longer durations of activity. Fourth, we did not measure the participants' CO_2 levels or blood pressure during walking. Given that hyperventilation-induced hypocapnia and blood pressure fluctuations may contribute to the observed changes in prefrontal oxygenation, future studies should incorporate end-tidal CO_2 monitoring and continuous blood pressure measurement to better assess their impact.

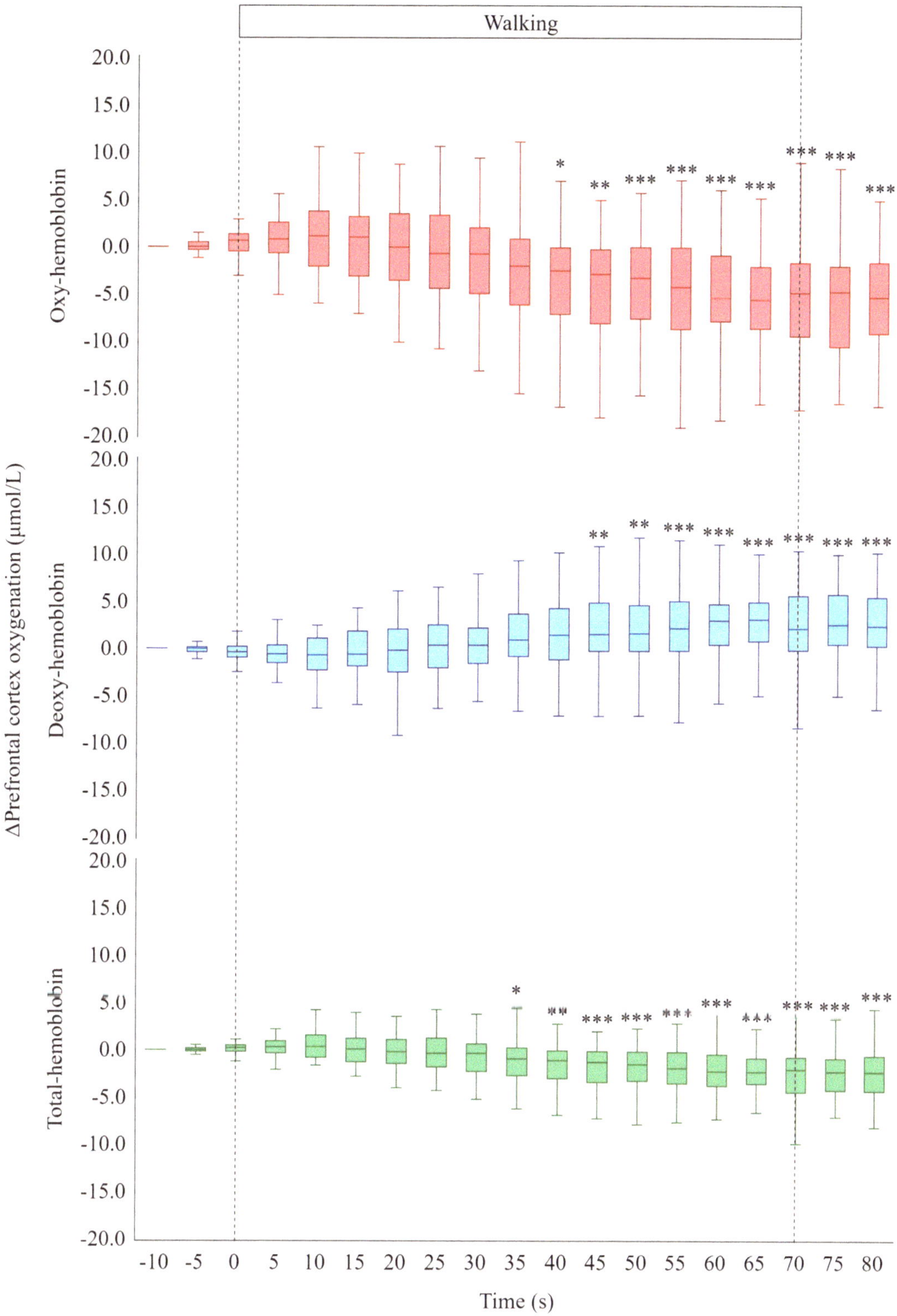

Fig. 26.1 Changes in the oxygenation dynamics of the prefrontal cortex, measured using near-infrared spectroscopy during walking. Boxplots represent temporal changes in oxy-hemoglobin, deoxy-hemoglobin, and total-hemoglobin concentrations, relative to the pre-task baseline (−10 s). Measurements were taken every 5 s during the 70-s walking task. Boxes indicate the interquartile range, lines inside the box denote the median, and whiskers represent the entire data range. $^{*}p < 0.05$ vs. −10 s. $^{**}p < 0.01$ vs. −10 s. $^{***}p < 0.001$ vs. −10 s

In conclusion, our study revealed decreased O_2Hb and THb levels and increased HHb levels in the PFC during walking in patients with acute stroke. This suggests the potential occurrence of cerebral tissue ischemia during early rehabilitation after stroke onset.

References

Bishnoi A, Holtzer R, Hernandez ME (2021) Brain activation changes while walking in adults with and without neurological disease: systematic review and meta-analysis of functional near-infrared spectroscopy studies. Brain Sci 11(3):291. https://doi.org/10.3390/brainsci11030291

Claassen JAHR, Thijssen DHJ, Panerai RB, Faraci FM (2021) Regulation of cerebral blood flow in humans: physiology and clinical implications of autoregulation. Physiol Rev 101:1487–1559. https://doi.org/10.1152/physrev.00022.2020

Herold F, Wiegel P, Scholkmann F, Müller NG (2018) Applications of functional near-infrared spectroscopy (fNIRS) neuroimaging in exercise¯cognition science: a systematic, methodology-focused review. J Clin Med 7:466. https://doi.org/10.3390/jcm7120466

Hillman EM (2014) Coupling mechanism and significance of the BOLD signal: a status report. Annu Rev Neurosci 37:161–181. https://doi.org/10.1146/annurev-neuro-071013-014111

Lim SB, Yang CL, Peters S, Liu-Ambrose T, Boyd LA, Eng JJ (2022) Phase-dependent brain activation of the frontal and parietal regions during walking after stroke - an fNIRS study. Front Neurol 19(13):904722. https://doi.org/10.3389/fneur.2022.904722

Liu Y, Yin JH, Lee JT, Peng GS, Yang FC (2022) Early rehabilitation after acute stroke: the golden recovery period. Acta Neurol Taiwan

Louie DR, Simpson LA, Mortenson WB, Field TS, Yao J, Eng JJ (2022) Prevalence of walking limitation after acute stroke and its impact on discharge to home. Phys Ther 102:pzab246. https://doi.org/10.1093/ptj/pzab246

Meng L, Gelb AW (2015) Regulation of cerebral autoregulation by carbon dioxide. Anesthesiology 122(1):196–205. https://doi.org/10.1097/ALN.0000000000000506

Nogueira RC, Aries M, Minhas JS, Petersen NH, Xiong L, Kainerstorfer JM, Castro P (2022) Review of studies on dynamic cerebral autoregulation in the acute phase of stroke and the relationship with clinical outcome. J Cereb Blood Flow Metab 42(3):430–453. https://doi.org/10.1177/0271678X211045222. Epub 2021 Sep 13

Salinet AS, Silva NC, Caldas J, de Azevedo DS, de-Lima-Oliveira M, Nogueira RC, Conforto AB, Texeira MJ, Robinson TG, Panerai RB, Bor-Seng-Shu E (2019) Impaired cerebral autoregulation and neurovascular coupling in middle cerebral artery stroke: influence of severity? J Cereb Blood Flow Metab 39:2277–2285. https://doi.org/10.1177/0271678X18794835

Truijen J, Rasmussen LS, Kim YS, Stam J, Stok WJ, Pott FC, van Lieshout JJ (2018) Cerebral autoregulatory performance and the cerebrovascular response to head-of-bed positioning in acute ischaemic stroke. Eur J Neurol 25:1365–e117. https://doi.org/10.1111/ene.13737

Yamada T, Umeyama S, Matsuda K (2012) Separation of fNIRS signals into functional and systemic components based on differences in hemodynamic modalities. PLoS One 7:e50271. https://doi.org/10.1371/journal.pone.0050271

Zhang Z, Guo Q, Wang E (2019) Hyperventilation in neurological patients: from physiology to outcome evidence. Curr Opin Anaesthesiol 32(5):568–573. https://doi.org/10.1097/ACO.0000000000000764

Cerebral Edema Depends on the Results of Large Vascular Occlusion Reperfusion Therapy

27

Alex O. Trofimov, Ekaterina Volkova, Edwin M. Nemoto, Olga A. Bragina, and Denis E. Bragin

Abstract

The objective was to quantify the effect of endovascular therapy (EVT) results of large-vessel occlusion (LVO) on brain tissue hydration, assessed by net water uptake (NWU). *Materials and Methods.* Ninety-three patients with EVT of acute ischemic stroke (AIS) due to acute LVO (men-44, women-49, median age 68.7) were included in this retrospective, non-randomized, single-center study. EVT was performed under general anesthesia with approved devices, including a stent retriever and aspiration catheters. Net water uptake (NWU) in ischemic zones was evaluated using non-contrasted computed tomography (CT). NWU was calculated using the well-known formula in the symmetrical zone of the brain, in admission and follow-up non-contrasted CT scans, 24 h after EVT. *Results.* Fifty-seven patients had successful arterial recanalization (Thrombolysis in Cerebral Infarction scale [TICI] 2b-3), while 36 patients had persistent or recurrent acute LVO (TICI 0-2a). The NWU median at admission was 3.4% in patients with successful EVT and 5.1% in patients with unsuccessful EVT ($p < 0.05$). The follow-up NWU median was 7.2% after successful EVT and 12.8% in patients with persistent/recurrent acute LVO ($p < 0.001$). *Conclusion.* Successful arterial recanalization at AIS due to LVO (TICI 2b – 3) led to a significant reduction in brain edema progression in cerebral ischemic foci.

Keywords

Net water uptake · Endovascular reperfusion treatment · Ischemic stroke

27.1 Introduction

Acute ischemic stroke (AIS) remains one of the leading causes of disability in cerebrovascular diseases, which places a significant burden on public health. Modern randomized clinical trials have convincingly demonstrated the advantages of endovascular reperfusion therapy (EVT) during the so-called "therapeutic window" compared

A. O. Trofimov (✉) · E. Volkova
Department of Neurological Diseases, Privolzhsky Research Medical University, Nizhny Novgorod, Russia

E. M. Nemoto
Department of Neurology, University of New Mexico School of Medicine, Albuquerque, NM, USA

O. A. Bragina
Lovelace Biomedical Research Institute, Albuquerque, NM, USA

D. E. Bragin
Department of Neurology, University of New Mexico School of Medicine, Albuquerque, NM, USA

Lovelace Biomedical Research Institute, Albuquerque, NM, USA

S. C. Pias et al. (eds.), *Oxygen Transport to Tissue XLVI*, Advances in Experimental Medicine and Biology 1498, https://doi.org/10.1007/978-3-032-10389-5_27

with a conservative approach to AIS treatment (Guo et al. 2024). Later interventions are usually accompanied by dangerous complications, which include post-reperfusion cerebral edema and hemorrhagic transformation, leading to brain swelling and herniation (Morsi et al. 2024).

Physiologically, it is based on cell destruction in the ischemic core (cytotoxic edema) and blood-brain barrier disruption (vasogenic edema). Although the distinction between these phenotypes is of great importance, in practice, their total effect is much more important (Yi et al. 2023). For the assessment of brain edema using non-contrasted computed tomography (CT), a new densitometric parameter (net water uptake—NWU) has been proposed (Broocks et al. 2022).

Since the edema dynamics in favorable and unfavorable EVT results subgroups remain poorly understood, we hypothesized that they may depend on recanalization success. The aim was to find out how endovascular therapy (EVT) in response to large vascular occlusion (LVO) influences brain tissue hydration measured by the net water uptake (NWU) method.

27.2 Methods

27.2.1 Design and Study Population

Ninety-three patients with EVT of AIS due to acute LVO (men-44, women-49, median age 68.7 [min 48; max 79]) were included in this retrospective, non-randomized, observational, single-center study that complied with Helsinki Declaration Standards. The protocol was reviewed and approved by the Local Ethical Committee.

Inclusion criteria were as follows: (1) AIS due to acute LVO of the M1–2 segments of the middle cerebral artery (MCA) or C6–7 segments of the internal carotid artery (ICA) (Bouthillier et al. 1996); (2) admission within 6 h after known AIS onset; (3) non-large AIS (Alberta Stroke Program Early CT Score (ASPECTS) more than 5 point); (4) absence of hemorrhagic transformation/contrast leakage on CT scans at follow-up CT; (5) EVT during first 6 h after AIS onset. Exclusion criteria were as follows: (1) unknown stroke onset or more than 6 h after known AIS onset; (2) age less than 18 years and more than 80 years; (3) occlusion of the C1–5 segments of ICA and/or tandem ICA/MCA occlusion; (4) ASPECTS ≤ 5; (5) any posterior circulation system lesions.

Imaging Analysis Net water uptake (NWU) in ischemic zones was calculated using non-contrasted CT. For establishing ischemic zones, all patients received multiphase perfusion CT (PCT) at admission on a 160-slice scanner (Canon Aquilion Prime SP, Canon Medical Systems, Otawara, Tochigi, Japan). The examinations were performed as non-contrasted CT with helical scanning using these parameters: 160×0.5 mm collimation, auto mA, 120kVp, 1.35 beam pitch, 0.5 s gantry rotation time, 512×512 matrix, and 280–370 mm field of view. The whole thin-section CT volume was then reconstructed with a back-projection filter provided by the hybrid iterative reconstruction (AIDR 3D, Canon Medical). All CT images were transferred to the PACS for further analysis. Anonymized PCT data were processed using the Vitrea workstation (Vitrea FX, Vital Images, USA). The ischemic thresholds were as follows: CBV <2.0 ml/100 g and MTT $>145\%$ compared with the contralateral hemisphere. NWU was calculated using the well-known formula, modified according to the study aim: NWU = (1-Dischemia/Dnormal) × 100%, where Dischemia—the ischemic focus density (Hounsfield unit—HU), Dnormal—the density of normal brain tissue in the symmetrical zone of the contralateral hemisphere in admission (1) and follow-up (2) non-contrasted CT scans during 24 h after EVT.

Endovascular Procedure All EVT procedures were performed on an ARTIS zee biplane angiography system (Siemens, Germany) under general anesthesia and neuromonitoring (blood pressure, heart rate, bispectral index, etc.) using the ADAPT method, using standard stent retrievers and reperfusion aspiration catheters

3MAX and aspiration pumps (Penumbra, USA). The number of maneuvers during the intervention was determined by the neurointerventionist depending on the current surgical situation. EVT was performed under general anesthesia. The EVT success for AIS was measured by the Thrombolysis in Cerebral Infarction (TICI) score (Table 27.1) (Zaidat et al. 2013).

Statistical Analysis Data were not normally distributed and expressed as median [interquartile range]. *U* test Mann-Whitney was used for continuous variables, and χ^2 test was used to compare dichotomous variables between EVT results groups. The level of significance was $p < 0.05$. The software package Statistica 12 (TIBCO Inc., USA) was used for data analysis.

27.3 Results

In our cohort, 57 patients (61.2%, men—27, women—30, median age 65.2 [43;72]) had successful arterial recanalization (TICI 2b-3), while 36 patients (38.8%, men—17, women—19, median age 68.7 [54; 78]) had persistent or recurrent acute LVO (TICI 0-2a) (Table 27.2). The NWU median at admission was 3.4% [1.2; 7.4] in patients with successful EVT and 5.1% [4.8; 6.1] in patients with unsuccessful EVT and persistent acute LVO ($p < 0.05$). The follow-up NWU median was 7.2% [5.8; 8.2] after successful EVT and 12.8% [10.3; 14.4] in patients with persistent/recurrent acute LVO ($p < 0.001$).

The median mRS in TICI 2b-3 patients was significantly higher ($p = 0.008$) than in the TICI

Table 27.1 Thrombolysis in Cerebral Infarction Classification (TICI)

Grade	Angiographic appearances
0	No perfusion
1	Penetration with minimal perfusion
2	Partial perfusion
2a	Only partial filling (less than two-thirds) of the entire vascular territory is visualized
2b	Complete filling of all of the expected vascular territory is visualized but the filling is slower than normal
3	Complete perfusion

Table 27.2 Comparison of the analyzed parameters

	TICI 0-2a ($n = 36$)	TICI 2b-3 ($n = 57$)	*p*
NWU at admission %	5.1% [4.8; 6.1]	3.4 [1.2; 7.4]	< 0.05[a]
Follow-up NWU %	12.8 [10.3; 14.4]	7.2 [5.8; 8.2]	< 0.05[a]
ASPECTS	7 [5;10]	8 [5;10]	0.822
Age, yo	74,2 [65;79]	71,3 [67;75]	0.785
NIHSS at admission	15 [13;18]	14 [11;16]	0.893
SBP (mmHg)	155 [123;177]	151 [132;169]	0.863
Male sex n (%)	17 (47.2)	27 (47.4)	ns
Hypertension n (%)	23 (64)	33 (57.9)	ns
Diabetes mellitus n (%)	22 (61)	25 (43.9)	ns
Coronary artery disease n (%)	13 (36)	15 (26.3)	ns
Previous stroke	11 (30.5)	13 (22.8)	ns
Thrombus in M1 MCA	22 (61.1)	38 (66.7)	ns
Thrombus in M2 MCA	10 (27.8)	10 (17.5)	<0.05[a]
Intracranial ICA	4 (11.1)	9 (15.8)	ns
mRS	4 [3;5]	5 [4;6]	0.008[a]

Values are presented as median [interquartile range] or n (%). *NIHSS* National Institutes of Health Stroke Scale, *SBP* systolic blood pressure, *ASPECTS* Alberta Stroke Program Early CT Score, *ICA* internal carotid artery, *EVT* endovascular treatment, *TICI* Thrombolysis in Cerebral Infarction score, *mRS* modified Rankin Scale, *ns* not significant difference, $p > 0.05$

[a]Significant difference, $p < 0.05$

0-2a patients (5 (Yi et al. 2023; Broocks et al. 2022; Bouthillier et al. 1996; Zaidat et al. 2013) vs. 4 (Morsi et al. 2024; Yi et al. 2023; Broocks et al. 2022; Bouthillier et al. 1996)). The mortality rate was significantly lower in the patients with favorable EVT results (14% (n = 8) vs. 41.7% (n =15), p <0.05, respectively).

27.4 Discussion

We studied the changes in brain tissue hydration using quantitative NWU changes compared to EVT results at acute LVO.

In the case of successful recanalization (TICI 2b-3), the significant increase was more than two-fold (3.4% versus 7.2%, p <0.05), while in case of unsuccessful recanalization (TICI 0-2a), the significant increase in NWU was almost three-fold compared to the level of NWU at admission (5.1% versus 12.8%, p <0.05).

There might be several reasons for such dynamics of the NWU. First, perfusion restoration does not mean the survival of all cells in the "at-risk" area. Some of them die, which leads to the development of cytotoxic edema. Second, although the blood-brain barrier damage reaches its maximum after 2–3 days, the glymphatic structures are damaged much faster, which leads to vasogenic edema (Zhu et al. 2024). In addition, LVO may recur even after successful arterial recanalization, and therefore, in some patients with TICI 2b-3, an NWU increase might indicate hidden (undetected) recurrent episodes of LVO. However, these mechanisms still remain poorly understood and unexplored (Meyer et al. 2021).

In the case of M2 segment occlusion, successful recanalization was more difficult than in its absence, which is consistent with previous studies. This results in an increase in the number of EVT maneuvers and intervention duration. Although these factors were not published and analyzed in this study, they may have had a greater impact on water homeostasis than expected (Menon et al. 2019). All this requires further research.

This study had some limitations. The first limitation arises from the study's retrospective design. Second, the study cohort was relatively small and not randomized; therefore, the results require validation in a multicenter randomized trial.

27.5 Conclusion

Successful arterial recanalization at AIS due to LVO (TICI 2b −3) led to a significant reduction in brain edema progression in cerebral ischemic foci.

Acknowledgments DB was supported by NIH R01 NS112808.

References

Bouthillier A, van Loveren H, Keller J (1996) Segments of the internal carotid artery: a new classification. Neurosurgery 38(3):425–433

Broocks G, Meyer L, Ruppert C et al (2022) Effect of intravenous alteplase on functional outcome and secondary injury volumes in stroke patients with complete endovascular recanalization. J Clin Med 11(6):1565. https://doi.org/10.3390/jcm11061565

Guo S, Qin S, Tan S, Su H, Chen X (2024) Endovascular thrombectomy without versus with different pre-intravenous thrombolysis in acute ischemic stroke: a network meta-analysis of randomized controlled trials. Front Neurol 15:1344961. https://doi.org/10.3389/fneur.2024.1344961

Menon BK, Hill MD, Davalos A et al (2019) Efficacy of endovascular thrombectomy in patients with M2 segment middle cerebral artery occlusions: meta-analysis of data from the HERMES Collaboration. J Neurointerv Surg 11(11):1065–1069. https://doi.org/10.1136/neurintsurg-2018-0146788

Meyer L, Schönfeld M, Bechstein M et al (2021) Ischemic lesion water homeostasis after thrombectomy for large vessel occlusion stroke within the anterior circulation: the impact of age. J Cereb Blood Flow Metab 41(1): 45–52. https://doi.org/10.1177/0271678X20915792

Morsi S, Ghozy S, Elfil M et al (2024) Endovascular thrombectomy for DAWN- and DEFUSE-3 ineligible acute ischemic stroke patients: a systematic review and meta-analysis. J Neurol 271(5):2230–2237. https://doi.org/10.1007/s00415-024-12198-3

Yi T, Li K, Lin XH et al (2023) Predictors of futile recanalization in basilar artery occlusion patients undergoing endovascular treatment: a post hoc analysis of the ATTENTION trial. Front Neurol 14:1308036. https://doi.org/10.3389/fneur.2023.1308036

Zaidat O, Yoo A, Khatri P et al (2013) Recommendations on angiographic revascularization grading standards for acute ischemic stroke: a consensus statement. Stroke 44(9):2650–2663. https://doi.org/10.1161/STROKEAHA.113.001972

Zhu J, Mo J, Liu K et al (2024) Glymphatic system impairment contributes to the formation of brain edema after ischemic stroke. Stroke 55(5):1393–1404. https://doi.org/10.1161/STROKEAHA.123.045941

Net Water Uptake and Cerebral Microcirculation in Perilesional Penumbra in TBI: CT Perfusion Study

28

Alex O. Trofimov, Ilya Kochsheev, George Kalentyev, Ekaterina Volkova, Edwin M. Nemoto, Olga Bragina, and Denis E. Bragin

Abstract

The aim of our study was to compare brain tissue net water uptake (NWU), cerebral oxygenation, and microcirculation changes in perilesional penumbra foci (PPF) in moderate isolated traumatic brain injury (moiTBI) patients. *Materials and Methods*: 77 moiTBI patients (women 35; men 42, age 37 years [34;39]) with unilateral frontal/frontotemporal lesions (Marshall II-III) were included in this retrospective, non-randomized, single-center study. Perfusion parameters were measured in PPF zones and determined using multiphase perfusion computed tomography (PCT) in the first 2 days after trauma and admission. NWU in hypoattenuated perilesional zones was calculated using non-contrast CT. Cerebral oxygenation ($SctO_2$) was simultaneously measured using near-infrared spectroscopy in the frontal lobes with PCT. Data are shown as a median [interquartile range]. Statistical analysis was performed using nonparametric statistics. $P < 0.05$ was considered statistically significant. *Results*: In PPF, $SctO_2$ was 64.03% [60.1;67.0], cerebral blood flow (CBF) was 30.6 ml/100 g/min [22.3; 38.7], cerebral blood volume (CBV) was 2 ml/100 g [1.6; 2.5], mean transit time (MTT) was 4.8 s [4.1;5.8], time to peak (TTP) was 21.8 s [20;24.2], NWU was 4.98% [2.21;7.39]. CBF, CBV, and $SctO_2$ in PPF were significantly lower; MTT was more prolonged than in non-damaged tissue ($p < 0.001$). Increased NWU values in PPF were significantly correlated with CBV ($r = 0.723$, $p < 0.001$); CBF ($r = -0.228$, $p = 0.048$), TTP (r 0.225, $p = 0.047$), and $SctO_2$ ($r = -0.608$, $p < 0.001$). There were no significant correlations between NWU_{PPF} and MTT. *Conclusions*: Brain edema in PPF was significantly associated with cerebral microcirculation and cerebral O_2 saturation disturbances in moiTBI.

Keywords

Net water uptake · Cerebral oxygenation · Perilesional penumbra · Perfusion CT

A. O. Trofimov (✉) · I. Kochsheev · G. Kalentyev · E. Volkova
Department of Neurological Diseases, Privolzhsky Research Medical University, Nizhny Novgorod, Russia

E. M. Nemoto
Department of Neurology, University of New Mexico School of Medicine, Albuquerque, NM, USA

O. Bragina
Lovelace Biomedical Research Institute, Albuquerque, NM, USA

D. E. Bragin
Department of Neurology, University of New Mexico School of Medicine, Albuquerque, NM, USA

Lovelace Biomedical Research Institute, Albuquerque, NM, USA

S. C. Pias et al. (eds.), *Oxygen Transport to Tissue XLVI*, Advances in Experimental Medicine and Biology 1498, https://doi.org/10.1007/978-3-032-10389-5_28

28.1 Introduction

Traumatic brain injury (TBI) remains one of the leading causes of mortality and disability in the world (Yue et al. 2024). This is mainly due to the high heterogeneity of its forms and specific brain reactions to parenchymal damage, such as the formation of an edematous zone around intracerebral hemorrhage, which is designated as the perilesional penumbra zone (by analogy with the penumbra of stroke) (Sun et al. 2024). However, the processes occurring in the posttraumatic penumbra have significant differences from the ischemic one (Vlodavsky et al. 2005). It has been shown that its formation occurs against the background of higher cerebral blood flow (CBF) and is associated not only with cell damage but also with endothelial disorders due to blood-brain barrier (BBB) disruption (Trabold et al. 2010). As a result, posttraumatic edema has a mixed nature (both cytotoxic and vasogenic), which has been confirmed in many studies (Stoffel et al. 1997). Although ischemia in TBI turned out to be much more common than previously thought, and its characteristics have significant differences from stroke, the perfusion parameters in the perilesional penumbra focus (PPF) remain insufficiently studied. The use of the net water uptake parameter made it possible to establish the basic patterns of the brain edema dynamics during ischemic stroke. Still, its use for studying posttraumatic edema is limited to single studies (Schwarzmaier et al. 2010; van Horn et al. 2022). Meanwhile, the uncoupling of cerebral oxygenation and microcirculation is interconnected with brain hydration dynamics in this area and is of great importance for understanding the characteristics of intracerebral hemorrhage pathomorphosis. The aim of this study was to compare brain net water uptake (NWU), cerebral oxygenation, and microcirculation changes in PPF in moderate isolated traumatic brain injury (moiTBI) patients.

28.2 Methods

28.2.1 Design and Study Population

The protocol of this retrospective, nonrandomized, single-center study was approved by the Local Ethics Committee. Seventy-seven adult patients of Regional Trauma Center Level I with moderate isolated TBI (moiTBI) were included in this study (women 35; men 42, median age 37 years [34;39]). All patients were conscious upon admission; thus, informed consent was obtained from all of them.

Inclusion criteria were acute moderate isolated TBI; Glasgow Coma Score (GCS) more than 10 and less than 13 points; perfusion computed tomography (PCT) performed in the first 2 days after trauma and admission; unilateral frontal/frontotemporal lesions with midline shift of 0 to 5 mm, basal cisterns remain visible, no high or mixed density lesions >25 cm^3, and swelling (Marshall Classification II-III).

The exclusion criteria were age >60 years, GCS more the 12 and less than 11 points, and concomitant TBI.

28.2.2 Image/Data Acquisitions

All patients received multiphase PCT on a 160-slice scanner (Canon Aquilion Prime SP, Canon Medical Imaging, Japan). After native CT, the bolus (30–50 mL) of the contrast agent (Ultravist 370, Schering AG, Germany) was administered using an automatic injector (Stellant, Medrad, USA) into a peripheral vein of the right hand. The scanning parameters were: 160 kVp, 160 mA, 70 mAs, 512 × 512. PCT data were analyzed using Vitrea software (Vitrea FX, Vital Images, USA). The calculation of cerebral blood flow (CBF), cerebral blood volume (CBV), mean transit time (MTT), and time to peak (TTP) was performed on a voxel-wise basis in frontotemporal hypoattenuating perilesional PPF. The voxels with cerebral blood flow >100 mL/100 g/min and/or cerebral blood volume >8 mL/100 g were removed from the perfusion map because they might belong to large brain vessels.

Brain hydration was calculated using net water uptake on baseline CT images in the same hypoattenuated perilesional focus using the modified well-known eq. (Steffen et al. 2024):

$$NWU_{PPF} = (1 - D_{PPF}/D_{\text{normal}}) \times 100\%$$

where NWU_{PPF}—net water uptake in PPF, D_{PPF}—the PPF density, D_{normal}—the density of normal brain tissue in the symmetrical zone of the contralateral hemisphere (Hounsfield Unit).

Cerebral oxygen saturation ($SctO_2$) was measured using near-infrared spectroscopy (NIRS) (Fore-Sight 2030 MS (CAS Medical Systems Inc., Branford, CT, USA)) in the frontal lobes simultaneously with PCT or immediately after.

28.2.3 Statistical Analysis

Data are shown as a median [interquartile range]. The parametric and nonparametric statistics methods were used for analysis. Pairwise correlations between studied parameters were tested. The level of significance was $p < 0.05$. The software Statistica 12 (TIBCO Inc., Palo Alto, CA, USA) was used for data analysis.

28.3 Results

In PPF, $SctO_2$ was 64.03% [60.1;67.0], CBF was 30.6 ml/100 g/min [22.3; 38.7], cerebral blood volume (CBV) was 2 ml/100 g [1.6; 2.5], mean transit time (MTT) was 4.8 s [4.1;5.8], time to peak (TTP) was 21.8 s [20;24.2], NWU was 4.98% [2.21;7.39]. In non-damaged tissue, $SctO_2$ was 73.5% [67.3;84.7], CBF was 38.2 ml/100 g/min [35.1; 39.2], cerebral blood volume (CBV) was 2.7 ml/100 g [2.3; 3.0], mean transit time (MTT) was 4.2 s [3.8;4.6], time to peak (TTP) was 20.2 s [18.7;22.2].

CBF, CBV, and $SctO_2$ in PPF were significantly lower; MTT was substantially more prolonged than in non-damaged tissues (Table 28.1). Increased NWU_{PPF} values were significantly correlated with CBV ($r = 0.723$, $p < 0.001$); CBF ($r = -0.228$, $p = 0.048$), TTP (r 0.225, $p = 0.047$), and $SctO_2$ ($r = -0.608$, $p < 0.001$). There were no significant correlations between NWU_{PPF} and TTP, MTT ($p > 0.05$). Moreover, no significant correlations existed between NWU level and Glasgow Coma Scale score or Glasgow Outcome Scale ($p > 0.05$). However, we found significant correlations between $SctO_2$ and the Glasgow Coma Scale score ($R^2 = 0.018$, $p < 0.05$).

28.4 Discussion

As Brooks et al. show, the brain's density in the X-ray spectrum is linearly correlated with the degree of hydration of the brain tissue. Thus, measuring the network water absorption of the brain is similar to a direct volumetric measurement by measuring the density (Broocks et al. 2018). This was confirmed in an experimental study by West et al. (1976). In addition, it was shown that edema formed after 15 min of ischemia or more, while hydration increased proportionally to its duration and, upon reaching 10%,

Table 28.1 Analyzed cohort data

	In PPF	In non-damaged tissue	P
CBF ml/100 g/min)	30.6[22.3; 8.7]	38.2 [35.1; 39.2]	<0.01
CBV (ml/100 g)	2.0 [1.6; 2.5]	2.7 [2.3; 3.0]	<0.01
MTT (s)	4.8 [4.1;5.8]	4.2 [3.8;4.6]	<0.01
TTP (s)	21.8 [20;24.2]	20.2 [18.7;22.2]	0.121
$SctO_2$ (%)	64.03 [60.1;67.0]	73.5 [67.3;84.7]	<0.001

caused irreversible electrolyte disturbances due to cell death, i.e., the edema became cytotoxic. At the same time, it was shown that edema of less than 10% was potentially reversible (Watanabe et al. 1977). Although ischemic stroke and TBI differ significantly in pathophysiology, it is evident that cerebral edema (both vasogenic and cytotoxic) plays a significant role in both. It should be noted that cerebral edema can occur both in the case of primary injury, when mitochondrial dysfunction leads to intracellular sodium accumulation, and in the case of secondary injury, leading to dysfunction of the blood-brain barrier and intercompartmental movement of water into the intercellular space (Kenda et al. 2024). Our results demonstrate that in the PPF zone, CBF, CBV, and $SctO_2$ remained at the lower limit of normal values and did not reach the "ischemic border." In addition, the NWU_{PPF} did not exceed the threshold of 10% in any case. Although this rather confirms that the PPF is apparently not only ischemic, the presence of a high correlation between oxygen saturation and NWU_{PPF} does not exclude the cytotoxic component in PPF. It may be nonlinear since there was no correlation between NWU_{PPF} and GCS severity, while the relationship between $SctO_2$ and GCS was significant. In this study, we could not answer the question of how perfusion and hydration in the PPF will change with cerebral oxygen saturation with an increase in the partial pressure of oxygen in the inspired air (e.g., during normobaric or hyperbaric hyperoxia). In other words, will hyperoxia lead to decreased hydration in the PPF, and if so, to what extent? In addition, it is possible that in the PPF zone, tissue oxygenation could be as "mosaic" and heterogeneous as pial cerebral blood flow (Bellapart et al. 2016). This is indicated by the difference between the median MTT_{PPF} values and those calculated by CT, which is not observed in intact tissues. It is likely that rapidly changing local oxygenation in the PPF zone can cause such rapid lumen microvessel changes that microcirculation indices can change during the PCT scan itself. This assumption requires further study. However, this may open new perspectives in the treatment of patients with parenchymal lesions.

28.5 Conclusion

Brain edema in PPF was significantly associated with altered cerebral microcirculation and cerebral O_2 saturation in moiTBI.

Acknowledgments DB was supported by NIH R01 NS112808.

References

Bellapart J, Abi-Fares C, Cuthbertson K et al (2016) Cerebral microcirculation during mild head injury after a contusion and acceleration experimental model in sheep. Brain Inj 30(13–14):1542–1551. https://doi.org/10.1080/02699052.2016.1199894

Broocks G, Flottmann F, Ernst M et al (2018) Computed tomography-based imaging of voxel-wise lesion water uptake in ischemic brain: relationship between density and direct volumetry. Investig Radiol 53:207–213

Kenda M, Lang M, Nee J, Hinrichs C et al (2024) Regional brain net water uptake in computed tomography after cardiac arrest - a novel biomarker for neuroprognostication. Resuscitation 200:110243. https://doi.org/10.1016/j.resuscitation.2024.110243

Schwarzmaier SM, Kim SW, Trabold R, Plesnila N (2010) Temporal profile of thrombogenesis in the cerebral microcirculation after traumatic brain injury in mice. J Neurotrauma 27(1):121–130. https://doi.org/10.1089/neu.2009.1114

Steffen P, Winkelmeier L, Kniep H et al (2024) Quantification of ischemic brain edema after mechanical thrombectomy using dual-energy computed tomography in patients with ischemic stroke. Sci Rep 14(1):4148. https://doi.org/10.1038/s41598-024-54600-0

Stoffel M, Eriskat J, Plesnila M, Aggarwal N, Baethmann A (1997) The penumbra zone of a traumatic cortical lesion: a microdialysis study of excitatory amino acid release. Acta Neurochir Suppl 70:91–93. https://doi.org/10.1007/978-3-7091-6837-0_28

Sun A, Cao Y, Jia Z, Zhao L, Shi H, Liu S (2024) Prognostic value of CBV index in patients with acute ischemic stroke treated with endovascular thrombectomy in the late therapeutic window. Front Neurol 14:1282159. https://doi.org/10.3389/fneur.2023.1282159

Trabold R, Erös C, Zweckberger K et al (2010) The role of bradykinin B(1) and B(2) receptors for secondary brain damage after traumatic brain injury in mice. J Cereb Blood Flow Metab 30(1):130–139. https://doi.org/10.1038/jcbfm.2009.196

van Horn N, Broocks G, Kabiri R et al (2022) Cerebral hypoperfusion intensity ratio is linked to progressive early edema formation. J Clin Med 11(9):2373. https://doi.org/10.3390/jcm11092373

Vlodavsky E, Palzur E, Feinsod M, Soustiel JF (2005) Evaluation of the apoptosis-related proteins of the BCL-2 family in the traumatic penumbra area of the rat model of cerebral contusion, treated by hyperbaric oxygen therapy: a quantitative immunohistochemical study. Acta Neuropathol 110(2):120–126. https://doi.org/10.1007/s00401-004-0946-8

Watanabe O, Bremer AM, West CR (1977) Experimental regional cerebral ischemia in the middle cerebral artery territory in primates. Part 1: angio-anatomy and description of an experimental model with selective embolization of the internal carotid artery bifurcation. Stroke 8:61–70

West CR, Hong CI, Bremer AM, Bourke RS, Tower DB (1976) Hemispheral ischemic cerebral cortical edema in a primate (M. mulatta). In: Pappius HM, Feindel W (eds) Dynamics of brain edema. Springer, Berlin, Heidelberg, pp 244–253

Yue JK, Etemad LL, Zafonte RD (2024) Prior traumatic brain injury is a risk factor for in-hospital mortality in moderate to severe traumatic brain injury: a TRACK-TBI cohort study. Trauma Surg Acute Care Open 9(1): e001501. https://doi.org/10.1136/tsaco-2024-001501

Uncontrolled Orthostatic Hypotension and Multi-channel Cerebral Hemodynamic Changes in a Young Male Participant: A Case Report

29

Atsuhiro Tsubaki, Sho Kojima, Hajime Tamiya, Shinichiro Morishita, Soshi Samejima, and Kentaro Iwata

Abstract

We aimed to determine the changes in mean arterial pressure (MAP) due to uncontrolled orthostatic hypotension and the oxygenated hemoglobin(O_2Hb), deoxygenated hemoglobin (HHb), and total hemoglobin (THb) changes in a young male participant. The participant (23 y.o. male, no medication) stood in a quiet room before the main experiment. Changes in O_2Hb, HHb, and THb levels were detected using NIRS optodes comprising 12 light sources and 12 detectors, providing 24-channel simultaneous recording. A 30-mm inter-optode distance was set for measuring cortical tissue hemodynamics. The Cz position of the international 10–20 system was used to ensure consistent optode placement. The regions of interest were the right (R-) and left (L-) prefrontal cortex (PFC), supplementary motor area (SMA), and sensorimotor cortex (SMC). We also measured MAP and skin blood flow (SBF) from the forehead and hemodynamic changes in the left vastus lateralis (VL) muscle. The participant stood for 600 s, including baseline measurements. The MAP decreased rapidly at approximately 500 s. A decrease was noted in the O_2Hb and THb of the SMA from 420 s and L- and R-PFC from 450 s, but was more pronounced after 500 s. The HHb signal showed a change opposite to that of O_2Hb and THb. The SBF decreased momentarily at approximately 500 s but increased immediately after and did not recover until the end of the measurement at 540 s. The O_2Hb and THb of the VL increased from 500 to 540 s. NIRS may detect uncontrolled orthostatic hypotension before a decrease in the MAP.

Keywords

Orthostatic hypotension · Near-infrared spectroscopy · Multichannel · Mean arterial pressure

A. Tsubaki (✉) · S. Kojima · H. Tamiya
Institute for Human Movement and Medical Sciences, Niigata University of Health and Welfare, Niigata, Japan
e-mail: tsubaki@nuhw.ac.jp

S. Morishita
Department of Physical Therapy, School of Health Sciences, Fukushima Medical University, Fukushima, Japan

S. Samejima
Division of Physical Therapy, Department of Rehabilitation Medicine, University of Washington, Seattle, WA, USA

K. Iwata
Department of Rehabilitation, Kobe City Medical Center General Hospital, Kobe, Japan

S. C. Pias et al. (eds.), *Oxygen Transport to Tissue XLVI*, Advances in Experimental Medicine and Biology 1498, https://doi.org/10.1007/978-3-032-10389-5_29

29.1 Introduction

Orthostatic hypotension (OH) is an unusually large decrease in blood pressure when standing, mostly occurring among older adults and young people (Wieling et al. 2022). Blood pressure and cortical signal changes due to posture change (i.e., supine to standing) are measured using near-infrared spectroscopy (NIRS) to detect OH and related symptoms. It is reasonable to reproduce these results in clinical or experimental settings. However, OH develops in the same posture, and the relationship between blood pressure and cortical signal changes remains unclear.

Herein, we aimed to report OH-related blood pressure and cortical signal changes in the standing posture.

29.2 Methods

29.2.1 Participant

The participant was a 23-year-old male university student with no history of cardiovascular, neurodegenerative, or cerebrovascular disease, and no previous OH history. He was not taking any medications. Before measuring the cortical signal changes during the walking task, we allowed the participant to stand for 10 min as a baseline measurement. During this measurement, OH occurred spontaneously, and we successfully recorded it. He received oral and written explanations of the objectives, measurement techniques, and risks and benefits of the investigation and consented to participate in the study. We also received informed consent from him to publish this case report. This study was approved by the Ethics Committee of Niigata University of Health and Welfare and was conducted in accordance with the Declaration of Helsinki.

29.2.2 IRS Measurements

To monitor cortical hemodynamics, a multichannel NIRS imaging system (OMM-3000; Shimadzu Co., Kyoto, Japan) with multiple continuous wavelengths (780, 805, and 830 nm) was used to detect changes in oxyhemoglobin (O_2Hb_C), deoxyhemoglobin (HHb_C), and total hemoglobin (THb_C) levels at a sampling rate of 130 ms based on the modified Beer-Lambert law (Boas et al. 2001). The NIRS optodes consisted of 12 light-source fibers and 12 detectors, which provided 24-channel simultaneous recording. The optodes were arranged in a 4 × 6 pattern in a multichannel probe holder, and a 30-mm inter-optode distance was set to measure cortical tissue oxygenation. The Cz position of the international 10–20 system was used to ensure consistent optode placement for the participant. The regions of interest (ROIs) included the right (R-) and left (L-) prefrontal cortex (PFC), sensorimotor area (SMA), and primary motor cortex (M1) (Fig. 29.1).

A NIR spectrometer (BOM-L1TRW, Omega Wave Co., Ltd., Tokyo, Japan) was placed on the left vastus lateralis muscle to measure the muscle oxygenation profile. Changes in oxyhemoglobin (O_2Hb_{VL}), deoxyhemoglobin (HHb_{VL}), and total

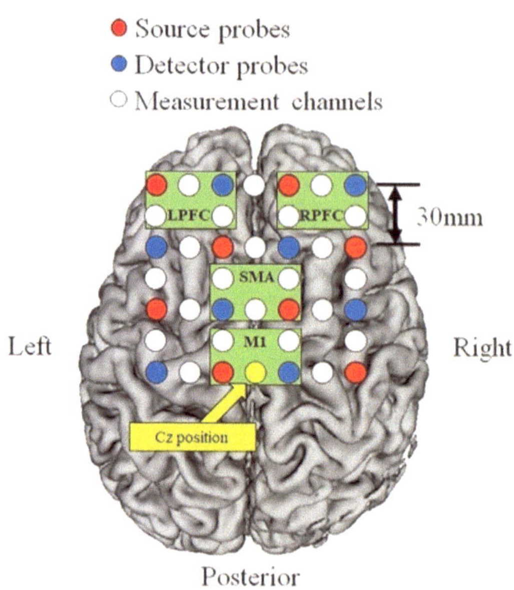

Fig. 29.1 Locations of the light-source fibers, detectors, and regions of interest. R-PFC right prefrontal cortex, L-PFC left prefrontal cortex, SMA sensorimotor area, M1 primary motor cortex

hemoglobin (THb_{VL}) levels were measured. The NIRS system was connected to a personal computer via a Power Lab (AD Instruments, Colorado Springs, CO, USA), and data were acquired at a sampling frequency of 1000 Hz.

29.2.3 Central Hemodynamics and Skin Blood Flow Measurements

Beat-to-beat mean arterial pressure (MAP) was recorded on the left middle finger using the volume-clamping method with a finger photoplethysmograph (Finometer; Finapres Medical Systems, Amsterdam, Netherlands). Heart rate (HR), stroke volume (SV), and cardiac output (CO) were measured using a non-invasive impedance cardiac output meter (PhysioFlow Q-Link, Manatec Biomedical, France). Changes in skin blood flow (SBF) were measured at the forehead using a laser Doppler blood flowmeter (Omegaflow FLO-CI; Omegawave Inc., Osaka, Japan). Analog data were converted into digital data using an A/D converter (PowerLab; AD Instruments, Australia) at a sampling rate of 1000 Hz.

29.3 Results

The participant was scheduled to stand for 600 s, including baseline measurements. During standing, the MAP decreased rapidly at approximately 500 s (Fig. 29.2). The O_2Hb_C and THb_C of the SMA and SMC decreased from 420 s, and L- and

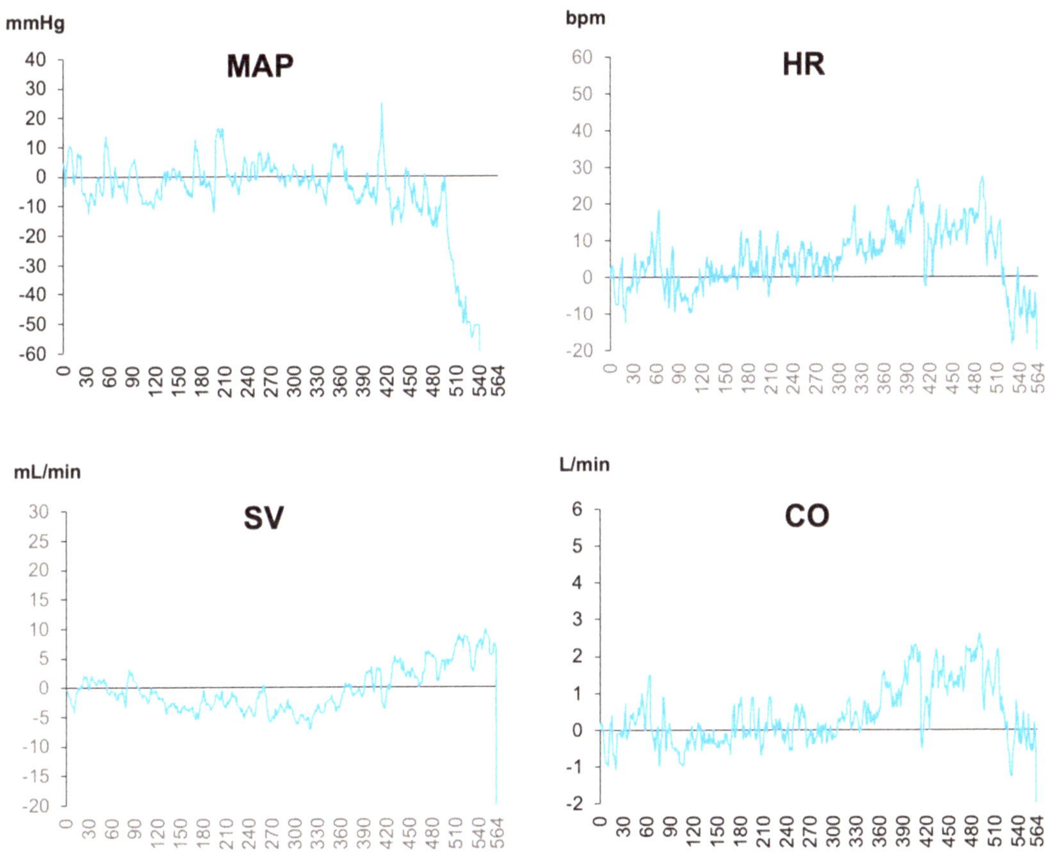

Fig. 29.2 Temporal changes in systemic circulatory parameters. MAP mean arterial pressure, HR heart rate, SV stroke volume, CO cardiac output

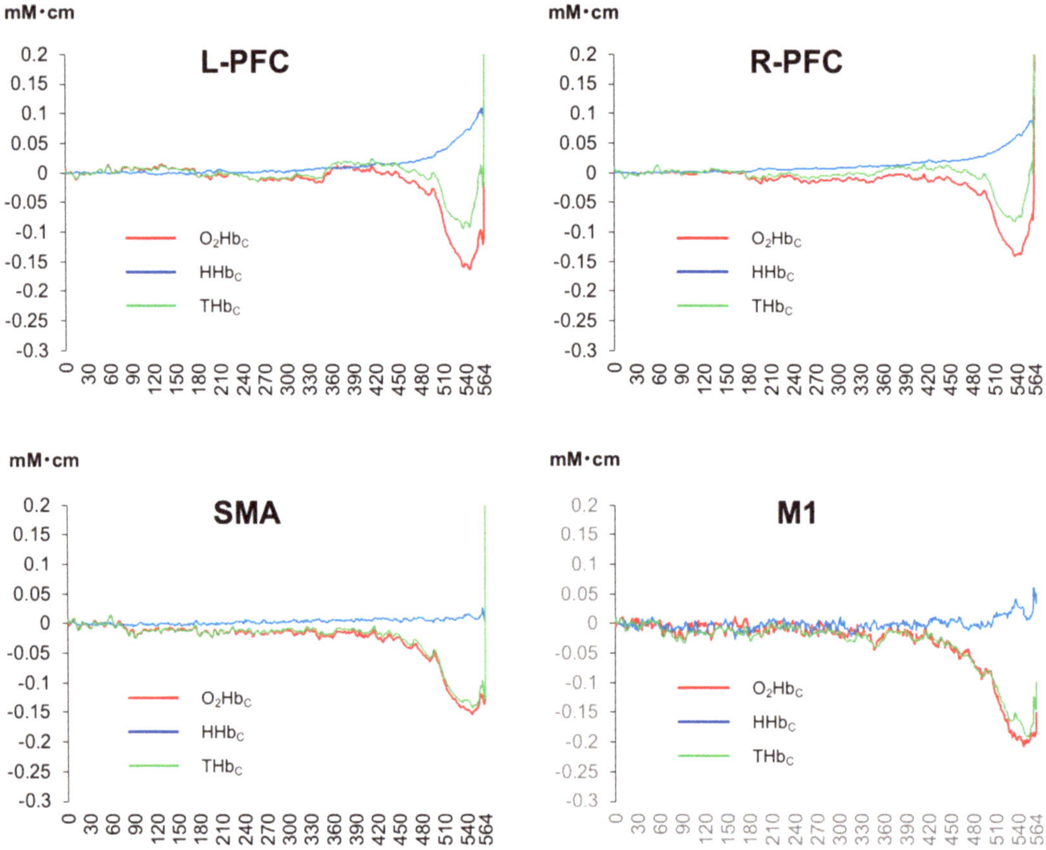

Fig. 29.3 Temporal changes in O_2Hb_C, HHb_C, and THb_C at four regions. R-PFC right prefrontal cortex, L-PFC left prefrontal cortex, SMA sensorimotor area, M1 primary motor cortex, O_2Hb_C cortical oxyhemoglobin, HHb_C cortical deoxyhemoglobin, THb_C cortical total hemoglobin

R-PFC decreased from 450 s; the decrease was more pronounced after 500 s. The HHb_C signal showed an opposite trend from that of O_2Hb_c and THb_C (Fig. 29.3). The SBF decreased momentarily at approximately 500 s but began to rise immediately and did not recover until the end of the measurement at 540 s. The O_2Hb_{VL} and THb_{VL} increased from 500 to 540 s (Fig. 29.4).

29.4 Discussion

This study shows the OH-related blood pressure and critical signal changes during standing in a young male university student. A decline in O_2Hb_C and an increase in HHb_C were observed in some regions before the MAP drop while standing. This information is critical for preventing or detecting OH and OH-related symptoms. In addition, we could prevent fainting or falling unsupported during OH episode by monitoring these changes and discontinuing the measurement.

The bedside standing test examines the basic physiology and is essential for diagnosing and classifying OH. It measures blood pressure changes from sitting or lying to standing with or without a tilt table (Finucane et al. 2019; Cheshire et al. 2021). However, symptoms of OH (e.g., headache, lightheadedness, and visual disturbances) are associated with cerebral hypoperfusion (Kim and Farrell 2022). Therefore,

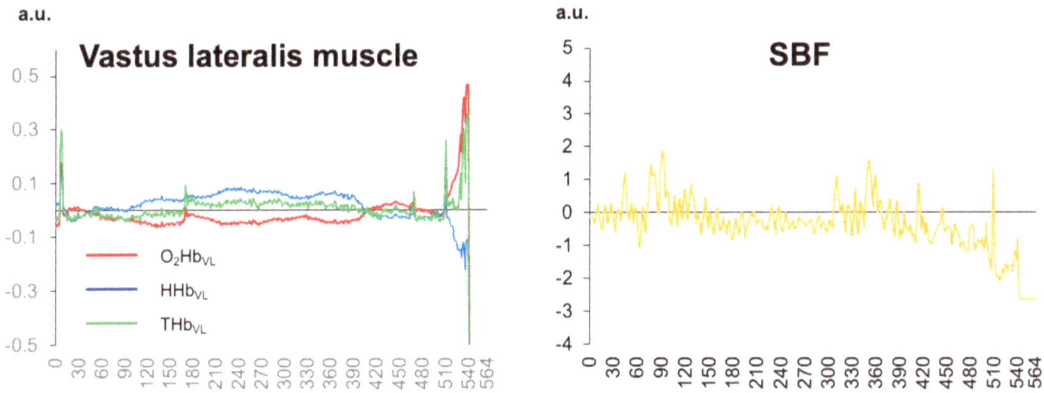

Fig. 29.4 Temporal changes in O_2Hb_{VL}, HHb_{VL}, THb_{VL}, and SBF. O_2Hb_{VL} oxyhemoglobin of the vastus lateralis, HHb_{VL} deoxyhemoglobin of the vastus lateralis, THb_{VL} total hemoglobin of the vastus lateralis, SBF skin blood flow

monitoring cerebral hypoperfusion is mandatory, and NIRS is the most suitable tool for this purpose (Kim et al. 2023).

The participant was instructed to maintain a standing posture for 10 min as the baseline measurement. MAP and SBF decreased, and O_2Hb_{VL} and THb_{VL} increased after 500 s. We measured OH levels at this time point. The O_2Hb_C and THb_C of the SMA and SMC decreased from 420 s, and the L- and R-PFC decreased from 450 s. These results indicate that cerebral hypoperfusion occurs before MAP drops, and multichannel NIRS can sensitively detect OH-related cerebral hemodynamic changes. However, the mechanism(s) underlying this time lag remain unclear. Therefore, further pathophysiological studies are warranted.

In conclusion, NIRS might detect uncontrolled OH before MAP drop.

Acknowledgments This study was supported by a Grant-in-Aid for Scientific Research (B) from the Japan Society for the Promotion of Science and a Grant-in-Aid for Exploratory Research from the Niigata University of Health and Welfare.

References

Boas DA, Gaudette T, Strangman G, Cheng X, Marota JJ, Mandeville JB (2001) The accuracy of near infrared spectroscopy and imaging during focal changes in cerebral hemodynamics. Neuroimage 13:76–90. https://doi.org/10.1006/nimg.2000.0674

Cheshire WP, Freeman R, Gibbons CH, Cortelli P, Wenning GK, Hilz MJ et al (2021) Electrodiagnostic assessment of the autonomic nervous system: a consensus statement endorsed by the American Autonomic Society, American Academy of Neurology, and the International Federation of Clinical Neurophysiology. Clin Neurophysiol 132:666–682. https://doi.org/10.1016/j.clinph.2020.11.024

Finucane C, van Wijnen VK, Fan CW, Soraghan C, Byrne L, Westerhof BE et al (2019) A practical guide to active stand testing and analysis using continuous beat-to-beat non-invasive blood pressure monitoring. Clin Auton Res 29:427–441. https://doi.org/10.1007/s10286-019-00606-y

Kim MJ, Farrell J (2022) Orthostatic hypotension: a practical approach. Am Fam Physician 105:39–49

Kim JM, Park KI, Choi SY, Park HE, Lee H, Bae HM (2023) Dynamic alterations in cerebral hemodynamics measured by portable near-infrared spectroscopy in orthostatic hypotension and intolerance. Am J Hypertens 36:385–393. https://doi.org/10.1093/ajh/hpad025

Wieling W, Kaufmann H, Claydon VE, van Wijnen VK, Harms MPM, Juraschek SP et al (2022) Diagnosis and treatment of orthostatic hypotension. Lancet Neurol 21:735–746. https://doi.org/10.1016/S1474-4422(22)00169-7

30 Dynamic Changes in Cerebral Oxygenation and Arterial Pressure During Repeated Rapid Hypotension

Saya Tokuda, Hyuga Kojima, Hiroki Baba, Hajime Tamiya, and Atsuhiro Tsubaki

Abstract

This study aimed to determine the changes in the time to nadir of mean arterial pressure (MAP) and oxygenated hemoglobin (O_2Hb) in the right and left prefrontal cortices (R-PFC and L-PFC, respectively) during repeated episodes of rapid hypotension. Four cycles of blood pressure reduction were induced via bilateral thigh occlusion at 250 mmHg followed by rapid deflation. Sixteen healthy male university students participated. MAP was recorded beat by beat, and O_2Hb was continuously monitored using near-infrared spectroscopy. The time from deflation to nadir was measured and compared across each deflation cycle. MAP and O_2Hb in the R-PFC and L-PFC reached their nadir values approximately 7 to 8 s post-deflation, with no significant changes observed across repetitions. These findings indicate that the time to nadir of MAP and O_2Hb in the R-PFC and L-PFC remains stable during repeated rapid hypotension.

Keywords

Acute hypotension · Oxygenated hemoglobin · Near-infrared spectroscopy · Prefrontal cortex · Mean arterial pressure

30.1 Introduction

The brain is an important organ, accounting for 15–20% of cardiac output (CO) (Ainslie and Brassard 2014) and consuming approximately 20% of oxygen (Claassen et al. 2021) for 2% of body weight. Lassen (1959) reported that cerebral blood flow remains constant even when mean arterial pressure (MAP) varies between 60 and 150 mmHg. However, standing up from a sitting position causes a rapid decrease in MAP to 24.0 ± 9.8 mmHg and a decrease in oxygenated hemoglobin (O_2Hb) (Mol et al. 2019). In addition, a study on orthostatic loading found that O_2Hb—a proxy for cerebral blood flow—remains stable in the right and left prefrontal cortices (R-PFC and L-PFC, respectively) despite postural blood pressure changes. However, another study reported differential O_2Hb responses to blood pressure changes between the R-PFC and L-PFC (Ishii et al. 2020). This study aimed to investigate

S. Tokuda · H. Kojima · H. Baba
Department of Physical Therapy, Faculty of Rehabilitation, Niigata University of Health and Welfare, Niigata, Japan

H. Tamiya · A. Tsubaki (✉)
Department of Physical Therapy, Faculty of Rehabilitation, Niigata University of Health and Welfare, Niigata, Japan

Institute for Human Movement and Medical Sciences, Niigata University of Health and Welfare, Niigata, Japan
e-mail: tsubaki@nuhw.ac.jp

S. C. Pias et al. (eds.), *Oxygen Transport to Tissue XLVI*, Advances in Experimental Medicine and Biology 1498, https://doi.org/10.1007/978-3-032-10389-5_30

changes in the time to nadir of MAP and O_2Hb in the R-PFC and L-PFC during repeated episodes of rapid hypotension.

30.2 Materials and Methods

30.2.1 Participants

Sixteen healthy male university students (mean ± standard deviation: age, 20.9 ± 0.3 years; height, 171.8 ± 5.6 cm; weight, 62.6 ± 7.5 kg; body mass index 21.1 ± 1.6 kg/m^2; resting MAP, 98.2 ± 1.9 mmHg) participated in this study. The study was approved by the Ethics Committee of Niigata University of Health and Welfare (approval number: 18649–210,618) and was conducted in accordance with the Declaration of Helsinki. All participants received verbal and written explanations of the study and provided informed consent prior to participation.

30.2.1.1 Experimental Procedures

Consumption of alcohol and caffeine and engaging in strenuous exercise were prohibited on the day before the experiment. During the experiment, the participants sat on a recumbent chair in a quiet laboratory with a 105 × 750 mm thigh cuff wrapped around both proximal thighs and were instructed to rest for 5 min. After resting, a digital air tourniquet (motorized digital air tourniquet MT-960; Mizuho Corporation, Tokyo, Japan) was used for occluding the thigh cuff at a pressure of 250 mmHg for 2 min, and it was then rapidly deflated to facilitate a decrease in MAP. The operation was repeated four times to establish a 5-min rest period until the next cycle. During the experiment, we measured the MAP and O_2Hb of the R-PFC and L-PFC as indicators of regional cerebral blood flow.

30.2.1.2 Measurements

O_2Hb was measured over time as an index of regional cerebral blood flow using a functional optical brain imaging system (LABNIRS; Shimadzu Corporation, Kyoto, Japan) in the frontal lobe. Probes were placed at a 30-mm inter-optode distance based on the international 10–20 method, and the source and detector fibers were placed as shown in Fig. 30.1. MAP was measured using continuous blood pressure and hemodynamic monitoring device (Finapres NOVA; Finapres Medical Systems, Enschede, the Netherlands) equipped with an arm cuff on the upper left arm and a finger cuff on the left third finger to detect blood pressure at each pulse beat.

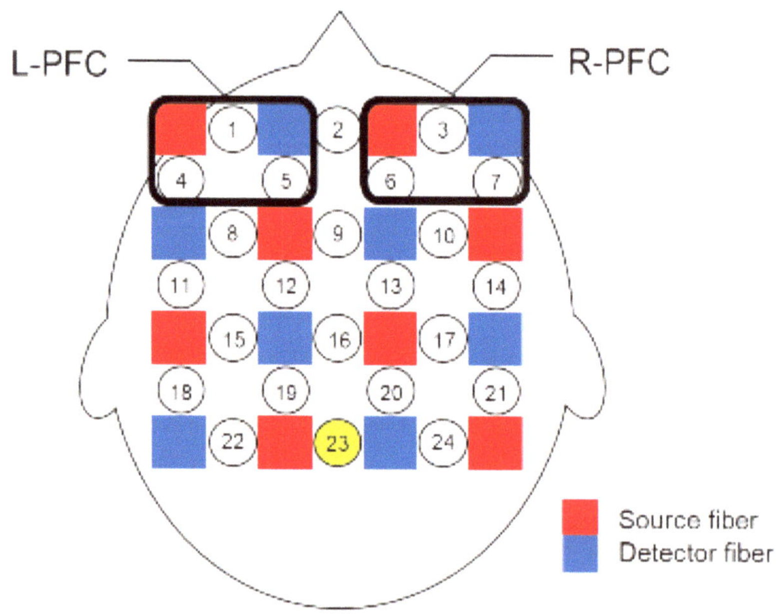

Fig. 30.1 Location of the light source and detector fibers and the ROI. The areas circled in black indicate the R-PFC and L-PFC, which are the ROIs. ROI region of interest; R-PFC right prefrontal cortex; L-PFC left prefrontal cortex

30.2.1.3 Analyses

The change in O_2Hb of the MAP, R-PFC, and L-PFC over time was calculated from the average value during 5 min of rest (Ishii et al. 2020), and the average value of each second was used as the analyzed value. An average value of 30 s before release was used as the baseline, and the time from 0 s after release to the time to nadir—the lowest values of MAP and O_2Hb in the R-PFC and L-PFC—was calculated. When normality of values was confirmed, repeated measures one-way analysis of variance was performed. GraphPad Prism 8 (GraphPad Software, San Diego, CA, USA) was used for the statistical analysis, and the significance level was set at 5%.

30.3 Results

The O_2Hb of the MAP, R-PFC, and L-PFC decreased immediately after deflation (Fig. 30.2).

No significant changes were observed in the time to nadir of MAP across the four cycles of thigh-cuff deflation. O_2Hb in the R-PFC and L-PFC showed no significant changes in the time to nadir between cycles (Fig. 30.3).

30.4 Discussion

In the present study, we examined the change in the time to nadir of O_2Hb of the MAP, R-PFC, and L-PFC following repeated rapid blood pressure reductions and found no significant change in the time to nadir of O_2Hb of the MAP, R-PFC, and L-PFC at each deflation. To the best of our knowledge, this is the first study to measure changes in MAP and O_2Hb time to nadir in response to repeated rapid hypotension maneuvers. These results suggest that the response of O_2Hb in the R-PFC and L-PFC to hypotension is not altered in healthy young individuals.

The time to nadir in MAP did not significantly change across the four rapid hypotension cycles, indicating that this maneuver did not affect autonomic nerve activity. It has been reported that remote preconditioning with repeated de-stimulation increases parasympathetic indicators after 2 weeks of intervention, 4 times a day, 6 times a week (Gardner et al. 2020). However, the intervention in this study was conducted for only 1 day; thus, it is possible that no change occurred in autonomic nervous activity immediately after the rapid hypotension.

Additionally, we observed that the time to nadir in O_2Hb did not significantly change across repeated hypotension cycles, suggesting that repeated hypotension did not affect vascular tone. A previous study reported that the speed of response to brain autoregulatory capacity is highly dependent on vascular tone (Aaslid et al. 1989). Thus, no change in vascular tone or in the speed of response of cerebral autoregulatory capacity was observed.

A limitation of this study is that we did not measure sympathetic nerve activity or evaluate arterial partial pressure of carbon dioxide; hence, we were unable to measure other factors that affect cerebral blood flow during hypotension. Therefore, these factors that regulate cerebral blood flow should be investigated in the future.

In conclusion, our findings suggest that repeated rapid hypotension does not alter the time to nadir of MAP or O_2Hb in the R-PFC and L-PFC in healthy young individuals.

Acknowledgments This study was supported by a Grant-in-Aid for Scientific Research (B) from the Japan Society for the Promotion of Science (A.T.) and a Grant-in-Aid for Exploratory Research from Niigata University of Health and Welfare (A.T.).

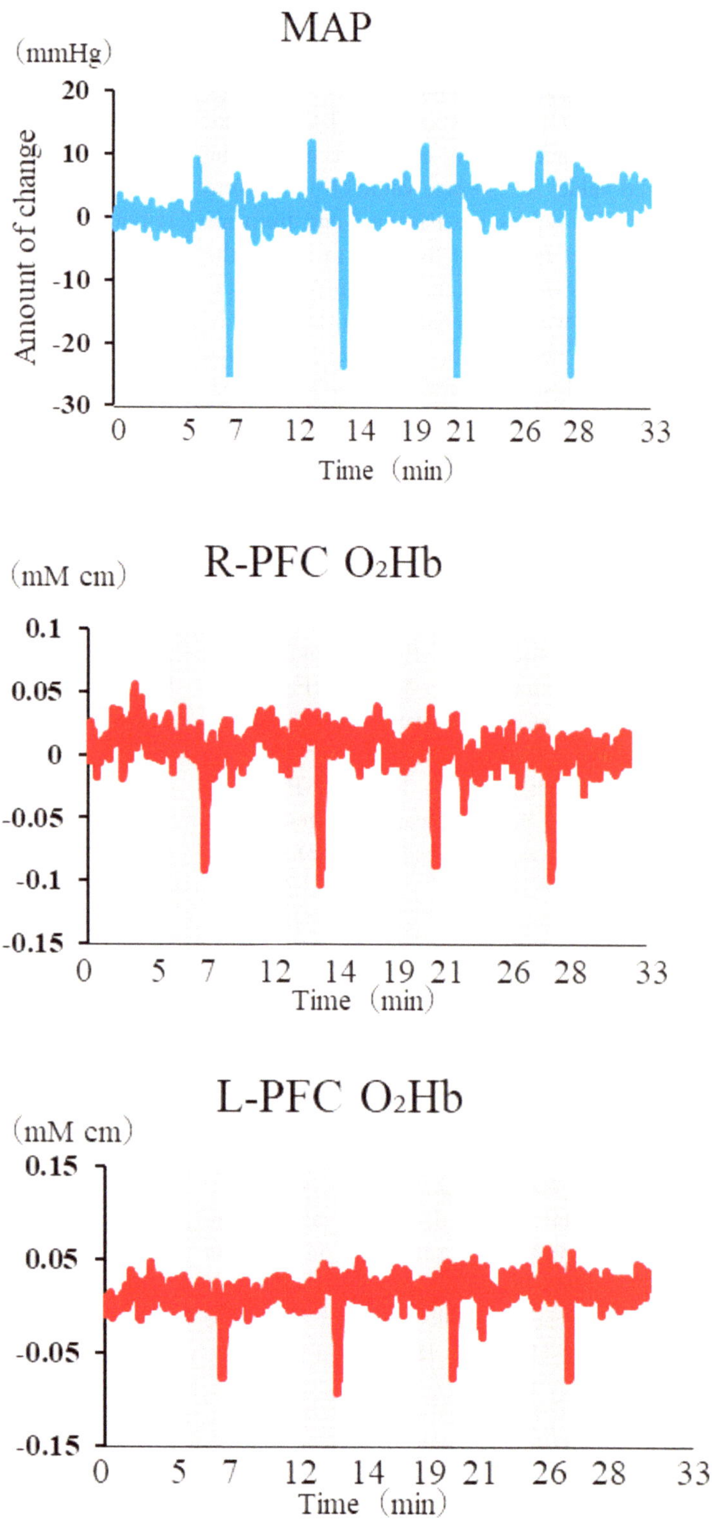

Fig. 30.2 Temporal changes in MAP, R-PFC O_2Hb, and F-PFC O_2Hb over time. The gray area indicates deflation. MAP, mean arterial pressure; R-PFC O_2Hb, right prefrontal cortex oxygenated hemoglobin; L-PFC O_2Hb, left prefrontal cortex oxygenated hemoglobin

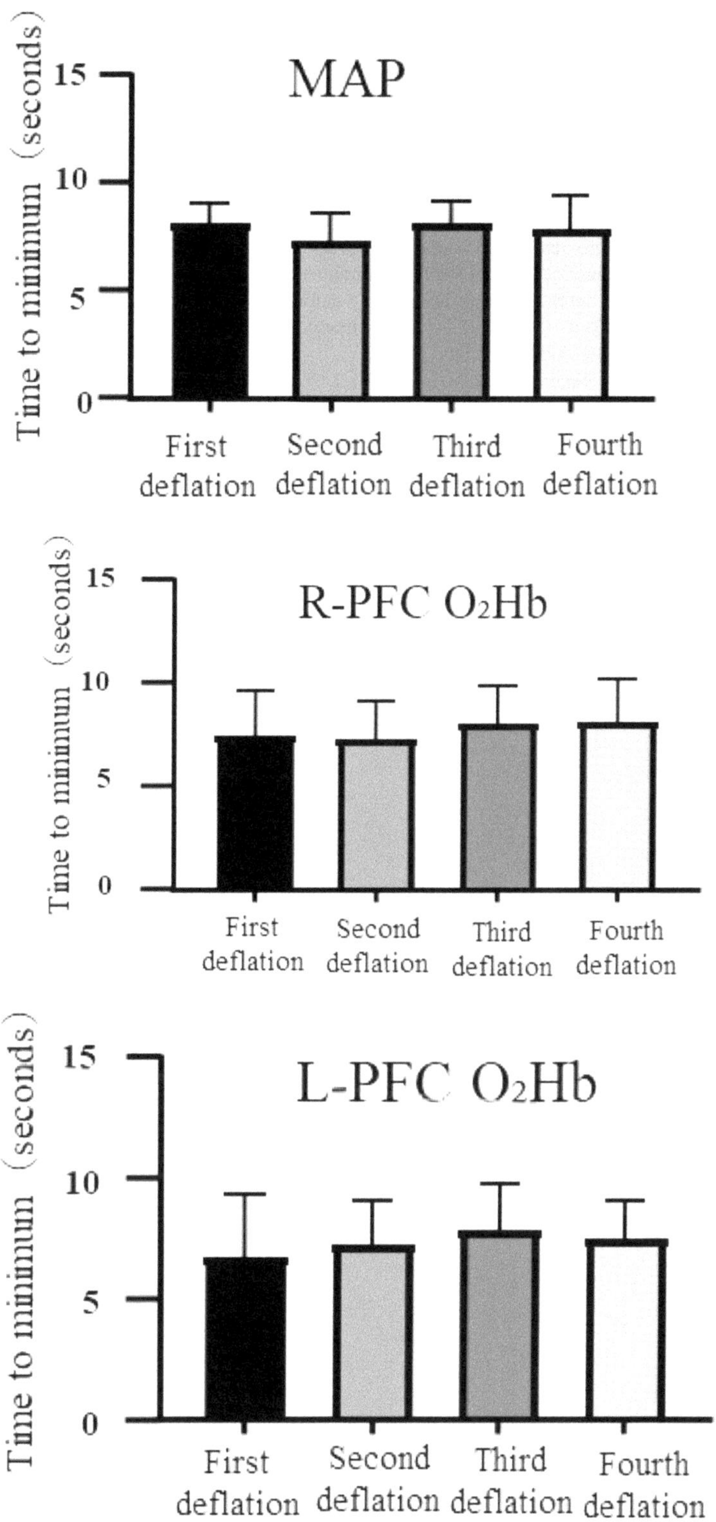

Fig. 30.3 Time to nadir of MAP, R-PFC O_2Hb, and L-PFC O_2Hb for each deflation. There was no significant change in the time to nadir of O_2Hb minimum for each deflation for MAP, R-PFC, and L-PFC. O_2Hb, oxygenated hemoglobin; MAP mean arterial pressure; R-PFC right prefrontal cortex; L-PFC left prefrontal cortex

References

Aaslid R, Lindegaard KF, Sorteberg W, Nornes H (1989) Cerebral autoregulation dynamics in humans. Stroke 20:45–52

Ainslie PN, Brassard P (2014) Why is the neural control of cerebral autoregulation so controversial? F1000Prime Rep 6:14. https://doi.org/10.12703/p6-14

Claassen J, Thijssen DHJ, Panerai RB, Faraci FM (2021) Regulation of cerebral blood flow in humans: physiology and clinical implications of autoregulation. Physiol Rev 101(4):1487–1559. https://doi.org/10.1152/physrev.00022.2020

Gardner RN, Sabino-Carvalho JL, Kim J, Vianna LC, Lang JA (2020) Two weeks of remote ischaemic preconditioning alters sympathovagal balance in healthy humans. Exp Physiol 105:1500–1506. https://doi.org/10.1113/EP088789

Ishii W, Fujita Y, Kawaguchi T, Kimura K, Fukuda A, Fuchigami T, Morioka I (2020) Cerebral autoregulation during active standing test in juvenile patients with instantaneous orthostatic hypotension. Adv Exp Med Biol 1232:77–83. https://doi.org/10.1007/978-3-030-34461-011

Lassen NA (1959) Cerebral blood flow and oxygen consumption in man. Physiol Rev 39(2):183–238. https://doi.org/10.1152/physrev.1959.39.2.183

Mol A, Woltering JHH, Colier W, Maier AB, Meskers CGM, van Wezel RJA (2019) Sensitivity and reliability of cerebral oxygenation responses to postural changes measured with near-infrared spectroscopy. Eur J Appl Physiol 119(5):1117–1125. https://doi.org/10.1007/s00421-019-04101-0

31 Patients Admitted to the ICU Who Cannot Be Mobilized for the First Time Within 72 Hours Have Low Regional Cerebral Oxygen Saturation

Ryota Imai, Atsuhiro Tsubaki, Takafumi Abe, and Seigo Yamaguchi

Abstract

Mobilization of critically ill patients within 72 hours of admission is associated with improved outcomes. Recently, the predictive value of regional cerebral oxygen saturation (rSO_2) measured via near-infrared spectroscopy (NIRS) has been emphasized. This study aimed to evaluate the differences in rSO_2 values depending on the availability of mobilization within 72 hours in the ICU. Eighty patients admitted to the emergency center between June 2020 and December 2022 were analyzed. Patients were assessed based on whether they could be mobilized within 72 hours (early mobilization group, EM) or later (non-early mobilization group, non-EM). During mobilization, prefrontal rSO_2 values were monitored. Regarding patient background, significant differences were noted between the groups, including delayed release in non-EM patients. rSO_2 values varied significantly, with the lowest values in the first half of end-sitting in both groups (non-EM 56.5 $\pm$ 4.8%, EM 58.6 $\pm$ 4.3%, $p < 0.05$). The rSO_2 value was also lower in the non-EM group than in the EM group ($P < 0.014$). A weak correlation was observed between rSO_2 and the number of days to first mobilization ($r = -0.251$, $p = 0.025$). The rSO_2 value may serve as a potential marker to guide the timing of mobilization in ICU patients.

Keywords

NIRS · Regional cerebral oxygen saturation · ICU · Rehabilitation · Early mobilization

31.1 Introduction

Early rehabilitation of critically ill patients improves functional outcomes upon discharge, shortens the duration of delirium, and increases the number of ventilator-free days (Schweickert et al. 2009). Additionally, early mobilization for these patients is based on the criteria for stopping or proceeding depending on respiratory and circulatory status, consciousness, auto-induced symptoms, and mobilization, especially within 72 hours of admission, which may significantly improve patient outcomes (Ruo Yu et al. 2024). However, there is no objective index of the brain

R. Imai · T. Abe
Department of Rehabilitation, Uonuma Kikan Hospital, Minamiuonuma, Japan

A. Tsubaki (✉)
Institute for Human Movement and Medical Sciences, Niigata University of Health and Welfare, Niigata, Japan
e-mail: tsubaki@nuhw.ac.jp

S. Yamaguchi
Department of Emergency and Critical Care, Uonuma Kikan Hospital, Minamiuonuma, Japan

S. C. Pias et al. (eds.), *Oxygen Transport to Tissue XLVI*, Advances in Experimental Medicine and Biology 1498, https://doi.org/10.1007/978-3-032-10389-5_31

for markers to be considered during mobilization. Previous studies have explored the use of near-infrared spectroscopy (NIRS) for monitoring cerebral oxygenation in critically ill patients, primarily focusing on its relationship with shock states, neurological outcomes, and mortality risk (Sun et al. 2014; Peng et al. 2021). However, the role of NIRS in guiding early rehabilitation and mobilization decisions remains unclear. This study aimed to fill this gap by investigating whether regional cerebral oxygen saturation (rSO_2) can serve as an indicator for the timing of early mobilization in ICU patients. Identifying a physiological marker for mobilization readiness could contribute to personalized rehabilitation strategies and improve patient outcomes.

31.2 Methods

31.2.1 Participants

Between June 2020 and December 2022, 132 patients were admitted to the emergency center and prescribed physiotherapy, with all 132 patients deciding to initiate mobilization according to the criteria of the Japanese Society of Intensive Care Medicine (Ad Hoc Committee for Early Rehabilitation, The Japanese Society of Intensive Care Medicine 2017). Excluded patients were as follows: 17 patients with an incompatible NIRS probe, 34 patients whose probes slipped during bed release, and 1 patient who discontinued treatment due to hypotension. Finally, 80 patients were included in the analysis. The patients were divided into two groups: EM patients who could be mobilized within 72 hours of admission ($n = 50$) and non-EM patients who were mobilized after 72 hours ($n = 30$). Basic patient information was collected from medical records. Data were collected after informed consent was obtained from the patients or their families. Ethical approval was obtained from the Ethics Committee of Uonuma Kikan Hospital (approval number 02-006).

31.2.2 Measurements

In this study, "mobilization" was defined as rehabilitation at the level of sitting on the edge of a bed (SEB) or higher (Liu et al. 2022). The procedure sequence was as follows: rest (supine), elevation of the head of the bed (EHOB) to 45°, SEB, and supine, each lasting approximately 3 minutes. Patients relaxed in bed for 5 minutes prior to the exercise, and swaying of the head and trunk was minimized during the exercise.

Wearable brain NIRS (Brain Activity Monitor Hb133; Astem, Kanagawa, Japan) was used to measure rSO_2. A dedicated head mount was attached such that the probes were in close contact with FP1 and FP2 using the international 10–20 method, and rSO_2 was continuously measured throughout the experiments. The sampling rate was 0.1 second. The FP1 and FP2 data measured were averaged, and the average values for each 1-minute period were used: the last minute of rest (supine), the first and second halves of EHOB 45°, the first and second halves of SEB, and the first minute of the supine position after SEB. Simultaneously, heart rate (HR), percutaneous oxygen saturation (SpO_2), and mean arterial pressure (MAP) were measured using a biometric monitor (IntelliVue X2; Philips, Amsterdam, the Netherlands) and recorded at six time points within 1 minute before and after the postural change.

31.2.3 Statistical Analysis

Changes in rSO_2, HR, SpO_2, and MAP over time during mobilization were compared for each group using two-way repeated-measures analysis of variance. Post hoc tests (Bonferroni method) were performed for indices presenting an interaction effect. Correlations between items that significantly differed in patient background and resting rSO_2 were examined using Pearson's or Spearman's rank correlation coefficients. All statistical analyses were performed using EZR in R

Commander version 1.54 (R Foundation for Statistical Computing, Vienna, Austria). Statistical significance was set at $p < 0.05$.

31.3 Results

Patient characteristics are presented in Table 31.1. No significant differences were found in the mean age, sex distribution, BMI, catecholamine use history, oxygenation, or Hb levels between the EM and non-EM groups. The severity of illness was significantly higher in the non-EM group ($p = 0.015$), with significantly more patients receiving ventilator management ($p = 0.017$) and a longer time to first mobilization ($p < 0.001$). The most common illness on admission was acute heart failure (30%), followed by sepsis (24%) in EM patients, and sepsis was the most common illness (30%), followed by respiratory failure (17%) in non-EM patients. Figure 31.1 illustrates the comparison of rSO_2 over time. During mobilization, rSO_2 showed significant temporal changes in both groups ($p < 0.001$). rSO_2 was lowest in the first half of SEB (EM, 58.6 ± 4.3%; non-EM, 56.5 ± 4.8%) and increased from the second half of SEB to the end of SEB. An interaction effect was observed between these changes ($P = 0.018$). rSO_2 was significantly lower in the non-EM group ($p = 0.014$). HR and MAP showed significant changes over time in both groups ($p < 0.001$ and $p = 0.005$, respectively) and were significantly higher in the non EM group ($p = 0.047$ and $p = 0.026$, respectively) (Fig. 31.2). No correlation was found between rSO_2 and APACHE II score ($r = -0.142$, $p = 0.21$); however, there was a weak correlation with days to first mobilization ($r = -0.251$, $p = 0.025$).

31.4 Discussion

Previous studies have explored rSO_2 as a prognostic marker in critically ill patients, focusing on shock states, mortality, and neurological outcomes (Sun et al. 2014; Peng et al. 2021). However, its role in guiding early mobilization remains unclear. Our study is among the first to investigate rSO_2 as a potential indicator for mobilization timing, providing a new perspective on ICU rehabilitation.

Unlike prior studies that linked rSO_2 to mortality or systemic oxygenation (Bertini et al. 2023), we focused on functional recovery and mobilization status. Although rSO_2 has been associated with systemic oxygen delivery markers (Al Tayar et al. 2017; Paarmann et al. 2012), our findings suggest its potential application in mobilization decision-making, although further validation is needed.

This study evaluated whether rSO_2 measured using NIRS during initial mobilization in critically ill patients differed depending on the availability of early mobilization within the first 72 hours in the ICU and changes in general vital

Table 31.1 Patient characteristics ($n = 80$)

	EM($n = 50$)	Non-EM($n = 30$)	P-value
Age, years	78.8 ± 11.4	75.5 ± 12.0	0.23
Sex, female, n (%)	14 (28%)	10 (33%)	0.624
APACHE II at ICU admission	16.5 ± 6.4	20.2 ± 6.7	0.015
BMI (kg/m^2)	22.4 ± 3.6	22.7 ± 3.2	0.687
Intubation history, n (%)	14 (28%)	17 (57%)	0.017
Catecholamine use history, n (%)	21 (42%)	14 (47%)	0.816
Oxygen administration, n (%)	29 (58%)	20 (67%)	0.485
Hb (g/dL)	11.1 ± 1.9	10.5 ± 1.6	0.154
Days to first mobilization (day)	2 (1–2)	6 (4–8)	<0.001

Values are presented as mean ± standard deviation, n (%), or median (interquartile range)
APACHEII Acute Physiology and Chronic Health Disease Classification System II; *BMI* body mass index; *ICU* intensive care unit; *Hb* hemoglobin

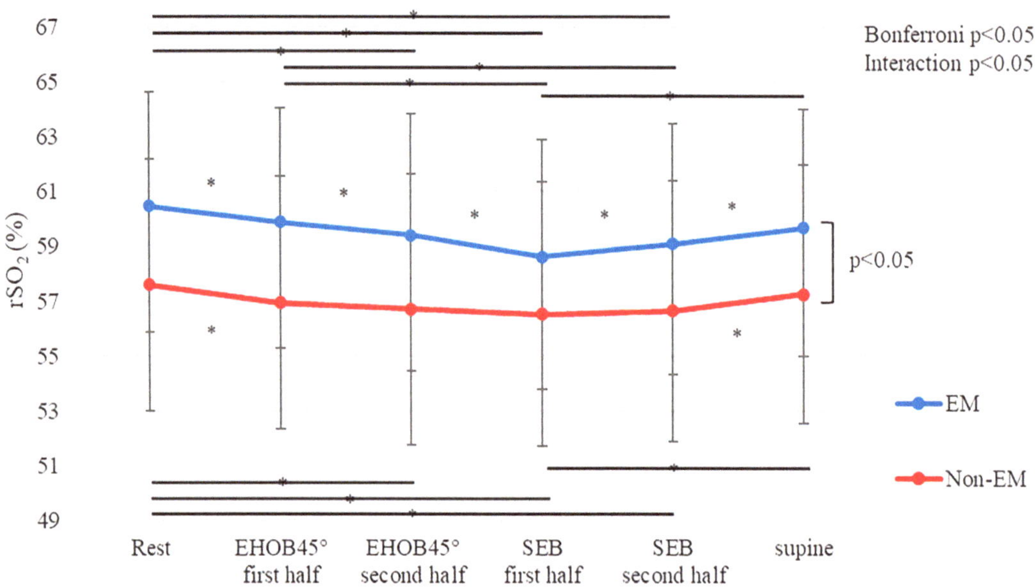

Fig. 31.1 Comparison of rSO2 temporal changes between the groups. rSO_2 during mobilization revealed significant temporal changes in both groups ($p < 0.001$), decreasing in the first half and increasing from the second half to the end of SEB. An interaction effect was observed ($p = 0.018$). This was significantly lower in the non-EM group than in the EM group ($p = 0.014$). Error bars represent the standard deviation. * $p < 0.05$, post hoc test. *EM* early mobilization; *rSO_2* regional cerebral oxygen saturation; *EHOB* elevation to the head of the bed; *SEB* sitting on the edge of the bed.

signs. The rSO_2 at mobilization revealed significant changes over time in both groups, with an interaction between the changes, and was significantly lower in the non-EM group. A weak negative correlation was found between rSO_2 and the number of days to first mobilization.

The observed reduction in rSO_2 in the delayed mobilization group may be attributed to reduced cerebral perfusion from prolonged immobility (Zhang et al. 2019), increased systemic oxygen demand, as indicated by correlations with lactate levels and hemodynamic instability (Gündüz et al. 2024; Bojan et al. 2019), and inflammatory responses in critically ill patients, which may impair mitochondrial function and oxygen extraction efficiency (Peng et al. 2021).

Our results indicate that rSO_2 is significantly lower in patients with delayed mobilization, but the optimal threshold has not been undetermined. Prior studies suggest that rSO_2 <55% may indicate critical hypoxia (Sun et al. 2014). Our data suggest that persistently <56% values may be associated with delayed mobilization, but further research is required.

Clinically, patients with low rSO_2 may benefit from targeted interventions such as oxygen therapy, circulatory support, or pre-mobilization optimization. Future studies should explore whether rSO_2-guided rehabilitation strategies improve patient outcomes.

Critically ill patients are treated to maintain adequate circulatory and respiratory status and are mobilizable according to their general vital signs. However, even patients deemed mobilizable are not typically in a stable general condition. rSO_2 may reflect instability in the patient's condition and may be indicative of the patient's prognosis. However, rSO_2 may vary depending on the treatment status and disease specificity of the patient. Further investigations are warranted to clarify these effects.

This study had several limitations. First, NIRS measurements are influenced by extracranial blood flow, which makes it difficult to isolate

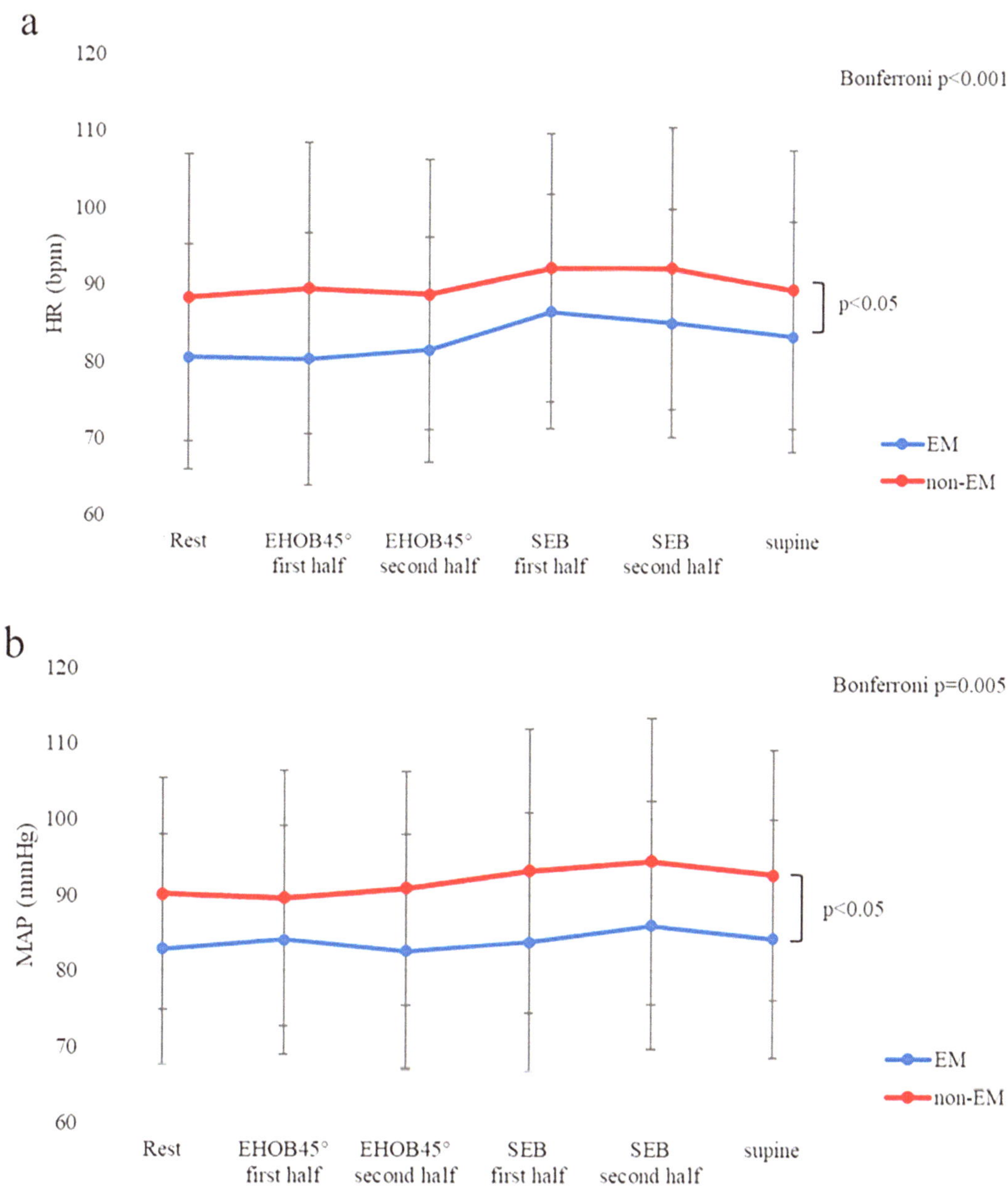

Fig. 31.2 Comparison of temporal changes in vital signs between the groups. HR (**a**) and MAP (**b**) show significant gradual changes in both groups ($p < 0.001$ and $p = 0.005$, respectively) and were significantly higher in the non-EM group ($p = 0.047$ and $p = 0.026$, respectively). No differences in the temporal changes in SpO_2 (**c**) were observed between the groups. Error bars represent the standard deviation. *EM*, early mobilization; *EHOB* elevation to the head of the bed; *SEB* sitting on the edge of the bed; *HR*, heart rate; *MAP* mean arterial pressure; *SpO_2* percutaneous oxygen saturation

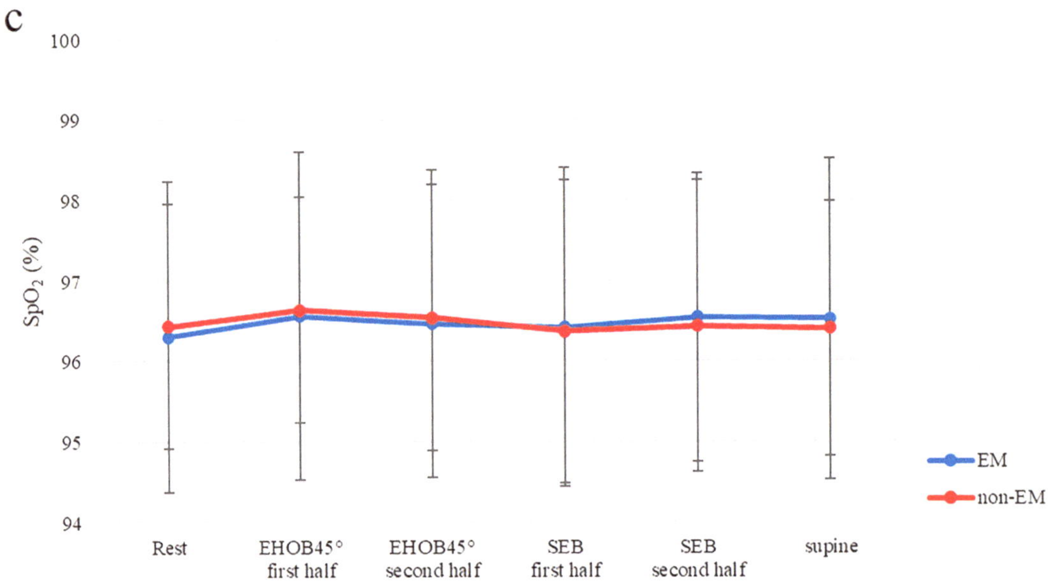

Fig. 31.2 (continued)

cerebral oxygenation changes (Takahashi et al. 2011). Additionally, patient movement, cooperation, and probe placement may have introduced variability in rSO_2 readings, although we made every effort to minimize these effects. Second, we did not perform a multivariate regression analysis to adjust for confounders, such as the APACHE II score or ventilator status, owing to the limited sample size and risk of overfitting. While our findings suggest an association between rSO_2 and mobilization timing, future larger-scale, multi-center studies should incorporate multivariate analyses to confirm whether rSO_2 is an independent predictor of early mobilization.

31.5 Conclusion

The change in rSO_2 during the first mobilization was influenced by the timing of mobilization and tended to be lower in patients mobilized after 72 hours. This suggests that rSO_2 may serve as a marker for guiding ICU mobilization, with lower values indicating a higher risk of delayed mobilization. Although further validation is required, rSO_2 could be considered an adjunct marker in ICU mobilization protocols to assess cerebral oxygenation before mobilization. Future research should explore the feasibility of developing an rSO_2-based risk stratification model and conduct prospective studies to evaluate its clinical utility in ICU rehabilitation.

References

Ad Hoc Committee for Early Rehabilitation, The Japanese Society of Intensive Care Medicine (2017) The japanese society of intensive care medicine: evidence based expert consensus for early rehabilitation in the intensive care unit. J Jpn Soc Intensive Care Med 24:255–303

Al Tayar A, Abouelela A, Mohiuddeen K (2017) Can the cerebral regional oxygen saturation be a perfusion parameter in shock? J Crit Care 38:164–167

Bertini P, Marabotti A, Paternoster G et al (2023) Regional cerebral oxygen saturation to predict favorable outcome in extracorporeal cardiopulmonary resuscitation: a systematic review and meta-analysis. J Cardiothorac Vasc Anesth 37:1265–1272

Bojan M, Bonaveglio E, Dolcino A, Mirabile C (2019) Somatic and cerebral near infrared spectroscopy for the monitoring of perfusion during neonatal cardiopulmonary bypass. Interact Cardiovasc Thorac Surg 29:955–959

Gündüz E, Simsek A, Temel H, Arici G, Titiz T (2024) Investigation of the effect of high-dose peripheral noradrenaline infusion in acute hemorrhage on the brain in rabbits and its correlation with cerebral oximetry and S100B protein. J Cardiothorac Vasc Anesth 38:54–55

Liu K, Shibata J, Fukuchi K, Takahashi K, Sonoo T, Ogura T, Goto T (2022) Optimal timing of introducing mobilization therapy for ICU patients with sepsis. J Intensive Care 10:22

Paarmann H, Heringlake M, Heinze H, Hanke T, Sier H, Karsten J, Schön J (2012) Non-invasive cerebral oxygenation reflects mixed venous oxygen saturation during the varying haemodynamic conditions in patients undergoing transapical transcatheter aortic valve implantation. Interact Cardiovasc Thorac Surg 14:268–272

Peng Q, Zhang L, Ai M, Huang L, Ai Y (2021) Clinical values of cerebral oxygen saturation monitoring in patients with septic shock. Zhong Nan Da Xue Xue Bao Yi Xue Ban 46:1212–1219

Ruo Yu L, Jia Jia W, Meng Tian W, Tian Cha H, Ji Yong J (2024) Optimal timing for early mobilization initiatives in intensive care unit patients: a systematic review and network meta-analysis. Intensive Crit Care Nurs 82: 103607

Schweickert WD, Pohlman MC, Pohlman AS et al (2009) Early physical and occupational therapy in mechanically ventilated, critically ill patients: a randomised controlled trial. Lancet 373:1874–1882

Sun X, Ellis J, Corso PJ, Hill PC, Lowery R, Chen F, Lindsay J (2014) Mortality predicted by preinduction cerebral oxygen saturation after cardiac operation. Ann Thorac Surg 98:91–96

Takahashi T, Takikawa Y, Kawagoe R, Shibuya S, Iwano T, Kitazawa S (2011) Influence of skin blood flow on near-infrared spectroscopy signals measured on the forehead during a verbal fluency task. NeuroImage 57:991–1002

Zhang L, Hu W, Cai Z et al (2019) Early mobilization of critically ill patients in the intensive care unit: a systematic review and meta-analysis. PLoS One 14: e0223185

Part VIII

Peripheral Oxygen Monitoring

NIRS-Derived Muscle Deoxygenation During Arterial Occlusion at Rest Before and After Body Weight Resistance Training with Slow Movement and Tonic Force Generation

32

Shun Takagi, Yuya Watanabe, and Ryotaro Kime

Abstract

The aim of this study was to investigate the effects of body weight resistance training with slow movement and tonic force generation on muscle deoxygenation during arterial occlusion at rest. Eleven untrained healthy young male subjects performed Bulgarian split squat training for 3 days per week for 8 weeks. Each training session consisted of 4 sets of 10 repetitions (or until volitional failure). Before and after training, with the subject at rest in supine position, a pneumatic cuff was placed at the proximal site of the thigh and inflated to 250 mmHg to induce the occlusion of arterial blood flow. During arterial occlusion, in the thigh distal to the cuff, muscle O_2 saturation (SmO_2) was continuously monitored at the vastus lateralis (VL) and rectus femoris (RF) muscles using spatially resolved near-infrared spectroscopy. During occlusion, the slope of desaturation and area above the curve (AAC) from baseline values were evaluated. At VL, a significantly shallower desaturation slope was found after training, compared to before. As a result, AAC from baseline was significantly lower after training. In contrast, no significant changes in the desaturation slope or AAC were observed at RF. Body weight resistance training for 8 weeks may cause a shallow desaturation slope during arterial occlusion at rest, though the effects of resistance training on desaturation response during arterial occlusion may be regional in quadriceps muscles.

Keywords

Near-infrared spectroscopy · Quadriceps muscles · Regional difference · Slow training · Squat exercise

S. Takagi (✉)
Faculty of Education and Welfare, Biwako-Gakuin University, Higashiomi, Japan

Department of Sports Medicine for Health Promotion, Tokyo Medical University, Shinjuku, Japan
e-mail: s-takagi@biwakogakuin.ac.jp

Y. Watanabe
Faculty of Sport Study, Biwako Seikei Sport College, Otsu, Japan

R. Kime
Department of Sports Medicine for Health Promotion, Tokyo Medical University, Shinjuku, Japan

32.1 Introduction

Muscle oxidative metabolism can be indirectly evaluated by measuring muscle deoxygenation using near-infrared spectroscopy (NIRS) with arterial occlusion (McLay et al. 2016). This occlusion maneuver at rest may be useful to assess muscle oxidative metabolism because it is more time efficient and less costly, without any

S. C. Pias et al. (eds.), *Oxygen Transport to Tissue XLVI*, Advances in Experimental Medicine and Biology 1498, https://doi.org/10.1007/978-3-032-10389-5_32

type of exercise or electrical stimulation needed (Koutlas et al. 2024). Difference in muscle deoxygenation during arterial occlusion at rest between aerobically trained and untrained subjects has been reported in some cross-sectional studies (McLay et al. 2016; Koutlas et al. 2024; Soares et al. 2017). However, a longitudinal study was needed to clarify the effects of training on muscle deoxygenation during arterial occlusion, and to date, few data are available regarding the effect of resistance training on muscle deoxygenation during arterial occlusion. Moreover, muscle deoxygenation was shown to be heterogeneous between muscles during exercise (Takagi et al. 2013); however, regional differences in deoxygenation during arterial occlusion are not fully understood.

Body weight resistance training is a popular type of exercise because no equipment is required. Although exercise intensity of body weight resistance training may generally be insufficient in young men (Yamauchi et al. 2009), body weight resistance with slow movement and tonic force generation may potentially be effective for improving muscle strength and hypertrophy (Tanimoto and Ishii 2006). Therefore, the purpose of this pilot study was to investigate the effects of body weight resistance training with slow movement and tonic force generation on muscle deoxygenation during arterial occlusion at rest in young men.

32.2 Methods

32.2.1 Subjects

Eleven untrained, healthy young males (age 20 ± 1 years; height 171.8 ± 5.1 cm; body weight 62.4 ± 8.0 kg, mean $\pm$ SD) participated in the study. None had been involved in any type of resistance training for at least 12 months before enrollment. This study was approved by the Institutional Ethics Committee. All subjects were informed of the purpose of the study and written informed consent was received.

32.2.2 Experimental Design

The subjects performed Bulgarian split squat training for 3 days per week for 8 weeks. Each training session consisted of 4 sets of 10 repetitions (or until volitional failure), with 1 min rest between sets. The subjects were asked to exercise under slow movement and tonic force generation (Tanimoto and Ishii 2006) (i.e., 5-sec eccentric, 5-sec concentric, and 1-sec isometric actions with no rest between each repetition). Conditions during exercise, such as posture, start (finish), and bottom position, and stride distance were normalized and specified according to a previous study (Takagi et al. 2024).

Before and after 8 weeks of training, with the subject at rest in supine position, a pneumatic cuff was placed at the proximal site of the thigh and inflated to 250 mmHg for 5 min to induce the occlusion of arterial blood flow. In the thigh distal to the cuff, muscle O_2 saturation (SmO_2) and relative changes from rest in oxygenated hemoglobin/myoglobin concentration (oxy-Hb/Mb), deoxygenated hemoglobin/myoglobin concentration (deoxy-Hb/Mb), and total hemoglobin/myoglobin concentration were evaluated at the vastus lateralis (VL) and rectus femoris (RF) muscles at the right leg by using spatial resolved near infrared spectroscopy (SR-NIRS) oximetry (Hb12–4, Astem Co., Japan). The oximetry system utilized two wavelengths: 770 nm and 830 nm. The probe consisted of two light sources and one photodiode detector, with the distances from source to detector at 20 mm and 30 mm, respectively. Data were monitored at a sampling rate of 1 Hz. Fat layer thickness at the measurement site was checked by using an ultrasound device (Vscan Dual Probe, GE HealthCare, Japan), and the effects of light scattering in the fat layer on NIRS data were corrected by normalizing measurement sensitivity (Niwayama et al. 2012). During occlusion, the slope, minimal value, and area above the curve (AAC) from baseline values were calculated for SmO_2 (McLay et al. 2016). Though SmO_2, which is less sensitive to changes in blood volume than oxy-Hb/Mb (Takagi 2016), was considered as an indicator of the balance between O_2 supply and utilization, the raw NIRS signal may have been partly affected, not only by muscle metabolism but also by changes in blood volume during arterial occlusion. Therefore, after correcting for the effects of blood volume fluctuations, deoxy-Hb/Mb was also calculated during arterial occlusion, as in previous studies (Ryan et al. 2012), and slope, maximal value, and area under the curve (AUC) during occlusion were also evaluated for deoxy-Hb/Mb (Fig. 32.1) (Koutlas et al. 2024; Soares et al. 2017). In addition, before and after training, Bulgarian split squat exercise with slow movement and tonic force generation was performed until volitional failure to assess the number of repetitions for squats.

32.2.3 Statistics

All data are represented as means ± standard deviation (SD). For differences in NIRS variables and fat layer thickness, a 2-way repeated-measures analysis of variance (ANOVA) was performed, with time (before vs after) and muscle (VL vs. RF) as factors. Where appropriate, the

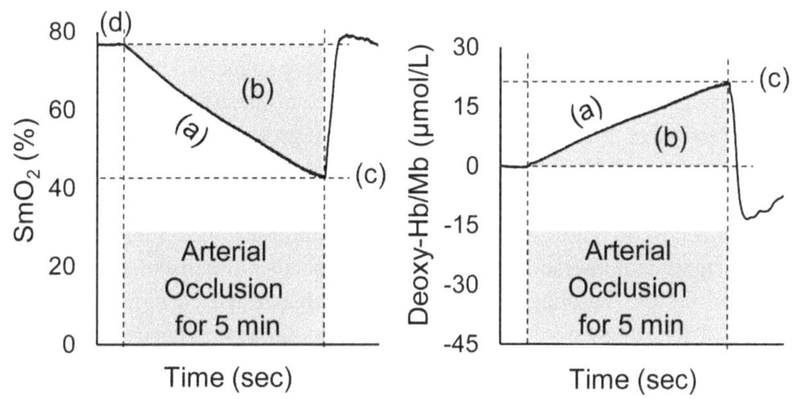

Fig. 32.1 Representative profile of muscle O_2 saturation (SmO_2: left panel) and deoxygenated hemoglobin/myoglobin concentration (deoxy-Hb/Mb: right panel) signal analysis during arterial occlusion. (**a**) slope, (**b**) area above the curve for SmO_2 or area under the curve for deoxy-Hb/Mb, (**c**) Minimum value for SmO_2 or maximum value for deoxy-Hb/Mb, (**d**) baseline value

Table 32.1 NIRS variables during arterial occlusion

	VL Before RT	VL After RT	RF Before RT	RF After RT
SmO_2				
Slope (%·sec^{-1}) #	−0.14 ± 0.04	−0.12 ± 0.03**	−0.12 ± 0.05	−0.12 ± 0.04
AAC for 5 min (%·sec) #	6749 ± 1625	5685 ± 1519**	5931 ± 2135	5916 ± 2028
Minimum value (%) #	27.4 ± 9.3	31.2 ± 6.1†	38.9 ± 12.9	40.3 ± 10.3
Baseline value (%)	70.5 ± 4.2	70.2 ± 4.0	75.5 ± 3.3	76.9 ± 3.2
Deoxy-Hb/Mb				
Slope (μmol/L·sec^{-1}) #	0.09 ± 0.03	0.07 ± 0.02*	0.07 ± 0.03	0.08 ± 0.04
AUC for 5 min (μmol/L·sec) #	4101 ± 1196	3491 ± 802†	3959 ± 1237	4360 ± 2437
Maximum value (μmol/L) #	24.6 ± 6.8	21.9 ± 4.3†	22.4 ± 7.3	24.0 ± 12.2

All data are given as mean ± SD. Significantly different from before training (*: $p < 0.05$, **: $p < 0.01$, †: $p < 0.1$). There was a significant interaction (#: $p < 0.05$). RT: resistance training, AAC: area above the curve, AUC: area under the curve. Symbols indicating a significant difference between VL and RF have been omitted for the sake of clarity

Bonferroni post hoc test was used to test specific significant differences. Pearson's correlation coefficient was employed to determine the relationship between variables. Body weight and the number of maximal repetitions (nMR) for squats were compared between before and after training by paired t-tests. For all statistical analyses, significance was accepted at $p < 0.05$.

32.3 Results

Table 32.1 represents the changes in NIRS variables during arterial occlusion at rest before and after squat exercise for 8 weeks. For SmO_2, there were significant interactions for slope, AAC, and minimal value during arterial occlusion ($p < 0.05$). At VL, a significantly shallower slope of SmO_2 was observed after training than before ($p < 0.01$), and as a result, AAC for SmO_2 was significantly reduced after training ($p < 0.01$). Minimum values tended to increase after training ($p = 0.06$), and no significant interaction or main effect was observed in the baseline value for SmO_2. In contrast, at RF, there were no significant changes in SmO_2 variables.

Similar to the results for SmO_2, significant interactions for slope, AUC, and maximal value during arterial occlusion were also found in deoxy-Hb/Mb ($p < 0.05$). At VL, the slope during arterial occlusion was significantly shallowed after training ($p < 0.05$). AUC ($p = 0.08$) and maximal values ($p = 0.06$) for deoxy-Hb/Mb tended to decrease after training. At RF, no significant differences in deoxy-Hb/Mb were observed between before and after training.

The number of maximal repetitions (nMR) for squats was significantly increased after training (11 ± 4 vs. 19 ± 5 rep, $p < 0.05$). Change in nMR for squats was significantly related to change in slope of SmO_2 ($r = 0.64$, $p < 0.05$), while change in nMR tended to be related to change in slope of deoxy-Hb/Mb ($r = -0.58$, $p = 0.06$), change in AAC for SmO_2 ($r = -0.57$, $p = 0.07$), and change in AUC for deoxy-Hb/Mb ($r = -0.55$, $p = 0.08$). However, no significant relationships were observed between NIRS variables at RF. No significant interactions or main effects were observed in fat layer thickness, and no significant difference was found in body weight.

32.4 Discussion

This study found that muscle deoxygenation during arterial occlusion at rest was diminished after body weight resistance training. Though there was no non-exercise control group, the data showed that change in deoxygenation variables was significantly related to change in the number of maximal repetitions for squats. Hence, these results reinforce the idea that change in deoxygenation variables during arterial occlusion may be caused by resistance training. Soares et al. have suggested that decreased muscle deoxygenation during arterial occlusion at rest in trained groups can in part be explained by a reduced reliance on oxidative metabolism (Soares et al. 2017). Resistance exercise with slow movement and sustained force generation causes an increase in intramuscular pressure, followed by somewhat restricted blood flow, and may therefore potentially enhance muscle deoxygenation during exercise (Tanimoto and Ishii 2006). Our findings may partly be explained by reduced reliance on oxidative metabolism, secondary to an enhancement of muscle deoxygenation during training. Because metabolic adaptation in skeletal muscle may not be affected by short-term or low-volume body weight resistance training with blood flow restriction (Jakobsgaard et al. 2018), our findings should be confirmed in long-term or higher-volume training, and with a non-exercise control group.

While a significantly shallower slope of deoxygenation was observed after training at VL, no significant changes were found in RF. These results indicate that the effects of resistance training on muscle deoxygenation during arterial occlusion at rest were regional. In our previous study (Takagi et al. 2024), deoxygenation during Bulgarian split squat exercise with slow movement and tonic force generation was more enhanced at VL than at RF, possibly due to a difference in intramuscular pressure (i.e., degree of blood flow restriction) (Kilgas et al. 2019) and/or change in muscle activities (Kacin and Strazar 2011). Although the detailed mechanisms are unclear, there may be a possibility that the regional differences in muscle deoxygenation during arterial occlusion at rest may be attributed to differences in these factors during resistance training.

In conclusion, body weight resistance training under slow movement and tonic force generation for 8 weeks may cause a shallow deoxygenation slope during arterial occlusion at rest, though the effects of resistance training on deoxygenation response during arterial occlusion may be regional in quadriceps muscles.

Acknowledgments The authors are grateful for revision of this manuscript by Andrea Hope. This work was supported in part by JSPS KAKENHI Grant Number 22 K11667 to Shun Takagi.

References

Jakobsgaard JE, Christiansen M, Sieljacks P et al (2018) Impact of blood flow-restricted bodyweight exercise on skeletal muscle adaptations. Clin Physiol Funct Imaging 38:965–975

Kacin A, Strazar K (2011) Frequent low-load ischemic resistance exercise to failure enhances muscle oxygen delivery and endurance capacity. Scand J Med Sci Sports 21(6):e231–e241

Kilgas MA, McDaniel J, Stavres J et al (2019) Limb blood flow and tissue perfusion during exercise with blood flow restriction. Eur J Appl Physiol 119(2):377–387

Koutlas A, Smilios I, Kokkinou EM et al (2024) NIRS-derived muscle-deoxygenation and microvascular reactivity during occlusion-reperfusion at rest are associated with whole-body aerobic fitness. Res Q Exerc Sport 95(1):127–139

McLay KM, Gilbertson JE, Pogliaghi S et al (2016) Vascular responsiveness measured by tissue oxygen saturation reperfusion slope is sensitive to different occlusion durations and training status. Exp Physiol 101(10):1309–1318

Niwayama M, Suzuki H, Yamashita T et al (2012) Error factors in oxygenation measurement using continuous wave and spatially resolved near-infrared spectroscopy. J Jpn Coll Angiol 52:211–215

Ryan TE, Erickson ML, Brizendine JT et al (2012) Noninvasive evaluation of skeletal muscle mitochondrial capacity with near-infrared spectroscopy: correcting for blood volume changes. J Appl Physiol 113(2): 175–183

Soares RN, McLay KM, George MA et al (2017) Differences in oxidative metabolism modulation induced by ischemia/reperfusion between trained and untrained individuals assessed by NIRS. Physiol Rep 5(19):e13384

Takagi S (2016) Skeletal muscle oxygen dynamics and peak aerobic capacity. J Phys Fitness Sports Med 5(5): 379–383

Takagi S, Kime R, Niwayama M et al (2013) Muscle oxygen saturation heterogeneity among leg muscles during ramp exercise. Adv Exp Med Biol 765:273–278

Takagi S, Watanabe Y, Kime R (2024) Regional differences in skeletal muscle O_2 dynamics during body weight resistance exercise with slow movement and tonic force generation. Adv Exp Med Biol 1463: 365-370

Tanimoto M, Ishii N (2006) Effects of low-intensity resistance exercise with slow movement and tonic force generation on muscular function in young men. J Appl Physiol (1985) 100(4):1150–1157

Yamauchi J, Nakayama S, Ishii N (2009) Effects of bodyweight-based exercise training on muscle functions of leg multi-joint movement in elderly individuals. Geriatr Gerontol Int 9(3):262–269

Characteristics of Muscle Oxygenation of the Upper and Lower Limb During Moderate-Intensity Upper Limb Crank Exercise: A Near-Infrared Spectroscopy Study

33

Yasuhiro Endo, Daichi Kataoka, Shinga Sagawa, Misaki Igarashi, Ryuya Ishibashi, Toshimi Sato, Atsuhiro Tsubaki, Hajime Tamiya, and Shinichiro Morishita

Abstract

[Purpose] To characterize oxygenated hemoglobin (O_2Hb), deoxygenated hemoglobin (HHb), total hemoglobin (THb), and muscle oxygen saturation (StO_2) changes in the upper and lower limb muscles during moderate-intensity upper limb crank exercise. [Methods] Fifteen healthy males (mean age, height, and weight: 21.1 ± 0.4 years, 172.2 ± 5.9 cm, and 63.3 ± 8.0 kg, respectively) were recruited. The experimental protocol consisted of a 4-min rest, 4-min warm-up, and 20-min constant-load exercise, followed by 10 min of post-exercise rest. O_2Hb, HHb, THb, and StO_2 of the triceps (TR), vastus lateralis (VL), and gastrocnemius muscles (GM) were measured using near-infrared spectroscopy. Changes from rest (ΔO_2Hb, ΔHHb, ΔTHb, ΔStO_2) were analyzed using a repeated-measures ANOVA. [Results] TR ΔO_2Hb significantly decreased at 1–3 min of warm-up and significantly increased from 5 min into exercise. The TR HHb significantly increased from 2 min after starting the warm-up to the end of exercise. VL ΔO_2Hb and ΔTHb significantly increased immediately after exercise onset. GM ΔO_2Hb and ΔTHb increased from 7 to 9 min after exercise onset. The VL ΔHHb increased from 14 to 15 min to the end of exercise, and the GM ΔHHb increased post-exercise. [Conclusion] This study elucidates the influence of systemic circulatory responses on oxygenation dynamics in non-active muscles during upper limb exercise, offering novel insights into whole-body hemodynamic adaptations to localized exercise.

Keywords

Near-infrared spectroscopy · Non-active muscle · Active muscle · Crank exercise · Constant-load exercise

Y. Endo (✉) · D. Kataoka · S. Sagawa · M. Igarashi · R. Ishibashi · T. Sato · S. Morishita
Department of Physical Therapy, School of Health Sciences, Fukushima Medical University, Fukushima, Japan
e-mail: ysendo@fmu.ac.jp

A. Tsubaki · H. Tamiya
Department of Physical Therapy, Faculty of Rehabilitation, Niigata University of Health and Welfare, Niigata, Japan

33.1 Introduction

Near-infrared spectroscopy (NIRS) evaluates intramuscular oxygen saturation (StO_2) and oxygen consumption (VO_2) in real time, facilitating the assessment of oxygen utilization efficiency during exercise and rehabilitation. Oxygenated

S. C. Pias et al. (eds.), *Oxygen Transport to Tissue XLVI*, Advances in Experimental Medicine and Biology 1498, https://doi.org/10.1007/978-3-032-10389-5_33

hemoglobin (O_2Hb) levels in the vastus lateralis (VL) muscle reportedly decrease during bicycle ergometer exercise (Racinais et al. 2014), and exercise-induced increases in endothelium-dependent vasodilation have been observed in non-primary working muscles (DeSouza et al. 2000). Investigations into blood flow in inactive limbs during progressive upper and lower extremity exercise reveal increased blood flow in the forearm, whose muscles are inactive during bicycle ergometer exercise, due to endothelium-dependent vasodilation (Ogata et al. 2004; Tanaka et al. 2006). Similarly, blood flow in inactive lower extremities increases during single-leg or arm exercise, with vascular responses being more pronounced in the forearm than in the leg (Ahlborg et al. 1975; Newcomer et al. 2004). This study aimed to clarify the oxygen dynamics of the inactive lower limb muscles during upper limb exercise at the anaerobic threshold (AT). We hypothesized that O_2Hb, deoxygenated hemoglobin (HHb), and total hemoglobin (THb) levels in the inactive lower extremity muscles would increase during constant-load upper limb exercise based on previous findings, showing increased blood flow and shear stress in non-working limbs during localized exercise, driven by systemic circulatory responses and endothelium-dependent vasodilation.

33.2 Methods

33.2.1 Participants

Fifteen healthy male volunteers (mean age, 21.1 ± 0.4 years; mean height, 172.2 ± 5.9 cm; mean body weight, 63.3 ± 8.0 kg) participated in this study. None of the participants had any symptoms of neurological or cardiovascular disease, and all were medication-free at the time of enrollment. Each participant provided written informed consent after being thoroughly informed of the potential risks, study objectives, measurement procedures, and benefits of the study.

33.2.2 Experimental Procedure

To determine the AT, participants first performed a cardiopulmonary exercise test using a lower limb cycle ergometer (STRENGTH ERGO 8; Mitsubishi Electric Engineering, Tokyo, Japan) with a ramp protocol (20 W/min increments). This was accompanied by electrocardiography (ECG; FCP-9800; Fukuda Denshi, Tokyo, Japan) and exhaled gas analysis (AE-310S; Minato Medical Science, Osaka, Japan). Based on the AT heart rate, participants then performed upper limb crank exercise at a constant load using an upper-body ergometer. The experimental protocol included a 4-min rest, a 4-min warm-up, 20 min of constant-load exercise, and a 10-min post-exercise recovery period.

33.2.3 NIRS Measurements

O_2Hb, HHb, THb, and StO_2 of the triceps (TR), VL, and gastrocnemius (GM) muscles were measured using an NIRS system with PortaLite MKII sensors (Artinis Medical Systems, Elst, Netherlands). The NIRS device incorporated three light-emitting diodes positioned at 30, 35, and 40 mm from a single receiver. The diodes transmitted light in two wavelengths (760 and 850 nm). The NIRS probe was located in the center of each muscle belly.

33.2.4 Statistical Analysis

The results were averaged every min, and changes from rest (ΔO_2Hb, ΔHHb, ΔTHb,

ΔStO_2) were analyzed using a repeated-measures ANOVA followed by Bonferroni post hoc testing. Statistical analyses were performed using SPSS (version 29.0; SPSS Japan Inc., Tokyo, Japan), with statistical significance set at $p < 0.05$.

33.3 Results

33.3.1 Triceps

O_2Hb levels significantly decreased at 2–4 min of the warm-up period and significantly increased from 5 to 20 min of constant-load exercise. A rapid increase was observed post-exercise. HHb significantly increased during the warm-up period, reaching a plateau at approximately 4 min of constant-load exercise. During the post-exercise period, values were not significantly different from those at rest. THb significantly increased during warm-up and continued to increase until the end of the exercise, showing a rapid increase post-exercise and plateauing at around 10 min (Fig. 33.1a). ΔStO_2 significantly decreased after the start of warm-up up to 11 min of constant load exercise and significantly increased after the end of exercise (Fig. 33.2).

33.3.2 Vastus Lateralis

O_2Hb levels significantly increased immediately after the start of exercise and continued to increase up to 20 min into and after exercise. HHb showed a significant increase 14–15 min into the constant-load exercise and increased again post-exercise. THb significantly increased from 1 min into the constant-load exercise and continued to rise until the end of the exercise period (Fig. 33.1b). ΔStO_2 significantly increased immediately after the start of the warm-up and continued to rise during the exercise period (Fig. 33.2).

33.3.3 Gastrocnemius

O_2Hb decreased during the warm-up, significantly increased 7 min into the exercise period, and continued to rise until the end of exercise, with a rapid post-exercise increase. HHb showed a temporary decrease immediately after the warm-up, followed by a lack of significant changes during exercise, and significantly increased from 6 min post-exercise. THb significantly decreased during the warm-up, significantly increased 9 min into the exercise period, and continued to increase until the end of the exercise period, with a rapid post-exercise increase before plateauing (Fig. 33.1c). ΔStO_2 significantly increased at the 12th min of the exercise phase and between the 4th and 5th min during the post-exercise phase (Fig. 33.2).

33.4 Discussion

In this study, the oxygenation dynamics of the triceps, VL, and GM were measured using NIRS during a constant-load exercise of the upper extremities based on the AT heart rate. O_2Hb and THb levels increased during constant-load exercise in both the VL and GM, which were inactive muscles. Previous research has demonstrated exercise-induced endothelium-dependent vasodilation in non-primary active muscles, such as those of the forearms during cycling or walking (DeSouza et al. 2000; Kingwell et al. 1997). Additionally, differences in vasodilatory responses between arm and leg vascular beds have been documented, with several-fold higher blood flow in the inactive lower extremities during upper arm cycling exercises (Ahlborg et al. 1975; Bangsbo et al. 1995). Thigh blood flow and shear stress have been reported to increase linearly during graded arm cycling exercise (Tanaka et al. 2006). We hypothesized that our study would similarly show increases in O_2Hb and THb levels in

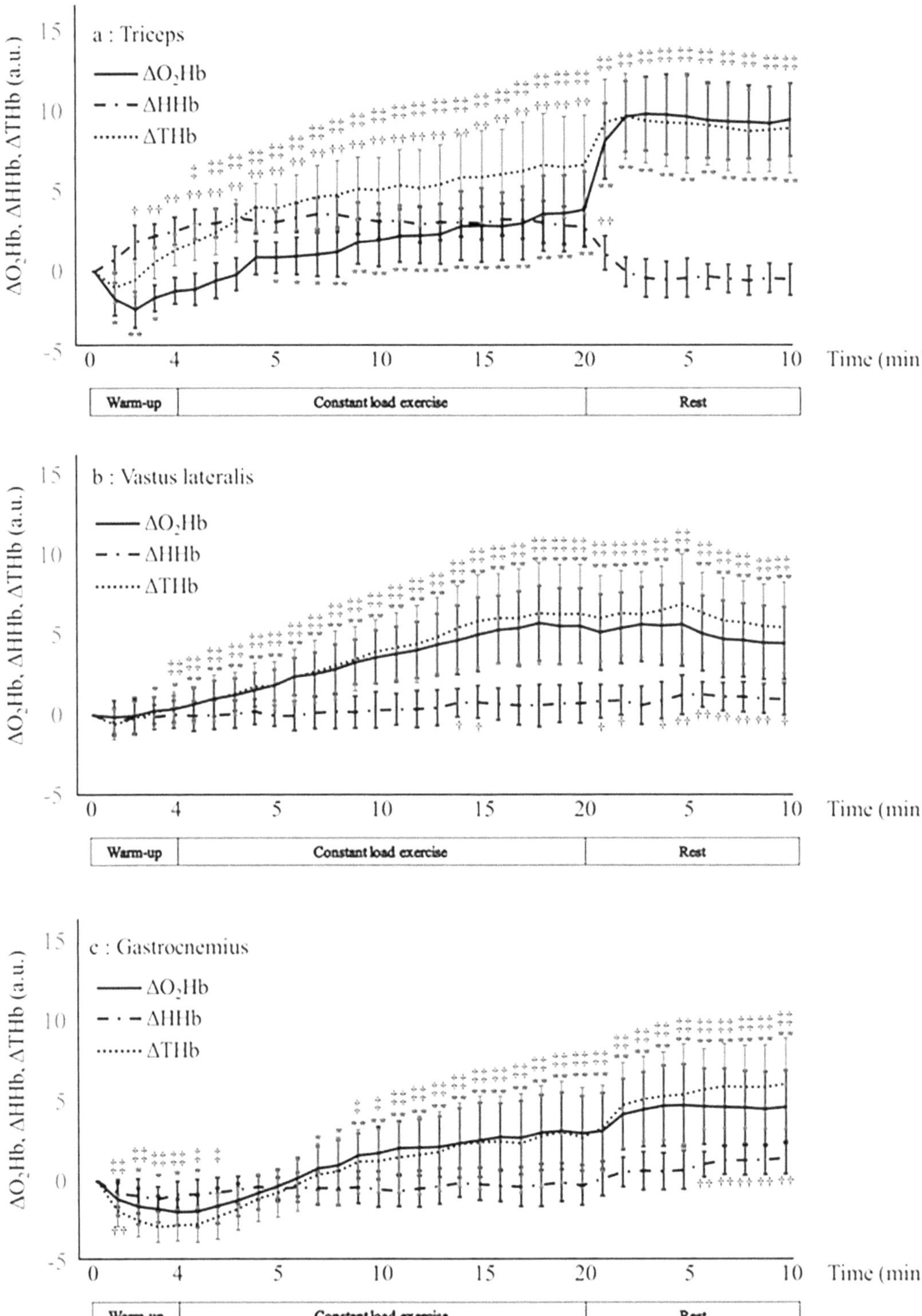

Fig. 33.1 Changes in oxygenated hemoglobin (O_2Hb), deoxygenated hemoglobin (HHb), and total hemoglobin (THb) during the task. (**a**) triceps, (**b**) vastus lateralis, (**c**) gastrocnemius, *: $p < 0.05$ (O_2Hb), **: $p < 0.01$ (O_2Hb), †: $p < 0.05$ (HHb), ††: $p < 0.01$ (HHb), ‡: $p < 0.05$ (THb), ‡‡: $p < 0.01$ (THb)

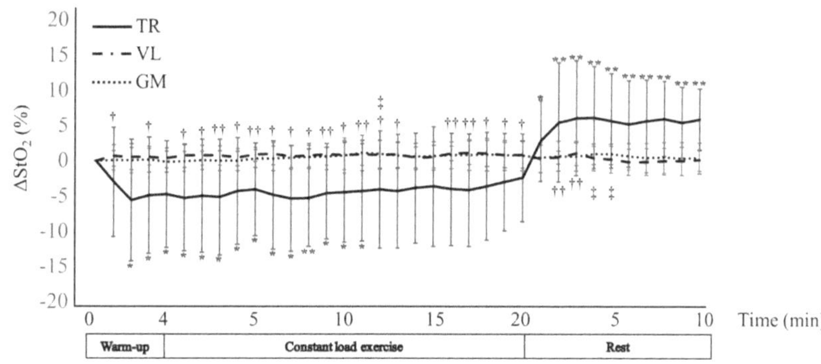

Fig. 33.2 Changes in intramuscular oxygen saturation during the task. *: $p < 0.05$ (triceps), **: $p < 0.01$ (triceps), †: $p < 0.05$ (vastus lateralis), ††: $p < 0.01$ (vastus lateralis), ‡: $p < 0.05$ (gastrocnemius), StO_2, tissue oxygen saturation

inactive muscles due to these mechanisms. Contrary to Ogata et al. (Ogata et al. 2004), who reported different findings at high intensity and ramp load exercises, our study used constant load exercises based on the AT heart rate. Our results suggest that exercise at the AT increases O_2Hb levels in inactive muscles, likely due to increased blood flow and shear stress in the non-exercising limbs (Tanaka et al. 2006). This may result from increased systemic blood circulation and inflow to inactive muscles. Comparing the VL and GM, we found that the response timing was more immediate in the VL, indicating possible differences in vascular responses between these muscles. Kime et al. (Kime et al. 2006) reported higher StO_2 proximally than distally within the same muscle at rest and during fatigue. Slower reoxygenation and lower muscle deoxygenation in distal regions may be due to differences in oxygen supply or muscle fiber composition between distal and proximal regions. There may be a time lag before increased blood flow affects oxygen dynamics in the distal GM.

This study has some limitations. The small sample size may limit the generalizability of the findings. In addition, the cardiopulmonary exercise test was conducted using a lower extremity ergometer, while the main experimental protocol used an upper extremity crank ergometer. This difference may reduce comparability with studies that used the same modality for both tests. Furthermore, noise in the NIRS signals, which may have resulted from motion artifacts and variations in probe placement, could have affected measurement accuracy. Future research with a larger sample, standardized protocols, and improved signal processing is needed to validate these findings.

In conclusion, this study demonstrated that O_2Hb and THb levels in VL and GM increased during upper limb crank exercise, despite these muscles being inactive. These findings suggest that systemic circulatory responses, including increased shear stress and blood flow, contribute to muscle oxygenation changes even in non-active muscles. This study highlights the importance of understanding whole-body hemodynamic responses to localized exercise, which may have implications for exercise prescription in rehabilitation and sports science.

Competing Interests There are no conflicts of interest related to this study.

References

Ahlborg G, Hagenfeldt L, Wahren J (1975) Substrate utilization by the inactive leg during one-leg or arm exercise. J Appl Physiol 39:718–723

Bangsbo J, Aagaard T, Olsen M, Kiens B, Turcotte LP, Richter EA (1995) Lactate and H+ uptake in inactive muscles during intense exercise in man. J Physiol 488: 219–229

DeSouza CA, Shapiro LF, Clevenger CM, Dinenno FA, Monahan KD, Tanaka H et al (2000) Regular aerobic exercise prevents and restores age-related declines in endothelium-dependent vasodilation in healthy men. Circulation 102:1351–1357

Kime R, Osada T, Shiroishi K, Ichimuri S, Kurosawa Y, Homma T et al (2006) Muscle oxygenation

heterogeneity in a single muscle at rest and during bicycle exercise. Jpn J Phys Fit Sport 55:S19–S22
Kingwell BA, Sherrard B, Jennings GL, Dart AM (1997) Four weeks of cycle training increases basal production of nitric oxide from the forearm. Am J Phys 272: H1070–H1077
Newcomer SC, Leuenberger UA, Hogeman CS, Handly BD, Proctor DN (2004) Different vasodilator responses of human arms and legs. J Physiol 556:1001–1011
Ogata H, Reyihan A, Yano T (2004) Kinetics of oxygenation in inactive forearm muscle during ramp leg cycling. J Physiol Anthropol Appl Hum Sci 23:7–17
Racinais S, Buchheit M, Girard O (2014) Breakpoints in ventilation, cerebral and muscle oxygenation, and muscle activity during an incremental cycling exercise. Front Physiol 5:142
Tanaka H, Shimizu S, Ohmori F, Muraoka Y, Kumagai M, Yoshizawa M et al (2006) Increases in blood flow and shear stress to nonworking limbs during incremental exercise. Med Sci Sports Exerc 38:81–85

34 Compensatory Oxygenation Changes in Non-active and Active Muscles During Incremental Exercise in Healthy Adults

Toshimi Sato, Shinga Sagawa, Daichi Kataoka, Misaki Igarashi, Ryuya Ishibashi, Yasuhiro Endo, Atsuhiro Tsubaki, Hajime Tamiya, and Shinichiro Morishita

Abstract

Purpose: We aimed to clarify the compensatory changes in oxygenation in non-active versus active muscles during incremental exercise. Methods: Fifteen male volunteers (age, 21.1 ± 0.5 years) underwent cardiopulmonary exercise testing (CPET) using a cycle ergometer ramp protocol, while maintaining the left upper extremity in a drooping position. We continuously recorded oxygenated hemoglobin (O_2Hb), deoxygenated hemoglobin (HHb), total hemoglobin (THb), and tissue oxygen saturation (StO_2) in the left vastus lateralis (active muscle) and triceps brachii (non-active muscle), as well as cardiopulmonary parameters, during the test. Results: There were significant interactions between time and muscles for all changes in O_2Hb ($p < 0.001$), HHb ($p < 0.001$), THb ($p < 0.001$), and StO_2 ($p < 0.001$). In the non-active muscle, O_2Hb and StO_2 remained higher than at rest until the 40% point of the test, unlike in the active muscle. HHb increased from the 80% point in the non-active muscle, but it increased immediately after the start of the test in the active muscle. THb showed no significant change in the non-active muscle, but it continued to increase immediately after the start of the test in the active muscle. Conclusion: During CPET, O_2Hb and StO_2 in the non-active muscle remained higher in the low- to moderate-intensity phases than in the rest phase and decreased in the high-intensity phase, unlike in the active muscle.

Keywords

Oxygenation · Cardiopulmonary exercise testing · Non-active muscle · Near-infrared spectroscopy · Oxygenated hemoglobin

34.1 Introduction

Determining the local oxygenation kinetics of skeletal muscle during exercise has an important role in improving exercise performance and health care. Several studies examining oxygenation kinetics in active muscle during exercise using near-infrared spectroscopy (NIRS) have shown that deoxygenated hemoglobin (HHb) and total hemoglobin (THb) increase while oxygenated hemoglobin (O_2Hb) and

T. Sato (✉) · S. Sagawa · D. Kataoka · M. Igarashi · R. Ishibashi · Y. Endo · S. Morishita
Department of Physical Therapy, School of Health Sciences, Fukushima Medical University, Fukushima City, Japan
e-mail: satot-pt@fmu.ac.jp

A. Tsubaki · H. Tamiya
Department of Physical Therapy, Faculty of Rehabilitation, Niigata University of Health and Welfare, Niigata, Japan

S. C. Pias et al. (eds.), *Oxygen Transport to Tissue XLVI*, Advances in Experimental Medicine and Biology 1498, https://doi.org/10.1007/978-3-032-10389-5_34

oxygen saturation (StO_2) decrease when the exercise load is increased (Perrey and Ferrari 2018; Boone et al. 2016). These responses reflect oxygen availability in active muscle associated with increased exercise load.

On the other hand, during exercise, local blood flow distribution and oxygenation responses are also modulated in other tissues to maintain oxygen supply to active muscle (Joyner and Casey 2015). It has also been shown that during high-intensity lower extremity cycling exercise, the tissue oxygenation index measured in the forearm muscle, a non-active muscle, was significantly reduced (Shiroishi et al. 2010). These findings suggest that changes in local oxygenation dynamics in non-active muscles may be important in maintaining oxygenation dynamics in active muscles. However, due to the lack of simultaneous monitoring of oxygenation dynamics in active and non-active muscles in previous studies, it is unclear how the oxygenation dynamics of both interact. In addition, because oxygen utilization in active muscles varies with exercise intensity, it is important to compare changes in the oxygenation kinetics of active and non-active muscles in response to detailed exercise intensity changes.

Therefore, the purpose of this study was to characterize the changes in oxygenation in non-active muscle compared with active muscle during cardiopulmonary exercise testing (CPET). Central changes, such as cardiac output (CO) and oxygen uptake (VO_2), were also examined as a secondary objective to understand the changes in oxygenation in each skeletal muscle location during CPET.

34.2 Methods

34.2.1 Participants

Fifteen healthy male volunteers (mean age, 21.1 ± 0.5 years; mean height, 172.9 ± 6.7 cm; mean body weight, 63.3 ± 8.3 kg) participated in this study. No participant exhibited symptoms of neurological or cardiovascular disease, and all were free from medication at the time of enrollment in the study. Each participant provided written informed consent after receiving information regarding the potential risks, study objectives, measurement procedures, and benefits associated with the study.

34.2.2 Experimental Procedure

The participants underwent CPET, engaging in 20 W/min cycle ergometer ramp exercise, focusing on the lower extremities with their left upper extremity in a drooping position. The CPET protocol consisted of a 4-min rest period, 4-min warm-up (WU) period, and cardiopulmonary exercise. A ramp program with an incremental increase in the workload of 20 W/min on a leg cycle ergometer (STRENGTH ERGO 8, Mitsubishi Electric Engineering, Tokyo, Japan) accompanied by electrocardiography (ECG; FCP-9800, Fukuda Denshi, Tokyo, Japan), an exhaled gas analyzer (AE-310S; Minato Medical Science, Osaka, Japan), and impedance cardiography (Physio Flow; Manatec Biomedical, Paris, France) were used. All participants were instructed to maintain a cadence of 50 rotations per minute (rpm) during CPET. Exhaustion was defined as follows: (1) a plateau in VO_2; (2) heart rate (HR) values near the age-predicted maximal HR, calculated as 220 − (0.65 × age); and (3) a decrease in the cycling cadence to <50 rpm, despite strong verbal encouragement (Rupp et al. 2008). The highest value obtained for VO_2 was considered the VO_2 peak. HR, CO, VO_2, and minute ventilation (VE) were continuously recorded, with average values calculated every 5 s during CPET.

34.2.3 NIRS Measurements

O_2Hb, HHb, THb, and StO_2 in the left vastus lateralis muscle, as an active muscle, and the left triceps brachii muscle, as a non-active muscle, were evaluated by NIRS using two PortaLite MKII sensors (Artinis Medical Systems, Einsteinweg, Netherlands). The triceps brachii was chosen because it is one of the muscles that can be relaxed most easily during CPET. These NIRS parameters were continuously recorded, with average values calculated every 5 s during CPET, as well as cardiopulmonary parameters. The total duration from the start to the end of the test for each participant was converted to 100%, and the values of each parameter for every 10% were presented as representative values.

34.2.4 Statistical Analysis

A generalized estimating equation was used to evaluate the differences in changes in O_2Hb, HHb, THb, and StO_2 over time between the active and non-active muscles. An unstructured working correlation structure was applied to adjust for within-participant correlation. Interaction terms between time and muscles were included in the generalized estimating equation model to investigate the difference of changes in the parameters during CPET between the muscles. Statistical analyses were performed using SPSS 29.0 (SPSS Japan Inc., Tokyo, Japan), with statistical significance set at $p < 0.05$.

34.3 Results

All participants completed CPET (Table 34.1). As shown in Fig. 34.1, respiratory and circulatory parameters, such as CO, HR, VO_2, and VE, showed a steady increase with a progressive increase in exercise intensity during CPET (Fig. 34.1).

There were significant interactions between time and muscles for all changes in O_2Hb ($p < 0.001$), HHb ($p < 0.001$), THb ($p < 0.001$), and StO_2 ($p < 0.001$) (Fig. 34.2). In the post-hoc test results, O_2Hb and StO_2 in the non-active muscle remained higher than at rest until the 40% point, and significantly decreased during the high-intensity phase from the 80% point. Additionally, HHb increased from the 50% point in the active muscle, whereas it increased from the 80% point in the non-active muscle.

Table 34.1 Participant characteristics

Age, years	21.1 (0.5)
Height, cm	172.9 (6.7)
Body weight, kg	63.3 (8.3)
Cardio-pulmonary exercise testing	
At rest	
VO_2, mL/min	288.1 (46.6)
VE, L/min	9.5 (1.5)
HR, bpm	85.7 (11.3)
Cardiac output, L/min	7.6 (1.7)
At anaerobic threshold	
VO_2, mL/min	1227.2 (173)
VE, L/min	29.5 (3.5)
Work rate, watts	93.5 (13.2)
HR, bpm	127.5 (10)
Cardiac output, L/min	12.3 (2.5)
At peak	
VO_2, mL/min	2496.1 (336.4)
VE, L/min	82.7 (16.4)
Work rate, watts	229.5 (26.8)
HR, bpm	181.8 (8.6)
Cardiac output, L/min	20.2 (4.0)

Data are presented as mean (standard deviation)
HR heart rate, *VE* minute ventilation, VO_2 oxygen uptake

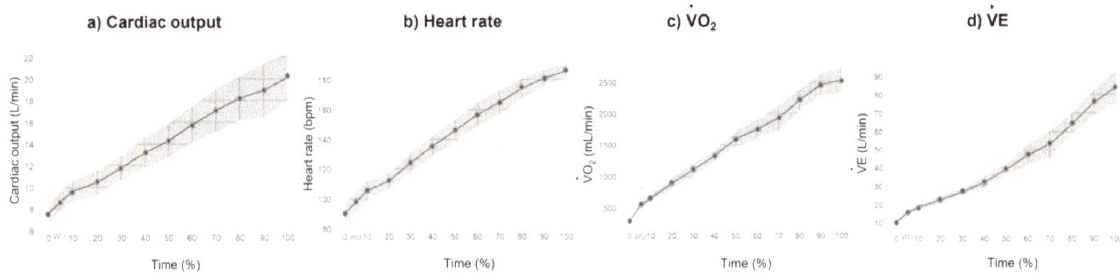

Fig. 34.1 Longitudinal changes in cardiopulmonary parameters during cardiopulmonary exercise testing. Data are presented with means and 95% confidence intervals. HR heart rate, VO_2 oxygen uptake, VE minute ventilation, WU warm-up

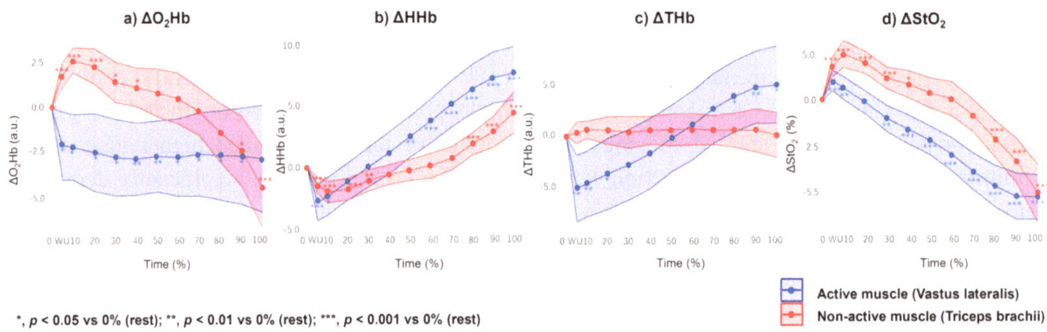

Fig. 34.2 Longitudinal changes in O_2Hb, HHb, THb, and StO_2 in active and non-active muscles during cardiopulmonary exercise testing. Data are presented with estimated marginal means and 95% confidence intervals calculated by the generalized estimating equation. HHb deoxygenated hemoglobin, O_2Hb oxygenated hemoglobin, StO_2 tissue oxygen saturation, THb total hemoglobin, WU warm-up

34.4 Discussion

The present study found that: (1) oxygenation of active and non-active muscles during CPET showed significantly different changes; (2) in the non-active muscle, O_2Hb and StO_2 remained higher than at rest until the 40% point, and significantly decreased at high loads beyond the 80% point, unlike in the active muscle; and (3) HHb increased from the 80% point in the non-active muscle, although it increased from the 50% point in the active muscle.

During CPET, the active muscle exhibited responses indicating increased local oxygen utilization with rising exercise intensity, such as decreased O_2Hb and StO_2, as well as increased HHb and THb, consistent with previous studies (Perrey and Ferrari 2018; Boone et al. 2016).

In contrast, O_2Hb and StO_2 increased in the non-active muscle during the low- to moderate-intensity phases during CPET. This response could be the result of increased oxygen transport not only in the active muscle but also in the non-active muscle due to increased CO and VO_2 responses associated with exercise initiation, peripheral vasodilation (Crecelius et al. 2014), and central commands (Ishii et al. 2012).

On the other hand, significant decreases in O_2Hb and StO_2 were observed not only in the active muscle but also in the non-active muscle during the high-intensity phase. The reason for these results, even though CO and VO_2 continued to increase until the end of the study, may be the result of the exercise pressor reflex function associated with autonomic alterations, a physical response that promotes blood flow distribution predominantly in active muscle (Secher and Volianitis 2006; Grotle et al. 2020). In actuality, the decrease in StO_2 in the active muscle was gradual from the 90% point in the present study (Fig. 34.2d).

In conclusion, during CPET, the non-active muscle was characterized by an increase in O_2Hb and StO_2 in the low- to moderate-intensity phases and a decrease in the high-intensity phase, showing oxygenation kinetics quite different from those of the active muscle.

References

Boone J, Vandekerckhove K, Coomans I, Prieur F, Bourgois JG (2016) An integrated view on the oxygenation responses to incremental exercise at the brain, the locomotor and respiratory muscles. Eur J Appl Physiol 116:2085–2102

Crecelius AR, Luckasen GJ, Larson DG, Dinenno FA (2014) KIR channel activation contributes to onset and steady-state exercise hyperemia in humans. Am J Physiol Heart Circ Physiol 307:H782–H791

Grotle AK, Macefield VG, Farquhar WB, O'Leary DS, Stone AJ (2020) Recent advances in exercise pressor reflex function in health and disease. Auton Neurosci 228:102698

Ishii K, Liang N, Oue A, Hirasawa A, Sato K, Sadamoto T, Matsukawa K (2012) Central command contributes to increased blood flow in the noncontracting muscle at the start of one-legged dynamic exercise in humans. J Appl Physiol 112:1961–1974

Joyner MJ, Casey DP (2015) Regulation of increased blood flow (hyperemia) to muscles during exercise: a hierarchy of competing physiological needs. Physiol Rev 95:549–601

Perrey S, Ferrari M (2018) Muscle oximetry in sports science: a systematic review. Sport Med 48:597–616

Rupp T, Thomas R, Perrey S, Stephane P (2008) Prefrontal cortex oxygenation and neuromuscular responses to exhaustive exercise. Eur J Appl Physiol 102:153–163

Secher NH, Volianitis S (2006) Are the arms and legs in competition for cardiac output? Med Sci Sports Exerc 38:1797–1803

Shiroishi K, Kime R, Osada T, Murase N, Shimomura K, Katsumura T (2010) Decreased muscle oxygenation and increased arterial blood flow in the non-exercising limb during leg exercise. Adv Exp Med Biol 662:379–384

Pulse Oximeters: Accuracy and Artifacts 35

Victor Ochoa-Gutierrez and Gennadi Saiko

Abstract

Oximetry is used to quantify the presence of oxygen in human blood within soft tissues of the human body. Among multiple implementations of this technology, pulsatile oxygen saturation (SpO_2) is a core medical technology and is being rapidly adopted in consumer health. However, despite its long history of clinical use, recent findings indicate that the accuracy of pulse oximetry may be affected by various factors and biases. For example, the COVID-19 pandemic showed that pulse oximeters exhibited flaws in accuracy due to the skin pigmentation of patients with darker skin. Thus, the future of this technology, particularly in consumer health devices, needs to be built on foundations that account for such biases. This chapter reviews the principles of pulse oximetry, sources of its artifacts, calibration methods, and the factors that may cause inaccuracy in pulse oximeters, particularly pertinent to two-wavelength pulse oximetry. Drawing upon recent research and clinical insights, we review the multifaceted nature of pulse oximetry biases, including motion artifacts, skin pigmentation, body mass index, environmental variables, device calibration, and nail polish, among others.

Keywords

Arterial oxygen saturation · Photoplethysmography · Signal artifacts · Skin pigmentation · Body mass index

35.1 Introduction

Pulse oximeters (POs) have been used in the last five decades in the clinical settings, and they have become known as the fifth vital sign (Torp et al. 2023). They were initially designed to measure the average of arterial oxygen present in the body (SaO_2), which in this context is known as SpO_2. According to the World Health Organization (WHO), a PO is an early-warning device of hypoxia detection that continuously measures the oxygen saturation in the arterial blood noninvasively to support practitioners' decision-making.

Over the last 50 years, POs have been rapidly evolving (Fig. 35.1) and the technological improvements have been beneficial for their broad use. However, the underlying principles have not been changed nor improved to bring a new

V. Ochoa-Gutierrez (✉)
mPulse Technologies, New York City, NY, USA

Jacobs Technion-Cornell Institute, Cornell Tech, New York City, NY, USA
e-mail: victorog@cornell.edu

G. Saiko
Department of Physics, Toronto Metropolitan University, Toronto, ON, Canada

S. C. Pias et al. (eds.), *Oxygen Transport to Tissue XLVI*, Advances in Experimental Medicine and Biology 1498, https://doi.org/10.1007/978-3-032-10389-5_35

framework for a future generation of pulse oximeters (Cabanas et al. 2022).

As a result, even though POs are widely used in hospitals, commonly with a finger probe, it has been demonstrated that certain implementations, including consumer health POs, are not accurate enough for hypoxia detection (Harskamp et al. 2021). In part, it can be explained by the fact that despite some sophisticated pulse oximetry technologies having been developed for clinical use, most clinical and consumer health devices still rely on two-wavelength technology. This chapter provides an overview of the principles of pulse oximetry technology, its sources of artifacts, and calibration methods toward development of new approaches. In particular, we look at the reported flaws that can be used to guide improvements or to redesign this medical device.

35.2 Principles of Pulse Oximetry

The typical PO probe assesses the capillary blood vessels in the finger. It usually consists of two optical parts placed in front of each other in a typical transmission configuration using two wavelengths, in red and infrared ranges of the spectrum. Commonly, it also has a broadband photodetector from 400 nm to 1000 nm. As the deoxygenated hemoglobin (HHb) has different absorption spectra than oxyhemoglobin (HbO_2), their relative fractions can be extracted from optical measurements. As such, the functional oxygen content (SaO_2) can be expressed as follows:

$$\mathrm{SaO}_{2(\mathrm{functional})} = \frac{[\mathrm{HbO}_2]}{[\mathrm{HbO}_2] + [\mathrm{HHb}]} \tag{35.1}$$

Pulse oximetry operates by assuming that only blood volume in the arterial compartment changes during the cardiac cycle (pulsatile volume) and ignoring or accounting for scattering (Beer-Lambert law or the Modified Beer-Lambert law (MBLL) (Delpy et al. 1988), respectively).

$$\begin{aligned} \mathrm{Attenuation(OD)} &= \log\left(\frac{I}{I_0}\right) \\ &= \varepsilon \cdot c \cdot l \cdot B + G \end{aligned} \tag{35.2}$$

Here, OD is the optical density, I_0 and I respectively denote the incident and transmitted light intensity, ε is the molar absorption extinction coefficient, c is the concentration of the hematocrit, the pathlength l is the physical distance between the light emitter and the photodetector, B is a "pathlength factor" that accounts for scattered light in tissue, and G a geometry factor related to the tissue.

Current pulse oximetry technology was developed by Japanese engineer Takuo Aoyagi in 1972 (Miyasaka et al. 2021). Pulse oximetry estimates the oxygen saturation based on a "ratio of ratios".

$$R = \frac{\mathrm{AC}_R/\mathrm{DC}_R}{\mathrm{AC}_{IR}/\mathrm{DC}_{IR}} \tag{35.3}$$

where the AC_λ component (with subscript R for "red" or IR for "infrared") is referred to as the "volume pulsatile component" and the DC_λ component is referred to as the "blood volume baseline" (Larsen et al. 1997). Both are detected via photoplethysmogram (PPG) signals, which are extracted from a fingertip using reflection or transmission principles. These signals contain abundant information about blood circulation: for example, about changes in heart rate, which can be obtained by applying the Fourier transformation in order to understand circulatory problems.

In the typical scenario, 660 nm (red light) and 940 nm (infrared light) wavelengths are used. Pulsatile oxygen saturation (SpO_2) can be obtained from R using look-up tables generated by taking arterial blood samples from healthy subjects, reducing the inhaled fraction of O_2, and then relating the R to a specific SpO_2 value. Most commercial pulse oximeters are calibrated in this way. However, some POs (e.g., Masimo) use other algorithms.

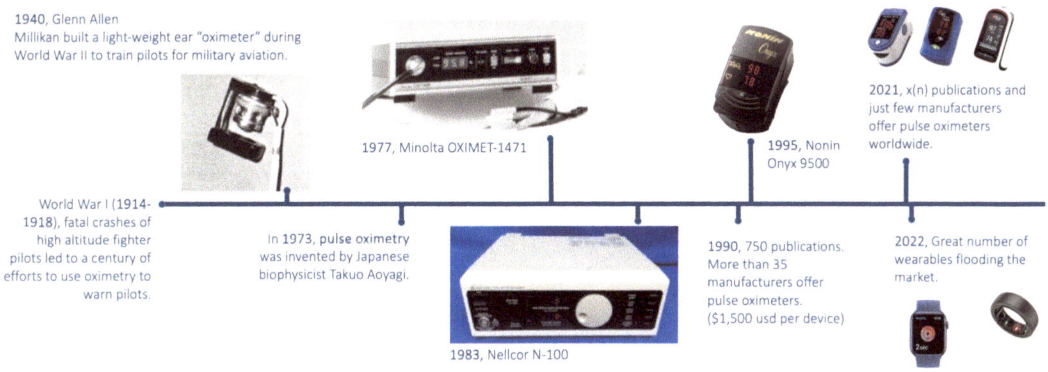

Fig. 35.1 Pulse oximeters and wearables timeline, 1940 until present time. Quick and steady evolution in size reduction, low power consumption upgrades, data transmission, and interconnectivity protocols improvements

35.3 Sources of Artifacts in Pulse Oximetry

Pulse oximeters have been studied intensively. During the last 50 years, multiple sources of error have been identified, including issues related to the different optical paths of the two wavelengths, the volume changes in the tissue during pulsation, changes in light scattering (since the blood has a higher scattering coefficient than the tissue), motion artifacts, repositioning changes, ambient light, low signal-to-noise ratio (SNR), non-standard manufactured LEDs or photodiodes. However, even if these artifacts are controlled, there are multiple other reasons for erroneous PO readings.

Pulse oximetry is used across populations, from neonates to geriatric patients, and across environments, from low temperatures to high-altitude settings—where overlooked hypoxia can cause serious consequences (Eltzschig and Carmeliet 2011; Nimmagadda et al. 2017). Thus, there are various issues associated with PO use. For example, with newborn babies, the sensor size is often inadequate (does not fit) (Hakemi and Bender 2005). The PO also can cause allergic reactions to adhesives of single-use sensors or pressure injury from clip sensors, in some cases (Torp et al. 2023).

In addition, there are several common pathways that result in incorrect SpO_2 readings. For example, artifacts may arise from discrepancies in the optical pathlength ratio between the two wavelengths during hypoxemia, which leads to overestimation of SpO_2 values (Elron et al. 2023). Artifacts can also be attributed to overlapping absorption of blood components (deoxygenated hemoglobin [HHb], methemoglobin [HbMet], or carboxyhemoglobin [HbCO]) and tissue components, such as melanin, in the same wavelength band (Fig. 35.2), reducing the capacity of red and infrared light to differentiate between biomarkers (Patterson et al. 2022).

Blood Components The presence of methemoglobin (HbMet) or carboxyhemoglobin (HbCO) (Fig. 35.2) can be a source of error, depending on their concentrations, when measuring oxygen in arterial blood. Other substances such as fetal hemoglobin (HbF), sulfhemoglobin, bilirubin (Pritišanac et al. 2021; Porath et al. 2001; Banerjee et al. 2023) and conditions (sickle cells, anemia) present can also interfere with the SpO_2 measurement (Cox et al. 2012; Pianosi et al. 1993).

Drugs Incorrect readings can also be caused by unintended effects of drugs used for pain relief or patient stabilization. Other substances, such as methylene blue, indigo carmine, indocyanine green, can also affect the readings (Yoo et al. 2015; Ralston et al. 1991; Chan et al. 2013).

Nail Polish Nail polish or artificial nails affect light transmission, causing inaccurate R values

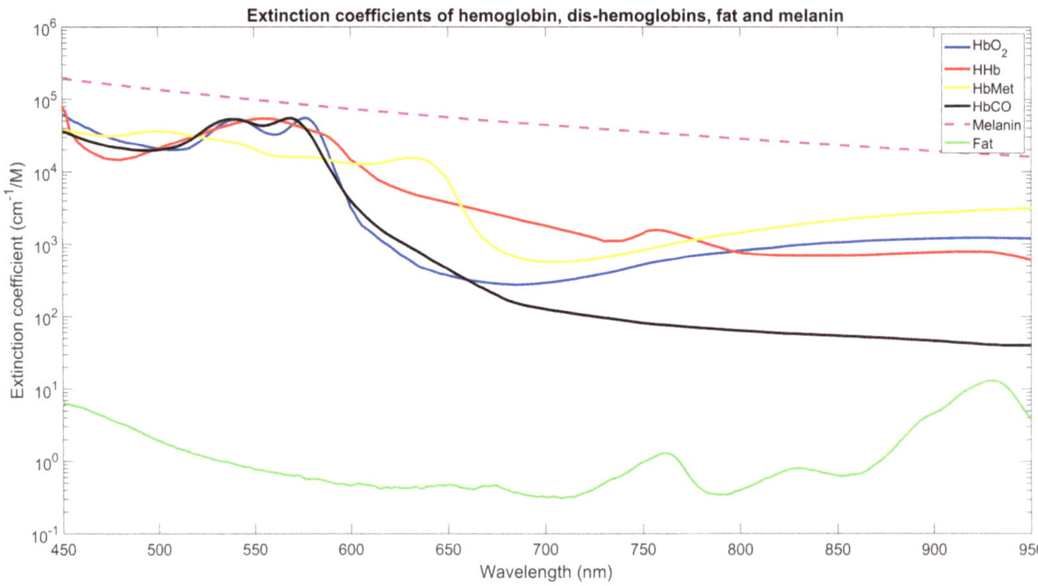

Fig. 35.2 Log-scale plot of absorbances of oxyhemoglobin, deoxyhemoglobin, methemoglobin, carboxyhemoglobin, fat, and melanin, in a wavelength range from 450 nm to 1000 nm. (Data compiled and adapted from Jacques 2013 and Zijlstra et al. 2000)

that affect the look-up table results (Hakverdioğlu Yönt et al. 2014), where the average errors were 3%, 5%, and 6% for the black, green, and blue dyes, respectively (Ralston et al. 1991).

BMI Recently, it has been noted that body mass index (BMI) (Rodriguez et al. 2022) also affects the light transmission through tissues, attenuating PPG signals and leading to inaccurate R values for SpO_2 calculations.

Skin Tone One of the biggest problems with pulse oximetry is the generic approach to measuring SpO_2 through the skin, which does not consider skin pigmentation. A 2020 study (Sjoding et al. 2020) showed that occult hypoxemia incidence was 11.7% in Black patients compared to 3.6% in White patients when SpO_2 values were compared against the blood gas analyser (BGA) gold standard. For more than 30 years, clinicians have mentioned that skin pigmentation has led to an overstated measurement of oxygen in the blood (Barker and Wilson 2023; Mardirossian and Schneider 1992; Davidson and Hosie 1993; Hutton and Clutton-Brock 1993; Sinex 1999; Hakemi and Bender 2005; DeMeulenaere 2007; Chan et al. 2013; Al-Halawani et al. 2023).

35.4 Calibration

Since continuous monitoring has become crucial, it is essential to ensure patient safety and accurate oxygen saturation for patient assessment. It is achieved by calibration and validation. There are multiple methods to validate whether a pulse oximeter is functioning properly. One method is the use of a "finger phantom" for oximeters that operates with two wavelengths. However, this method only provides calibration at 80%, 90%, and 100% blood oxygen saturation (Fisher et al. 1996). Even though clinicians typically know that SpO_2 values below 80% are too low and act accordingly, the approach still can be improved by including more reference data points.

Another common approach to calibration is to use healthy volunteers, following the ISO 80601-2-61 standard (ISO 2011). In this method, arterial blood samples are collected from healthy subjects while the inhaled fraction of oxygen (FiO_2) is gradually reduced. The corresponding measurements are then used to establish a relationship between the R value and a specific SpO_2. This calibration process could be improved by increasing the number of volunteers from various races and both genders, especially noting the effect in groups with high skin pigmentation, using a power calculator to determine the correct sample size. The approach can also be improved by including unhealthy groups, in addition to healthy groups (Wong et al. 2021).

35.5 Advances in Pulse Oximetry Technology

Recent advances in electronics have driven progress in pulse oximetry technology. For example, analog-to-digital-converter (ADC) advancements from 8-bit to 24-bit resolution conversion have significantly improved accuracy and extended the dynamic range. The development of low-power consumption solutions has also led to major advances in consumer-grade POs. In the last five years, there has been a noticeable increase in the implementation of pulse oximetry technology in wearables within unregulated markets (consumer health).

Multiple other approaches have been proposed. For example, pulse oximetry can be implemented as a noncontact modality (remote photoplethysmography, or rPPG) (Saiko et al. 2024). Solutions based on ambient light (without internal light sources) have also been proposed (Han et al. 2020). However, this approach is only practical under certain light conditions (daylight), which significantly impacts its applicability. For example, indoor lighting presents challenges for this method (Han et al. 2020).

Approaches using more than two wavelengths have been proposed and implemented in commercial devices already. They allow for the discrimination of more than two hemoglobin types. Particularly, fractional SaO_2 has been proposed to address the current shortcomings of the SaO_2 (Kleiser et al. 2018).

$$\mathrm{FrSaO}_{2(\mathrm{fractional})} = \frac{[\mathrm{HbO}_2]}{[\mathrm{HbO}_2] + [\mathrm{HHb}] + [\mathrm{MetHb}] + [\mathrm{COHb}]} \quad (35.4)$$

However, most POs used in the clinical settings are based on the two-wavelength approach. As such, the use of fractional SaO_2 is barely implemented in the clinical settings and is limited by the use of different sensors, depending on the biomarker to be measured, increasing the device cost.

35.6 Future Directions

With recently reported controversies, the credibility of pulse oximetry has been shaken. As such, it is important to regain trust and credibility by providing new solutions and improving existing ones. In particular, joint efforts among clinicians, industry, and academia are required to address the ongoing debates and controversies surrounding pulse oximetry.

It is important to mention that some of the previously noted artifacts can be solved by creating mathematical models that incorporate these variables and determine how to correct for them or how to decrease their impact on PO systems. New sensors need to be developed, to enhance the reliability and utility of POs in clinical practice and to prevent patient mistreatment. Algorithms must be improved, implemented, and tested to overcome the pitfalls already mentioned in this review. Although pulse oximetry is one of the most regulated areas in medical devices (see, for example, (ISO 2011)), changes to the standards are necessary. Notably, POs may benefit from the development of standardized calibration methods, validation protocols, and guidelines for oximetry use. Finally, clinical studies are required in which inclusion by race, age, and gender is properly established.

35.7 Conclusions

With SpO_2 errors of 6% when nails are polished and 10% in individuals with highly pigmented skin in clinical settings, pulse oximetry technology is ripe for change. Two-wavelength technology, the bare bones of clinical monitoring, is seductively simple. However, its broader use, particularly in consumer health devices, needs to be carefully managed. This review has discussed the primary artifacts that currently affect two-wavelength pulse oximetry technology. In particular, several understudied areas have emerged recently, including the impact of skin tone and BMI. In addition, novel approaches and solutions for newborns are required. We have also discussed future directions and alternative steps for interdisciplinary analysis, toward achieving better measurements in pulse oximetry technology for both lifestyle implementations and clinical pulse oximetry.

Acknowledgments V. Ochoa-Gutierrez acknowledges financial support from the OPTICA Pivot Fellowship and the Runway Program at Cornell Tech. G. Saiko acknowledges financial support from the personal Discovery Grant (NSERC).

References

Al-Halawani R, Charlton PH, Qassem M, Kyriacou PA (2023) A review of the effect of skin pigmentation on pulse oximeter accuracy. Physiol Meas 44:05TR01

Banerjee A et al (2023) Non-invasive estimation of hemoglobin, bilirubin and oxygen saturation of neonates simultaneously using whole optical spectrum analysis at point of care. Sci Rep 13:1–17

Barker SJ, Wilson WC (2023) Racial effects on Masimo pulse oximetry: a laboratory study. J Clin Monit Comput 37:567–574

Cabanas AM, Fuentes-Guajardo M, Latorre K, León D, Martín-Escudero P (2022) Skin pigmentation influence on pulse oximetry accuracy: a systematic review and bibliometric analysis. Sensors 22:3402

Chan ED, Chan MM, Chan MM (2013) Pulse oximetry: understanding its basic principles facilitates appreciation of its limitations. Respir Med 107:789–799

Cox SE et al (2012) Sickle cell anemia: iron availability and nocturnal oximetry. J Clin Sleep Med 8:541–545

Davidson JAH, Hosie HE (1993) Limitations of pulse oximetry: respiratory insufficiency – a failure of detection. Br Med J 307:372–373

Delpy DT et al (1988) Estimation of optical pathlength through tissue from direct time of flight measurement. Phys Med Biol 33:1433

DeMeulenaere S (2007) Pulse oximetry: uses and limitations. J Nurse Pract 3:312–317

Elron E et al (2023) Overestimation of oxygen saturation measured by pulse oximetry in hypoxemia. Part 1: effect of optical pathlengths-ratio increase. Sensors (Basel, Switzerland) 23:1434

Eltzschig HK, Carmeliet P (2011) Hypoxia and inflammation. N Engl J Med 364:656–665

Fisher JA, Martire T, Volgyesi GA (1996) Evaluation of a new pulse oximeter testing device. Can J Anaesth 43: 179–183

Hakemi A, Bender JA (2005) Understanding pulse oximetry, advantages, and limitations, vol 17, pp 416–418

Hakverdioğlu Yönt G, Akin Korhan E, Dizer B (2014) The effect of nail polish on pulse oximetry readings. Intensive Crit Care Nurs 30:111–115

Han D et al (2020) Pulse oximetry using organic optoelectronics under ambient light. Adv Mater Technol 5: 1901122

Harskamp RE et al (2021) Performance of popular pulse oximeters compared with simultaneous arterial oxygen saturation or clinical-grade pulse oximetry: a cross-sectional validation study in intensive care patients. BMJ Open Respir Res 8:e000939

Hutton P, Clutton-Brock T (1993) The benefits and pitfalls of pulse oximetry, vol 307, pp 457–458

ISO, in ISO 80601-2-61:2011 – Medical electrical equipment—Part 2-61: particular requirements for basic safety and essential performance of pulse oximeter equipment

Jacques SL (2013) Optical properties of biological tissues: a review. Phys Med Biol 58(11):R37. https://doi.org/10.1088/0031-9155/58/11/R37

Kleiser S et al (2018) Comparison of tissue oximeters on a liquid phantom with adjustable optical properties: an extension. Biomed Opt Express 9:86

Larsen PD, Harty M, Thiruchelvam M, Galletly DC (1997) Spectral analysis of AC and DC components of the pulse photoplethysmograph at rest and during induction of anaesthesia. Int J Clin Monit Comput 14:89–95

Mardirossian G, Schneider RE (1992) Limitations of pulse oximetry, vol 39. American Dental Society of Anesthesiology, pp 194–196

Miyasaka K et al (2021) Tribute to Dr. Takuo Aoyagi, inventor of pulse oximetry. J Anesth 35:671–709

Nimmagadda U, Salem MR, Crystal GJ (2017) Preoxygenation: physiologic basis, benefits, and potential risks. Anesth Analg 124:507–517

Patterson S, Sandercock N, Verhovsek M (2022) Understanding pulse oximetry in hematology patients: Hemoglobinopathies, racial differences, and beyond. Am J Hematol 97:1659–1663

Pianosi P, Charge TD, Esseltine DW, Coates AL (1993) Pulse oximetry in sickle cell disease. Arch Dis Child 68:735–738

Porath M, Sinha P, Dudenhausen JW, Luttkus AK (2001) Systematic instrumental errors between oxygen saturation analysers in fetal blood during deep hypoxemia. Clin Chim Acta 307:151–157

Pritišanac E, Urlesberger B, Schwaberger B, Pichler G (2021) Accuracy of pulse oximetry in the presence of fetal hemoglobin—a systematic. Children 8:361

Ralston AC, Webb RK, Runciman WB (1991) Potential errors in pulse oximetry III: effects of interference, dyes, dyshaemoglobins and other pigments. Anaesthesia 46:291–295

Rodriguez AJ et al (2022) Skin optical properties in the obese and their relation to body mass index: a review. J Biomed Opt 27:030902

Saiko G et al (2024) Observation of blood motion in the internal jugular vein by contact and contactless photoplethysmography during physiological testing: case studies. Biomed Opt Express 15(4):2578–2589

Sinex JE (1999) Pulse oximetry: principles and limitations, vol 17. W.B. Saunders, pp 59–66

Sjoding MW, Dickson RP, Iwashyna TJ, Gay SE, Valley TS (2020) Racial bias in pulse oximetry measurement. N Engl J Med 383:2477–2478

Torp KD, Modi P, Pollard EJ, Simon LV (2023) Pulse Oximetry. JAMA 298:8724

Wong AKI et al (2021) Analysis of discrepancies between pulse oximetry and arterial oxygen saturation measurements by race and ethnicity and association with organ dysfunction and mortality. JAMA Netw Open 4:e2131674

Yoo KY, Baek HY, Jeong S, Hallacoglu B, Lee JU (2015) Intravenously administered indocyanine green may cause falsely high near-infrared cerebral oximetry readings. J Neurosurg Anesthesiol 27:57–60

Zijlstra WG, Buursma A, van Assendelft OW, eds (2000) Visible and near-infrared absorption spectra of human and animal haemoglobin: determination and application (1st ed). CRC Press. https://doi.org/10.1201/9780429071096

Index

S. C. Pias et al. (eds.), *Oxygen Transport to Tissue XLVI*, Advances in Experimental Medicine and Biology 1498, https://doi.org/10.1007/978-3-032-10389-5

MIX
Papier aus verantwortungsvollen Quellen
Paper from responsible sources
FSC® C105338

If you have any concerns about our products,
you can contact us on
ProductSafety@springernature.com

In case Publisher is established outside the EU,
the EU authorized representative is:
Springer Nature Customer Service Center GmbH
Europaplatz 3, 69115 Heidelberg, Germany

Printed by Libri Plureos GmbH
in Hamburg, Germany